Handbook of Electromagnetism

Handbook of Electromagnetism

Edited by **Elliott Flanagan**

NY RESEARCH PRESS
New York

Published by NY Research Press,
23 West, 55th Street, Suite 816,
New York, NY 10019, USA
www.nyresearchpress.com

Handbook of Electromagnetism
Edited by Elliott Flanagan

International Standard Book Number: 978-1-63238-238-2 (Hardback)

Printed in the United States of America.

Contents

Preface

I am honored to present to you this unique book which encompasses the most up-to-date data in the field. I was extremely pleased to get this opportunity of editing the work of experts from across the globe. I have also written papers in this field and researched the various aspects revolving around the progress of the discipline. I have tried to unify my knowledge along with that of stalwarts from every corner of the world, to produce a text which not only benefits the readers but also facilitates the growth of the field.

The aim of this book is to introduce the readers with the science of electromagnetism. Among the various branches of classical physics, electromagnetism is the domain which experiences the most significant advancements, both in its practical and basic aspects. The quantum corrections which produce non-linear terms of the standard Maxwell equations, their particular form in curved spaces, whose predictions can be encountered with the cosmic polarization rotation, or the topological model of electromagnetism, constructed with electromagnetic knots, are important examples of current theoretical advancements. The book consists of in-depth information which makes it very easy for the readers to comprehend the concepts related to the science of electromagnetism. It also elucidates several industrial applications of electromagnetism, emphasizing on its significance. Therefore, this book serves as a valuable and interesting source of information for various specialists in this field.

Finally, I would like to thank all the contributing authors for their valuable time and contributions. This book would not have been possible without their efforts. I would also like to thank my friends and family for their constant support.

Editor

Part 1

Fundamentals

1

Reformulation of Electromagnetism with Differential Forms

Masao Kitano
Department of Electronic Science and Engineering,
Kyoto University,
Japan

1. Introduction

Classical electromagnetism is a well-established discipline. However, there remains some confusions and misunderstandings with respect to its basic structures and interpretations. For example, there is a long-lasting controversy on the choice of unit systems. There are also the intricate disputes over the so-called EH or EB formulations. In some textbooks, the authors respect the fields $\boldsymbol{E}$ and $\boldsymbol{B}$ as fundamental quantities and understate $\boldsymbol{D}$ and $\boldsymbol{H}$ as auxiliary quantities. Sometimes the roles of $\boldsymbol{D}$ and $\boldsymbol{H}$ in a vacuum are totally neglected.

These confusions mainly come from the conventional formalism of electromagnetism and also from the use of the old unit systems, in which distinction between $\boldsymbol{E}$ and $\boldsymbol{D}$, or $\boldsymbol{B}$ and $\boldsymbol{H}$ is blurred, especially in vacuum. The standard scalar-vector formalism, mainly due to Heaviside, greatly simplifies the electromagnetic (EM) theory compared with the original formalism developed by Maxwell. There, the field quantities are classified according to the number of components: vectors with three components and scalars with single component. But this classification is rather superficial. From a modern mathematical point of view, the field quantities must be classified according to the tensorial order. The field quantities $\boldsymbol{D}$ and $\boldsymbol{B}$ are the 2nd-order tensors (or 2 forms), while $\boldsymbol{E}$ and $\boldsymbol{H}$ are the 1st-order tensors (1 forms). (The anti-symmetric tensors of order n are called n-forms.)

The constitutive relations are usually considered as simple proportional relations between $\boldsymbol{E}$ and $\boldsymbol{D}$, and between $\boldsymbol{B}$ and $\boldsymbol{H}$. But in terms of differential forms, they associate the conversion of tensorial order, which is known as the Hodge dual operation. In spite of the simple appearance, the constitutive relations, even for the case of vacuum, are the non-trivial part of the EM theory. By introducing relativistic field variables and the vacuum impedance, the constitutive relation can be unified into a single equation.

The EM theory has the symmetry with respect to the space inversion, therefore, each field quantity has a definite parity, even or odd. In the conventional scalar-vector notation, the parity is assigned rather by hand not from the first principle: the odd vectors $\boldsymbol{E}$ and $\boldsymbol{D}$ are named the polar vectors and the even vectors $\boldsymbol{B}$ and $\boldsymbol{H}$ are named the axial vectors. With respect to differential forms, the parity is determined by the tensorial order and the pseudoness (twisted or untwisted). The pseudoness is flipped under the Hodge dual operation. The way of parity assignment in the framework of differential forms is quite natural in geometrical point of view.

It is well understood that the Maxwell equations can be formulated more naturally in the four dimensional spatio-temporal (Minkowski) space. However, the conventional expression with tensor components (with superscripts or subscripts) is somewhat abstract and hard to read out its geometrical or physical meaning. Here it will be shown that the four-dimensional differential forms are the most suitable method for expressing the structure of the EM theory. We introduce two fundamental, relativistic 2-forms, which are related by the four-dimensional Hodge's dual operation and the vacuum impedance.

In this book chapter, we reformulate the EM theory with the differential forms by taking care of physical perspective, the unit systems (physical dimensions), and geometric aspects, and thereby provide a unified and clear view of the solid and beautiful theory.

Here we introduce notation for dimensional consideration. When the ratio of two quantities X and Y is dimensionless (just a pure number), we write $X \overset{\mathrm{SI}}{\sim} Y$ and read "X and Y are dimensionally equivalent (in SI)." For example, we have $c_0 t \overset{\mathrm{SI}}{\sim} x$. If a quantity X can be measured in a unit u, we can write $X \overset{\mathrm{SI}}{\sim}$ u. For example, for $d = 2.5\,\mathrm{m}$ we can write $d \overset{\mathrm{SI}}{\sim} \mathrm{m}$.

2. The vacuum impedance as a fundamental constant

The vacuum impedance was first introduced explicitly in late 1930's (Schelkunoff (1938)) in the study of EM wave propagation. It is defined as the amplitude ratio of the electric and magnetic fields of plane waves in vacuum, $Z_0 = E/H$, which has the dimension of electrical resistance.

It is also called the characteristic impedance of vacuum or the wave resistance of vacuum. Due to the historical reasons, it has been recognized as a special parameter for engineers rather than a universal physical constant. Compared with the famous formula (Maxwell (1865)) representing the velocity of light c_0 in terms of the vacuum permittivity ε_0 and the vacuum permeability μ_0,

$$c_0 = \frac{1}{\sqrt{\mu_0 \varepsilon_0}}, \tag{1}$$

the expression for the vacuum impedance

$$Z_0 = \sqrt{\frac{\mu_0}{\varepsilon_0}}, \tag{2}$$

is used far less often. In fact the term is rarely found in index pages of textbooks on electromagnetism.

As we will see, the pair of constants (c_0, Z_0) can be conveniently used in stead of the pair (ε_0, μ_0) for many cases. However, conventionally the asymmetric pairs (c_0, μ_0) or (c_0, ε_0) are often used and SI equations become less memorable.

In this section, we reexamine the structure of electromagnetism in view of the SI system (The International System of Units) and find that Z_0 plays very important roles as a universal constant.

Recent development of new type of media called metamaterials demands the reconsideration of wave impedance. In metamaterials (Pendry & Smith (2004)), both permittivity ε and permeability μ can be varied from values for vacuum and thereby the phase velocity $v_{\rm ph} = 1/\sqrt{\varepsilon\mu}$ and the wave impedance $Z = \sqrt{\mu/\epsilon}$ can be adjusted independently. With the control of wave impedance the reflection at the interfaces of media can be reduced or suppressed.

2.1 Roles of the vacuum impedance

In this section, we show some examples for which Z_0 plays important roles (Kitano (2009)). The impedance (resistance) is a physical quantity by which voltage and current are related. In the SI system, the unit for voltage is $\mathrm{V}(= \mathrm{J/C})$ (volt) and the unit for current is $\mathrm{A}(= \mathrm{C/s})$ (ampere). We should note that the latter is proportional to and the former is inversely proportional to the unit of charge, C (coulomb). Basic quantities in electromagnetism can be classified into two categories as

$$\begin{aligned} &\phi, \quad A, \quad E, \quad B && \text{Force quantities} \quad \propto \mathrm{V}, \\ &D, \quad H, \quad P, \quad M, \quad \varrho, \quad J && \text{Source quantities} \quad \propto \mathrm{A}. \end{aligned} \tag{3}$$

The quantities in the former categories contain V in their units and are related to electromagnetic forces. On the other hand, the quantities in the latter contain A and are related to electromagnetic sources. The vacuum impedance Z_0 (or the vacuum admittance $Y_0 = 1/Z_0$) plays the role to connect the quantities of the two categories.

2.1.1 Constitutive relation

The constitutive relations for vacuum, $\boldsymbol{D} = \varepsilon_0 \boldsymbol{E}$ and $\boldsymbol{H} = \mu_0^{-1}\boldsymbol{B}$, can be simplified by using the relativistic pairs of variables as

$$\begin{bmatrix} \boldsymbol{E} \\ c_0\boldsymbol{B} \end{bmatrix} = Z_0 \begin{bmatrix} c_0\boldsymbol{D} \\ \boldsymbol{H} \end{bmatrix}. \tag{4}$$

The electric relation and magnetic relation are united under the sole parameter Z_0.

2.1.2 Source-field relation

We know that the scalar potential $\Delta\phi$ induced by a charge $\Delta q = \varrho\Delta v$ is

$$\Delta\phi = \frac{1}{4\pi\varepsilon_0}\frac{\varrho\Delta v}{r}, \tag{5}$$

where r is the distance between the source and the point of observation. The charge is presented as a product of charge density ϱ and a small volume Δv. Similarly a current moment (current times length) $\boldsymbol{J}\Delta v$ generates the vector potential

$$\Delta \boldsymbol{A} = \frac{\mu_0}{4\pi}\frac{\boldsymbol{J}\Delta v}{r}. \tag{6}$$

The relations (5) and (6) are unified as

$$\Delta \begin{bmatrix} \phi \\ c_0\boldsymbol{A} \end{bmatrix} = \frac{Z_0}{4\pi r} \begin{bmatrix} c_0\varrho \\ \boldsymbol{J} \end{bmatrix} \Delta v. \tag{7}$$

We see that the vacuum impedance Z_0 plays the role to relate the source $(\boldsymbol{J}, c_0\varrho)\Delta v$ and the resultant fields $\Delta(\phi, c_0\boldsymbol{A})$ in a unified manner.

2.1.3 Plane waves

We know that for linearly polarized plane waves propagating in one direction in vacuum, a simple relation $E = c_0 B$ holds. If we introduce $H\ (= \mu_0^{-1} B)$ instead of B, we have $E = Z_0 H$. This relation was introduced by Schelkunoff (Schelkunoff (1938)). The reason why H is used instead of B is as follows. A dispersive medium is characterized by its permittivity ε and and permeability μ. The monochromatic plane wave solution satisfies $E = vB$, $H = vD$, and $E/H = B/D = Z$, where $v = 1/\sqrt{\varepsilon\mu}$ and $Z = \sqrt{\mu/\varepsilon}$. The boundary conditions for magnetic fields at the interface of media 1 and 2 are $H_{1\mathrm{t}} = H_{2\mathrm{t}}$ (tangential) and $B_{1\mathrm{n}} = B_{2\mathrm{n}}$ (normal). For the case of normal incidence, which is most important practically, the latter condition becomes trivial and cannot be used. Therefore the pair of E and H is used more conveniently. The energy flow is easily derived from E and H with the Poynting vector $\boldsymbol{S} = \boldsymbol{E} \times \boldsymbol{H}$. In the problems of EM waves, the mixed use of the quantities (E and H) of the force and source quantities invites Z_0.

2.1.4 Magnetic monopole

Let us compare the force between electric charges q ($\overset{\mathrm{SI}}{\sim}$ A s = C) and that between magnetic monopoles g ($\overset{\mathrm{SI}}{\sim}$ V s = Wb). If these forces are the same for equal distance, r, i.e., $q^2/(4\pi\varepsilon_0 r^2) = g^2/(4\pi\mu_0 r^2)$, we have the relation $g = Z_0 q$. This means that a charge of 1 C corresponds to a magnetic charge of $Z_0 \times 1\,\mathrm{C} \sim 377\,\mathrm{Wb}$.

With this relation in mind, the Dirac monopole g_0 (Sakurai (1993)), whose quantization condition is $g_0 e = h$, can be beautifully expressed in terms of the elementary charge e as

$$g_0 = \frac{h}{e} = \frac{h}{Z_0 e^2}(Z_0 e) = \frac{Z_0 e}{2\alpha}, \tag{8}$$

where $h = 2\pi\hbar$ is Planck's constant. The dimensionless parameter $\alpha = Z_0 e^2/2h = e^2/4\pi\varepsilon_0\hbar c_0 \sim 1/137$ is called the fine-structure constant, whose value is independent of unit systems and characterizes the strength of electromagnetic interaction.

2.1.5 The fine-structure constant

We have seen that the fine-structure constant itself can be represented more simply with the use of Z_0. Further, by introducing the von Klitzing constant (the quantized Hall resistance) (Klitzing et al. (1980)) $R_\mathrm{K} = h/e^2$, the fine-structure constant can be expressed as $\alpha = Z_0/2R_\mathrm{K}$ (Hehl & Obukhov (2005)). We have learned that the use of Z_0 helps to keep SI-formulae in simple forms.

3. Dual space and differential forms

3.1 Covector and dual space

We represent a (tangential) vector at position $\boldsymbol{r}$ as

$$\boldsymbol{x} = x_1\boldsymbol{e}_1 + x_2\boldsymbol{e}_2 + x_3\boldsymbol{e}_3 \overset{\mathrm{SI}}{\sim} \mathrm{m}, \tag{9}$$

which represents a small spatial displacement from $\boldsymbol{r}$ to $\boldsymbol{r}+\boldsymbol{x}$. We have chosen an arbitrary orthonormal basis $\{\boldsymbol{e}_1, \boldsymbol{e}_2, \boldsymbol{e}_3\}$ with inner products $(\boldsymbol{e}_i, \boldsymbol{e}_j) = \delta_{ij}$, where

$$\delta_{ij} = \begin{cases} 1 & (i = j) \\ 0 & (i \neq j) \end{cases} \tag{10}$$

is Kronecker's delta. We note that $x_i \overset{\text{SI}}{\sim} \text{m}$ and $\boldsymbol{e}_i \overset{\text{SI}}{\sim} 1$.

Such vectors form a linear space which is called a tangential space at $\boldsymbol{r}$. The inner product of vectors $\boldsymbol{x}$ and $\boldsymbol{y}$ is $(\boldsymbol{x}, \boldsymbol{y}) = x_1y_1 + x_2y_2 + x_3y_3 \overset{\text{SI}}{\sim} \text{m}^2$.

We consider a linear function $\phi(\boldsymbol{x})$ on the tangential space. For any $c_1, c_2 \in \mathbb{R}$, and any vectors $\boldsymbol{x}_1$ and $\boldsymbol{x}_2$, $\phi(c_1\boldsymbol{x}_1 + c_2\boldsymbol{x}_2) = c_1\phi(\boldsymbol{x}_1) + c_1\phi(\boldsymbol{x}_2)$ is satisfied. Such linear functions form a linear space, because the (weighted) sum of two functions $d_1\phi_1 + d_2\phi_2$ with $d_1, d_2 \in \mathbb{R}$ defined with

$$(d_1\phi_1 + d_2\phi_2)(\boldsymbol{x}) = d_1\phi_1(\boldsymbol{x}) + d_2\phi_2(\boldsymbol{x}) \tag{11}$$

is also a linear function. This linear space is called a dual space. The dimension of the dual space is three. In general, the dimension of dual space is the same that for the original linear space. We can introduce a basis $\{\nu_1, \nu_2, \nu_3\}$, satisfying $\nu_i(\boldsymbol{e}_j) = \delta_{ij}$. Such a basis, which is dependent on the choice of the original basis, is called a dual basis. Using the dual basis, the action of a dual vector $\phi(\,) = a_1\nu_1(\,) + a_2\nu_2(\,) + a_3\nu_3(\,)$, $a_1, a_2, a_3 \in \mathbb{R}$ can be written simply as

$$\phi(\boldsymbol{x}) = (a_1\nu_1 + a_2\nu_2 + a_3\nu_3)(x_1\boldsymbol{e}_1 + x_2\boldsymbol{e}_2 + x_3\boldsymbol{e}_3) \tag{12}$$

$$= \sum_{i=1}^{3}\sum_{j=1}^{3} a_i x_j \nu_i(\boldsymbol{e}_j) = x_1a_1 + x_2a_2 + x_3a_3. \tag{13}$$

Here we designate an element of dual space with vector notation as $\boldsymbol{a}$ rather as a function $\phi(\,)$ in order to emphasize its vectorial nature, i.e.,

$$\boldsymbol{a} \cdot \boldsymbol{x} = \phi(\boldsymbol{x}). \tag{14}$$

We call $\boldsymbol{a}$ as a dual vector or a *covector*. The dual basis $\{\nu_1, \nu_2, \nu_3\}$ are rewritten as $\{\boldsymbol{n}_1, \boldsymbol{n}_2, \boldsymbol{n}_3\}$ with $\boldsymbol{n}_i \cdot \boldsymbol{e}_j = \delta_{ij}$. The dot product $\boldsymbol{a} \cdot \boldsymbol{x}$ and the inner product $(\boldsymbol{x}, \boldsymbol{y})$ should be distinguished. Here bold-face letters $\boldsymbol{x}$, $\boldsymbol{y}$, $\boldsymbol{z}$, and $\boldsymbol{e}$ represent tangential vectors and other bold-face letters represent covectors.

A covector $\boldsymbol{a}$ can be related to a vector $\boldsymbol{z}$ uniquely using the relation, $\boldsymbol{a} \cdot \boldsymbol{x} = (\boldsymbol{z}, \boldsymbol{x})$ for any $\boldsymbol{x}$. The vector $\boldsymbol{z}$ and the covector $\boldsymbol{a}$ are called conjugate each other and we write $\boldsymbol{z} = \boldsymbol{a}^\top$ and $\boldsymbol{a} = \boldsymbol{z}^\top$. In terms of components, namely for $\boldsymbol{a} = \sum_i a_i\boldsymbol{n}_i$ and $\boldsymbol{z} = \sum_i z_i\boldsymbol{e}_i$, $a_i = z_i$ $(i = 1, 2, 3)$ are satisfied

For the case of orthonormal basis, we note that $\boldsymbol{n}_i^\top = \boldsymbol{e}_i$, $\boldsymbol{e}_i^\top = \boldsymbol{n}_i$. Due to these incidental relations, we tend to identify $\boldsymbol{n}_i$ with $\boldsymbol{e}_i$. Thus a covector $\boldsymbol{a}$ is identified with its conjugate $\boldsymbol{a}^\top$ mostly. However, we should distinguish a covector as a different object from vectors since it bears different functions and geometrical presentation (Weinreich (1998)).

The inner product for covectors are defined with conjugates as $(\boldsymbol{a}, \boldsymbol{b}) = (\boldsymbol{a}^\top, \boldsymbol{b}^\top)$. We note the dual basis is also orthonormal, since $(\boldsymbol{n}_i, \boldsymbol{n}_j) = (\boldsymbol{n}_i^\top, \boldsymbol{n}_j^\top) = (\boldsymbol{e}_i, \boldsymbol{e}_j) = \delta_{ij}$.

A good example of covector is the electric field at a point $\boldsymbol{r}$. The electric field is determined through the gained work W when an electric test charge q at $\boldsymbol{r}$ is displaced by $\boldsymbol{x}$. A function $\phi : \boldsymbol{x} \mapsto W/q$ is linear with respect to $\boldsymbol{x}$ if $|\boldsymbol{x}|$ is small enough. Therefore $\phi(\,)$ is considered as a covector and normally written as $\boldsymbol{E}$, i.e., $\phi(\boldsymbol{x}) = \boldsymbol{E} \cdot \boldsymbol{x} \overset{\text{SI}}{\sim} \text{V}$. Thus the electric field vector can be understood as a covector rather than a vector. It should be expanded with the dual basis as

$$\boldsymbol{E} = E_1 \boldsymbol{n}_1 + E_2 \boldsymbol{n}_2 + E_3 \boldsymbol{n}_3 \overset{\text{SI}}{\sim} \text{V/m}. \tag{15}$$

The norm is given as $||\boldsymbol{E}|| = (\boldsymbol{E}, \boldsymbol{E})^{1/2} = \sqrt{E_1^2 + E_2^2 + E_3^2}$. We note that $\boldsymbol{n}_i \overset{\text{SI}}{\sim} 1$ and $E_i \overset{\text{SI}}{\sim} \text{V/m}$.

3.2 Higher order tensors

Now we introduce a tensor product of two covectors $\boldsymbol{a}$ and $\boldsymbol{b}$ as $T = \boldsymbol{ab}$, which acts on two vectors and yield a scalar as

$$T : \boldsymbol{xy} = (\boldsymbol{ab}) : \boldsymbol{xy} = (\boldsymbol{a} \cdot \boldsymbol{x})(\boldsymbol{b} \cdot \boldsymbol{y}). \tag{16}$$

It can be considered as a bi-linear functions of vectors, i.e., $T : \boldsymbol{xy} = \Phi(\boldsymbol{x}, \boldsymbol{y})$ with

$$\begin{aligned} \Phi(c_1 \boldsymbol{x}_1 + c_2 \boldsymbol{x}_2, \boldsymbol{y}) &= c_1 \Phi(\boldsymbol{x}_1, \boldsymbol{y}) + c_1 \Phi(\boldsymbol{x}_2, \boldsymbol{y}), \\ \Phi(\boldsymbol{x}, c_1 \boldsymbol{y} + c_2 \boldsymbol{y}_2) &= c_1 \Phi(\boldsymbol{x}, \boldsymbol{y}_1) + c_1 \Phi(\boldsymbol{x}, \boldsymbol{y}_2), \end{aligned} \tag{17}$$

where $c_1, c_2 \in \mathbb{R}$. We call it a bi-covector.

We can define a weighted sum of bi-covectors $T = d_1 T_1 + d_2 T_2$, $d_1, d_2 \in \mathbb{R}$, which is not necessarily written as a tensor product of two covectors but can be written as a sum of tensor products. Especially, it can be represented with the dual basis as

$$T = \sum_{i=1}^{3} \sum_{j=1}^{3} T_{ij} \boldsymbol{n}_i \boldsymbol{n}_j, \tag{18}$$

where $T_{ij} = T : \boldsymbol{e}_i \boldsymbol{e}_j$ is the (i, j)-component of T.

Similarly we can construct a tensor product of three covectors as $\mathcal{T} = \boldsymbol{abc}$, which acts on three vectors linearly as $\mathcal{T} : \boldsymbol{xyz}$. Weighted sums of such products form a linear space, an element of which is called a tri-covector. Using a tensor product of n covectors, a multi-covector or an n-covector is defined.

3.3 Anti-symmetric multi-covectors — n-forms

If a bicovector T satisfies $T : \boldsymbol{yx} = -T : \boldsymbol{xy}$ for any vectors $\boldsymbol{x}$ and $\boldsymbol{y}$, then it is called antisymmetric. Anti-symmetric bicovectors form a subspace of the bicovector space. Namely, a weighted sum of anti-symmetric bicovector is anti-symmetric. It contains an anti-symmetrized tensor product, $\boldsymbol{a} \wedge \boldsymbol{b} := \boldsymbol{ab} - \boldsymbol{ba}$, which is called a *wedge* product. In terms of basis, we have

$$\boldsymbol{a} \wedge \boldsymbol{b} = \sum_{i=1}^{3} a_i \boldsymbol{n}_i \wedge \sum_{j=1}^{3} b_j \boldsymbol{n}_j = \sum_{(i,j)} (a_i b_j - a_j b_i) \boldsymbol{n}_i \wedge \boldsymbol{n}_j, \tag{19}$$

where the last sum is taken for $(i,j) = (1,2),(2,3),(3,1)$. A general anti-symmetric bicovector can be written as

$$\mathsf{T} = \sum_{(i,j)} T_{ij}\boldsymbol{n}_i \wedge \boldsymbol{n}_j. \tag{20}$$

We see that the 2-form has three independent components; $T_{12} = -T_{21}$, $T_{23} = -T_{32}$, $T_{31} = -T_{13}$, and others are zero. The norm of T is $||\mathsf{T}|| = (\mathsf{T},\mathsf{T})^{1/2} = \sum_{(i,j)} T_{ij}T_{ij}$.

If a bicovector T satisfies $\mathsf{T} : \boldsymbol{xx} = 0$ for any $\boldsymbol{x}$, then it is anti-symmetric. It is easily seen from the relation: $0 = \mathsf{T} : (\boldsymbol{x}+\boldsymbol{y})(\boldsymbol{x}+\boldsymbol{y}) = \mathsf{T} : \boldsymbol{xx} + \mathsf{T} : \boldsymbol{xy} + \mathsf{T} : \boldsymbol{yx} + \mathsf{T} : \boldsymbol{yy}$.

An anti-symmetric multi-covector of order n are often called an n-form. A scalar and a covector are called a 0-form and a 1-form, respectively. The order n is bounded by the dimension of the vector space, $d = 3$, in our case. An n-form with $n > d$ vanishes due to the anti-symmetries.

Geometrical interpretations of n-forms are given in the articles (Misner et al. (1973); Weinreich (1998)).

3.4 Field quantities as n-forms

Field quantities in electromagnetism can be naturally represented as differential forms (Burke (1985); Deschamps (1981); Flanders (1989); Frankel (2004); Hehl & Obukhov (2003)). A good example of 2-form is the current density. Let us consider a distribution of current that flows through a parallelogram spanned by two tangential vectors $\boldsymbol{x}$ and $\boldsymbol{y}$ at $\boldsymbol{r}$. The current $I(\boldsymbol{x},\boldsymbol{y})$ is bilinearly dependent on $\boldsymbol{x}$ and $\boldsymbol{y}$. The antisymmetric relation $I(\boldsymbol{y},\boldsymbol{x}) = -I(\boldsymbol{x},\boldsymbol{y})$ can understood naturally considering the orientation of parallelograms with respect to the current flow. Thus the current density can be represented by a 2-form J as

$$\mathsf{J} : \boldsymbol{xy} = I(\boldsymbol{x},\boldsymbol{y}) \overset{\mathrm{SI}}{\sim} \mathrm{A}, \quad \mathsf{J} = \sum_{(i,j)} J_{ij}\boldsymbol{n}_i \wedge \boldsymbol{n}_j \overset{\mathrm{SI}}{\sim} \mathrm{A/m^2}. \tag{21}$$

The charge density can be represented by a 3-form $\mathcal{R}$. The charge Q contained in a parallelepipedon spanned by three tangential vectors $\boldsymbol{x}$, $\boldsymbol{y}$, and $\boldsymbol{z}$:

$$\mathcal{R} : \boldsymbol{xyz} = Q(\boldsymbol{x},\boldsymbol{y},\boldsymbol{z}) \overset{\mathrm{SI}}{\sim} \mathrm{C}, \quad \mathcal{R} = R_{123}\boldsymbol{n}_1 \wedge \boldsymbol{n}_2 \wedge \boldsymbol{n}_3 \overset{\mathrm{SI}}{\sim} \mathrm{C/m^3}. \tag{22}$$

Thus electromagnetic field quantities are represented as n-forms ($n = 0,1,2,3$) as shown in Table 1, while in the conventional formalism they are classified into two categories, scalars and vectors, according to the number of components. We notice that a quantity that is represented n-form contains physical dimension with m^{-n} in SI. An n-form takes n tangential vectors, each of which has dimension of length and is measured in m (meters).

In this article, 1-forms are represented by bold-face letters, 2-forms sans-serif letters, and 3-forms calligraphic letters as shown in Table 1.

3.5 Exterior derivative

The nabla operator $\boldsymbol{\nabla}$ can be considered as a kind of covector because a directional derivative $\boldsymbol{\nabla} \cdot \boldsymbol{u}$, which is a scalar, is derived with a tangential vector $\boldsymbol{u}$. It acts as a differential operator

rank	quantities (unit)	scalar/vector
0-form	ϕ (V)	scalar
1-form	$\boldsymbol{A}$ (Wb/m), $\boldsymbol{E}$ (V/m), $\boldsymbol{H}$ (A/m), $\boldsymbol{M}$ (A/m)	vector
2-form	B (Wb/m^2), D (C/m^2), P (C/m^2), J (A/m^2)	vector
3-form	$\mathcal{R}$ (C/m^3)	scalar

Table 1. Electromagnetic field quantities as n-forms

and also as a covector. Therefore it can be written as

$$\boldsymbol{\nabla} = \boldsymbol{n}_1 \frac{\partial}{\partial x_1} + \boldsymbol{n}_2 \frac{\partial}{\partial x_2} + \boldsymbol{n}_3 \frac{\partial}{\partial x_3} \overset{\text{SI}}{\sim} 1/\text{m}. \tag{23}$$

The wedge product of the nabla $\boldsymbol{\nabla}$ and a 1-form $\boldsymbol{E}$ yields a 2-form;

$$\boldsymbol{\nabla} \wedge \boldsymbol{E} = \sum_{(i,j)} \left(\frac{\partial E_j}{\partial x_i} - \frac{\partial E_i}{\partial x_j} \right) \boldsymbol{n}_i \wedge \boldsymbol{n}_j, \tag{24}$$

which corresponds to $\boldsymbol{\nabla} \times \boldsymbol{E}$ in the scalar-vector formalism. Similarly a 2-form J are transformed into a 3-form as

$$\boldsymbol{\nabla} \wedge J = \left(\frac{\partial J_{23}}{\partial x_1} + \frac{\partial J_{31}}{\partial x_2} + \frac{\partial J_{12}}{\partial x_3} \right) \boldsymbol{n}_1 \wedge \boldsymbol{n}_2 \wedge \boldsymbol{n}_3, \tag{25}$$

which corresponds to $\boldsymbol{\nabla} \cdot \boldsymbol{J}$.

3.6 Volume form and Hodge duality

We introduce a 3-form, called the volume form, as

$$\mathcal{E} = \boldsymbol{n}_1 \wedge \boldsymbol{n}_2 \wedge \boldsymbol{n}_3 = \sum_{i=1}^{3} \sum_{j=1}^{3} \sum_{k=1}^{3} \epsilon_{ijk} \boldsymbol{n}_i \boldsymbol{n}_j \boldsymbol{n}_k \overset{\text{SI}}{\sim} 1, \tag{26}$$

where

$$\epsilon_{ijk} = \begin{cases} 1 & (i,j,k : \text{cyclic}) \\ -1 & (\text{anti-cyclic}) \\ 0 & (\text{others}) \end{cases} . \tag{27}$$

It gives the volume of parallelepipedon spanned by $\boldsymbol{x}$, $\boldsymbol{y}$, and $\boldsymbol{z}$;

$$V(\boldsymbol{x}, \boldsymbol{y}, \boldsymbol{z}) = \mathcal{E} \vdots \boldsymbol{xyz} \quad \overset{\text{SI}}{\sim} \text{m}^3. \tag{28}$$

Using the volume form we can define a relation between n-forms and $(d-n)$-forms, which is call the Hodge dual relation. First we note that n-forms and $(d-n)$-forms have the same degrees of freedom (the number of independent components), ${}_dC_n = {}_dC_{d-n}$, and there could be a one-to-one correspondence between them. In our case of $d = 3$, there are two cases: $(n, d-n) = (0,3)$ and $(1,2)$. We consider the latter case. With a 1-form $\boldsymbol{E}$ and a 2-form D, we can make a 3-form $\boldsymbol{E} \wedge D = f(\boldsymbol{E}, D)\mathcal{E}$. The scalar factor $f(\boldsymbol{E}, D)$ is bilinearly dependent on $\boldsymbol{E}$ and D. Therefore, we can find a covector (a 1-form) $\boldsymbol{D}$ that satisfies $(\boldsymbol{E}, \boldsymbol{D}) = f(\boldsymbol{E}, D)$ for any $\boldsymbol{E}$. Then, $\boldsymbol{D}$ is called the Hodge dual of D and we write $\boldsymbol{D} = *D$ or $D = *\boldsymbol{D}$ using a unary

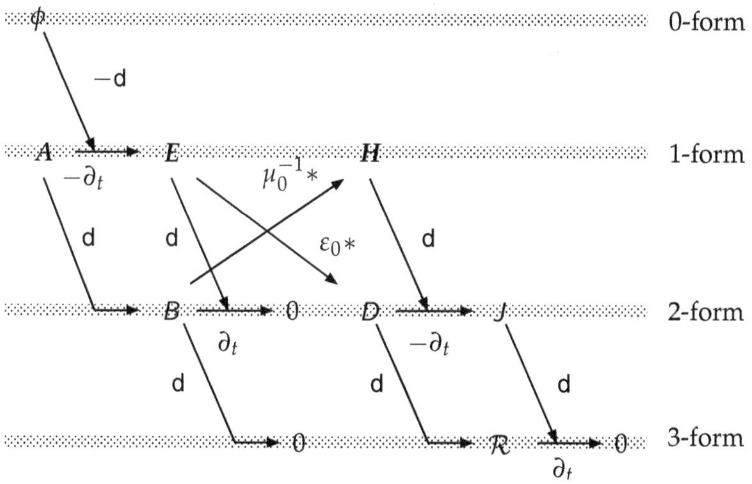

Fig. 1. Relations of electromagnetic field forms in three dimension

operator "$*$" called the Hodge star operator. In terms of components, $D_1 = D_{23}$, $D_2 = D_{31}$, $D_3 = D_{12}$ for $\boldsymbol{D} = *D$ with $\boldsymbol{D} = \sum_i D_i \boldsymbol{n}_i$ and $D = \sum_{(i,j)} D_{ij} \boldsymbol{n}_i \wedge \boldsymbol{n}_j$.

Physically, $(\boldsymbol{E}, \boldsymbol{D}) \overset{\text{SI}}{\sim} \text{J/m}^3$ corresponds to the energy density and can be represented by the 3-form $\mathcal{U} = \frac{1}{2}\boldsymbol{E} \wedge D = \frac{1}{2}(\boldsymbol{E}, \boldsymbol{D})\mathcal{E}$, because $\mathcal{U} \vdots \boldsymbol{xyz}$ is the energy contained in the parallelepipedon spanned by $\boldsymbol{x}$, $\boldsymbol{y}$, and $\boldsymbol{z}$.

The charge density form can be written as $\mathcal{R} = \varrho \boldsymbol{n}_1 \wedge \boldsymbol{n}_2 \wedge \boldsymbol{n}_3 = \varrho\mathcal{E}$ with the conventional scalar charge density ϱ. The relation can be expressed as $\mathcal{R} = *\varrho$ or $\varrho = *\mathcal{R}$. Similarly, we have $\mathcal{E} = *1$ and $1 = *\mathcal{E}$.

Equations (24) and (25) are related to the conventional notations; $*(\boldsymbol{\nabla} \wedge \boldsymbol{E}) = \boldsymbol{\nabla} \times \boldsymbol{E}$ and $*(\boldsymbol{\nabla} \wedge J) = \boldsymbol{\nabla} \cdot \boldsymbol{J}$, respectively.

3.7 The Hodge duality and the constitutive equation

In electromagnetism, the Hodge duality and the constitutive relations are closely related. We know that the electric field $\boldsymbol{E}$ and the electric flux density D are proportional. However we cannot compare them directly because they have different tensorial orders. Therefore we utilize the Hodge dual and write $D = \varepsilon_0(*\boldsymbol{E})$. Similarly, the magnetic relation can be written as $\boldsymbol{H} = \mu_0^{-1}(*B)$. Generally speaking, the constitutive relations in vacuum are considered to be trivial relations just describing proportionality. Especially in the Gaussian unit system, they tend to be considered redundant relations. But in the light of differential forms we understand that they are the keystones in electromagnetism.

With the polarization P and the magnetization $\boldsymbol{M}$, the constitutive relations in a medium are expressed as follows:

$$D = \varepsilon_0(*\boldsymbol{E}) + P, \quad \boldsymbol{H} = \mu_0^{-1}(*B) - \boldsymbol{M}. \tag{29}$$

4. The Maxwell equations in the differential forms

With differential forms, we can rewrite the Maxwell equations and the constitutive relations as,

$$\begin{aligned} &\nabla \wedge B = 0, \quad \nabla \wedge \boldsymbol{E} + \frac{\partial B}{\partial t} = 0, \\ &\nabla \wedge D = \mathcal{R}, \quad \nabla \wedge \boldsymbol{H} - \frac{\partial D}{\partial t} = J, \\ &D = \varepsilon_0 \mathcal{E} \cdot \boldsymbol{E} + P, \quad \boldsymbol{H} = \tfrac{1}{2}\mu_0^{-1}\mathcal{E} : B - \boldsymbol{M}. \end{aligned} \tag{30}$$

In the formalism of differential forms, the spatial derivative $\nabla \wedge \llcorner\!\lrcorner$ is simply denoted as $\mathrm{d}\llcorner\!\lrcorner$. Together with the Hodge operator "$*$", Eq. (30) is written in simpler forms;

$$\begin{aligned} &\mathrm{d}B = 0, \quad \mathrm{d}\boldsymbol{E} + \partial_t B = 0, \\ &\mathrm{d}D = \mathcal{R}, \quad \mathrm{d}\boldsymbol{H} - \partial_t D = J, \\ &D = \varepsilon_0(*\boldsymbol{E}) + P, \quad \boldsymbol{H} = \mu_0^{-1}(*B) - \boldsymbol{M}, \end{aligned} \tag{31}$$

where $\partial_t = \partial/\partial t$.

In Fig. 1, we show a diagram corresponding Eq. (31) and related equations (Deschamps (1981)). The field quantities are arranged according to their tensor order. The exterior derivative "d" connects a pair of quantities by increasing the tensor order by one, while time derivative ∂_t conserves the tensor order. $\boldsymbol{E}$ (B) is related to D ($\boldsymbol{H}$) with the Hodge star operator and the constant ε_0 (μ_0). The definitions of potentials and the charge conservation law

$$\boldsymbol{E} = -\mathrm{d}\phi - \partial_t \boldsymbol{A}, \quad B = \mathrm{d}\boldsymbol{A}, \quad \mathrm{d}J + \partial_t \mathcal{R} = 0 \tag{32}$$

are also shown in Fig. 1. We can see a well-organized, perfect structure. We will see the relativistic version later (Fig. 2).

5. Twisted forms and parity

5.1 Twist of volume form

We consider two bases $\Sigma = \{e_1, e_2, e_3\}$ and $\Sigma' = \{e'_1, e'_2, e'_3\}$. They can be related as $e'_i = \sum_j R_{ij} e_j$ by a matrix $R = [R_{ij}]$ with $R_{ij} = (e'_i, e_j)$. It is orthonormal and therefore $\det R = \pm 1$. In the case of $\det R = 1$, the two bases have the same orientation and they can be overlapped by a continuous transformation. On the other hand, for the case of $\det R = -1$, they have opposite orientation and an operation of reversal, for example, a diagonal matrix $\mathrm{diag}(-1, 1, 1)$ is needed to make them overlapped with rotations. Thus we can classify all the bases according to the orientation. We denote the two classes by C and C′, each of which contains all the bases with the same orientation. The two classes are symmetric and there are no *a priori* order of precedence, like for i and −i.

We consider a basis $\Sigma = \{e_1, e_2, e_3\} \in \mathrm{C}$ and the reversed basis $\Sigma' = \{e'_1, e'_2, e'_3\} = \{-e_1, -e_2, -e_3\}$, which belongs to C′. The volume form $\mathcal{E}$ in Σ is defined so as to satisfy $\mathcal{E} \,\vdots\, e_1 e_2 e_3 = +1$, i.e., the volume of the cube defined by e_1, e_2, and e_3 should be $+1$. Similarly, the volume form $\mathcal{E}'$ in Σ' is defined so as to satisfy $\mathcal{E}' \,\vdots\, e'_1 e'_2 e'_3 = +1$. We note that $\mathcal{E}' \,\vdots\, e_1 e_2 e_3 = -\mathcal{E}' \,\vdots\, e'_1 e'_2 e'_3 = -1$, namely, $\mathcal{E}' = -\mathcal{E}$. Thus we have two kinds of volume forms $\mathcal{E}$ and $\mathcal{E}' (= -\mathcal{E})$ and use either of them depending on the orientation of basis.

Assume that Alice adopts the basis $\Sigma \in C$ and Bob adopts $\Sigma' \in C'$. When we pose a parallelepipedon by specifying an ordered triple of vectors $(\boldsymbol{x}, \boldsymbol{y}, \boldsymbol{z})$ and ask each of them to measure its volume, their answers will always be different in the sign. It seems inconvenient but there is no principle to choose one over the other. It is only a customary practice to use the right-handed basis to avoid the confusion. Fleming's left-hand rule (or right-hand rule) seems to break the symmetry but it implicitly relies upon the use of the right-handed basis.

5.2 Twisted forms

Tensors (or forms) are independent of the choice of basis. For example, a second order tensor can be expressed in Σ and Σ' as

$$\mathsf{B} = \sum_i \sum_j B'_{ij} \boldsymbol{n}'_i \boldsymbol{n}'_j = \sum_k \sum_l B_{kl} \boldsymbol{n}_k \boldsymbol{n}_l, \tag{33}$$

with the change of components $B_{kl} = \sum_i \sum_j R_{ik} R_{jl} B'_{ij}$. We note the dual basis has been flipped as $\boldsymbol{n}'_i = -\boldsymbol{n}_i$.

Similarly, in the case of 3-forms, we have

$$\mathcal{T} = T'_{123} \boldsymbol{n}'_1 \wedge \boldsymbol{n}'_2 \wedge \boldsymbol{n}'_3 = T_{123} \boldsymbol{n}_1 \wedge \boldsymbol{n}_2 \wedge \boldsymbol{n}_3 \tag{34}$$

with $T_{123} = T'_{123}$. However, for the volume form the components must be changed as

$$\epsilon_{123} = (\det R) \epsilon'_{123}, \tag{35}$$

to have $\mathcal{E}' = -\mathcal{E}$ in the case of reverse of orientation. Therefore, the volume form is called a *pseudo* form in order to distinguish from an ordinary form. The pseudo (normal) form are also call a *twisted* (untwisted) form.

In electromagnetism, some quantities are defined in reference to the volume form or to the Hodge star operator. Therefore, they could be twisted or untwisted. First of all, ϕ, $\boldsymbol{A}$, $\boldsymbol{E}$, and B are not involved with the volume form, they are all untwisted forms. On the other hand, $\boldsymbol{H} = \mu_0^{-1}(*\mathsf{B})$ and $\mathsf{D} = \varepsilon_0(*\boldsymbol{E})$ are twisted forms. The Hodge operator transforms an untwisted (twisted) form to a twisted (untwisted) form.

The quantities $\boldsymbol{M}$, P, J, and $\mathcal{R}$ (charge density), which represent volume densities of electromagnetic sources, are also twisted as shown below. We have found that the force fields are untwisted while the source fields are twisted in general.

5.3 Source densities

Here we look into detail why the quantities representing source densities are represented by twisted forms. As examples, we deal with polarization and charge density. Other quantities can be treated in the same manner.

5.3.1 Polarization

We consider two tangential vectors $\mathrm{d}\boldsymbol{x}$, $\mathrm{d}\boldsymbol{y}$ at a point P. Together with the volume form $\mathcal{E}$, we can define

$$\mathrm{d}\mathsf{S} = \mathcal{E} : \mathrm{d}\boldsymbol{x}\mathrm{d}\boldsymbol{y} \quad \overset{\mathrm{SI}}{\sim} \mathrm{m}^2, \tag{36}$$

which is a pseudo 1-form. (Conventionally, it is written as $\mathrm{d}\boldsymbol{S} = \boldsymbol{x} \times \boldsymbol{y}$.) In fact, for a tangential vector $\boldsymbol{\zeta}$ at P, the volume $\mathrm{d}\boldsymbol{S} \cdot \boldsymbol{\zeta} = \mathcal{E} \,\vdots\, \mathrm{d}\boldsymbol{x}\mathrm{d}\boldsymbol{y}\,\boldsymbol{\zeta}$, spanned by the three vectors is a linear function of $\boldsymbol{\zeta}$. We choose $\mathrm{d}\boldsymbol{z}$ that is perpendicular to the plane spanned by $\mathrm{d}\boldsymbol{x}$ and $\mathrm{d}\boldsymbol{y}$, i.e., $(\mathrm{d}\boldsymbol{z}, \mathrm{d}\boldsymbol{x}) = (\mathrm{d}\boldsymbol{z}, \mathrm{d}\boldsymbol{y}) = 0$. We assume $|\mathrm{d}\boldsymbol{z}| \ll |\mathrm{d}\boldsymbol{x}|$ and $|\mathrm{d}\boldsymbol{z}| \ll |\mathrm{d}\boldsymbol{y}|$. $\mathrm{d}V = \mathrm{d}\boldsymbol{S} \cdot \mathrm{d}\boldsymbol{z}$ is the volume of thin parallelogram plate.

When a charge $+q$ is displaced by $\boldsymbol{l}$ from the other charge $-q$, they form an electric dipole $\boldsymbol{p} = q\boldsymbol{l}$. We consider an electric dipole moment at a point P in $\mathrm{d}V$. The displacement $\boldsymbol{l}$ can be considered as a tangential vector at P, to which $\mathrm{d}\boldsymbol{S}$ acts as $\mathrm{d}\boldsymbol{S} \cdot \boldsymbol{l} = q^{-1}\mathrm{d}\boldsymbol{S} \cdot \boldsymbol{p}$. Then

$$q' = \frac{\mathrm{d}\boldsymbol{S} \cdot \boldsymbol{p}}{\mathrm{d}\boldsymbol{S} \cdot \mathrm{d}\boldsymbol{z}} = q\frac{\mathrm{d}\boldsymbol{S} \cdot \boldsymbol{l}}{\mathrm{d}V} \tag{37}$$

is the surface charge that is contributed by $\boldsymbol{p}$. In the case of $\mathrm{d}\boldsymbol{S} \cdot \boldsymbol{p} = 0$, there are no surface charge associated with $\boldsymbol{p}$. If $\mathrm{d}\boldsymbol{z}$ and $\boldsymbol{p}$ are parallel, $q'\mathrm{d}\boldsymbol{z} = q\boldsymbol{l} = \boldsymbol{p}$ holds.

Next we consider the case where many electric dipoles $\boldsymbol{p}_i = q_i\boldsymbol{l}_i$ are spatially distributed. The total surface charge is given as

$$\begin{aligned} Q' &= \sum_{i\in\mathrm{d}V} q'_i = \sum_{i\in\mathrm{d}V} \frac{\mathrm{d}\boldsymbol{S} \cdot \boldsymbol{p}_i}{\mathrm{d}V} \\ &= (\mathrm{d}V)^{-1} \sum_{i\in\mathrm{d}V} \mathcal{E} \cdot \boldsymbol{p}_i : \mathrm{d}\boldsymbol{x}\mathrm{d}\boldsymbol{y} = \mathsf{P} : \mathrm{d}\boldsymbol{x}\mathrm{d}\boldsymbol{y}, \end{aligned} \tag{38}$$

where the sum is taken over the dipoles contained in $\mathrm{d}V$. The pseudo 2-form

$$\mathsf{P} := (\mathrm{d}V)^{-1} \sum_{i\in\mathrm{d}V} \mathcal{E} \cdot \boldsymbol{p}_i \quad \overset{\mathrm{SI}}{\sim} \mathrm{C/m^2} \tag{39}$$

corresponds to the polarization (the volume density of electric dipole moments).

5.3.2 Charge density

The volume $\mathrm{d}V$ spanned by tangential vectors $\mathrm{d}\boldsymbol{x}$, $\mathrm{d}\boldsymbol{y}$, $\mathrm{d}\boldsymbol{z}$ at P is

$$\mathrm{d}V = \mathcal{E} \,\vdots\, \mathrm{d}\boldsymbol{x}\mathrm{d}\boldsymbol{y}\mathrm{d}\boldsymbol{z} \quad \overset{\mathrm{SI}}{\sim} \mathrm{m}^3. \tag{40}$$

For distributed charges q_i, the total charge in $\mathrm{d}V$ is given as

$$\begin{aligned} Q &= \sum_{i\in\mathrm{d}V} q_i = \sum_{i\in\mathrm{d}V} \frac{q_i\mathrm{d}V}{\mathrm{d}V} \\ &= (\mathrm{d}V)^{-1} \sum_{i\in\mathrm{d}V} q_i\mathcal{E} \,\vdots\, \mathrm{d}\boldsymbol{x}\mathrm{d}\boldsymbol{y}\mathrm{d}\boldsymbol{z} = \mathcal{R} \,\vdots\, \mathrm{d}\boldsymbol{x}\mathrm{d}\boldsymbol{y}\mathrm{d}\boldsymbol{z}. \end{aligned} \tag{41}$$

The pseudo 3-form

$$\mathcal{R} := (\mathrm{d}V)^{-1} \sum_{i\in\mathrm{d}V} q_i\mathcal{E} \quad \overset{\mathrm{SI}}{\sim} \mathrm{C/m^3} \tag{42}$$

gives the charge density.

untwist/twist	rank	quantities	parity	polar/axial	scalar/vector
untwisted	0-form	ϕ	even	–	scalar
untwisted	1-form	A, $\boldsymbol{E}$	odd	polar	vector
untwisted	2-form	B	even	axial	vector
twisted	1-form	H, $\boldsymbol{M}$	even	axial	vector
twisted	2-from	D, P, J	odd	polar	vector
twisted	3-form	$\mathcal{R}$	even	–	scalar

Table 2. Electromagnetic field quantities as twisted and untwisted n-forms

5.4 Parity

Parity is the eigenvalues for a spatial inversion transformation. It takes $p = \pm 1$ depending on the types of quantities. The quantity with eigenvalue of $+1$ (-1) is called having even (odd) parity. In the three dimensional case, the spatial inversion can be provided by simply flipping the basis vectors; $\mathcal{P}\boldsymbol{e}_i = -\boldsymbol{e}_i$ $(i = 1, 2, 3)$. The dual basis covectors are also flipped; $\mathcal{P}\boldsymbol{n}_j = -\boldsymbol{n}_j$ $(j = 1, 2, 3)$.

A scalar (0-form) ϕ is even because $\mathcal{P}\phi = \phi$. The electric field $\boldsymbol{E}$ is a 1-form and transforms as

$$\mathcal{P}\boldsymbol{E} = \mathcal{P}(\sum_i E_i \boldsymbol{n}_i) = \sum_i E_i \mathcal{P}\boldsymbol{n}_i = -\sum_i E_i \boldsymbol{n}_i = -\boldsymbol{E}, \tag{43}$$

and, therefore, it is odd. The magnetic flux density B is a 2-form and even since it transforms as

$$\mathcal{P}B = \mathcal{P}(\sum_{(i,j)} B_{ij} \boldsymbol{n}_i \wedge \boldsymbol{n}_j) = \sum_{(i,j)} B_{ij} \mathcal{P}\boldsymbol{n}_i \wedge \mathcal{P}\boldsymbol{n}_j = B. \tag{44}$$

It is easy to see that the parity of an n-forms is $p = (-1)^n$.

The volume form is transformed as

$$\mathcal{P}\mathcal{E} = \mathcal{P}(V_{123}\boldsymbol{n}_1 \wedge \boldsymbol{n}_2 \wedge \boldsymbol{n}_3) = -V_{123}\mathcal{P}\boldsymbol{n}_1 \wedge \mathcal{P}\boldsymbol{n}_2 \wedge \mathcal{P}\boldsymbol{n}_3 = \mathcal{E}. \tag{45}$$

The additional minus sign is due to the change in the orientation of basis. If $\Sigma \in C$, then $\mathcal{P}\Sigma \in C'$, and *vice versa*. The twisted 3-form has even parity. In general, the parity of a twisted n-form is $p = (-1)^{(n+1)}$.

In the conventional vector-scalar formalism, the parity is introduced rather empirically. We have found that 1-forms and twisted 2-forms are unified as polar vectors, 2-forms and twisted 1-forms as axial vectors, and 0-forms and twisted 3-forms as scalars. Thus we have unveiled the real shapes of electromagnetic quantities as twisted and untwisted n-forms.

6. Relativistic formulae

6.1 Metric tensor and dual basis

Combining a three dimensional orthonormal basis $\{\boldsymbol{e}_1, \boldsymbol{e}_2, \boldsymbol{e}_3\}$ and a unit vector $\boldsymbol{e}_0$ representing the time axis, we have a four-dimensional basis $\{\boldsymbol{e}_0, \boldsymbol{e}_1, \boldsymbol{e}_2, \boldsymbol{e}_3\}$. With the basis, a four (tangential) vector can be written

$$\underline{x} = (c_0 t)\boldsymbol{e}_0 + x\boldsymbol{e}_2 + y\boldsymbol{e}_2 + z\boldsymbol{e}_3 = x^\alpha \boldsymbol{e}_\alpha, \tag{46}$$

where the summation operator $\sum_{\alpha=0}^{3}$ is omitted in the last expression according to the Einstein summation convention. The vector components are represented with variables with superscripts. The sum is taken with respect to the Greek index repeated once as superscript and once as subscript. With the four-dimensional basis, the Lorentz-type inner product can be represented as

$$(\underline{x}, \underline{x}) = -(c_0 t)^2 + x^2 + y^2 + z^2 = x^\alpha x^\beta (\boldsymbol{e}_\alpha, \boldsymbol{e}_\beta) = x^\alpha g_{\alpha\beta} x^\beta = x_\beta x^\beta, \tag{47}$$

where we set $x_\beta = x^\alpha g_{\alpha\beta}$ and $(\boldsymbol{e}_\alpha, \boldsymbol{e}_\beta) = g_{\alpha\beta}$ with $g_{\alpha\beta} = 0$ $(\alpha \neq \beta)$, $-g_{00} = g_{ii} = 1$ $(i = 1, 2, 3)$.

We introduce the corresponding dual basis as $\{\boldsymbol{e}^0, \boldsymbol{e}^1, \boldsymbol{e}^2, \boldsymbol{e}^3\}$ with $\boldsymbol{e}^\mu \cdot \boldsymbol{e}_\nu = \delta^\mu_\nu$, where $\delta^\mu_\nu = 0$ $(\mu \neq \nu)$, $\delta^0_0 = \delta^i_i = 1$ $(i = 1, 2, 3)$. The dual basis covector has a superscript, while the components have subscripts. A four covector can be expressed with the dual basis as

$$\underline{\boldsymbol{a}} = a_\alpha \boldsymbol{e}^\alpha. \tag{48}$$

Then the contraction (by dot product) can be expressed systematically as

$$\boldsymbol{a} \cdot \boldsymbol{x} = a_\alpha \boldsymbol{e}^\alpha \cdot x^\beta \boldsymbol{e}_\beta = a_\alpha x^\beta \, \boldsymbol{e}^\alpha \cdot \boldsymbol{e}_\beta = a_\alpha x^\beta \delta^\alpha_\beta = a_\alpha x^\alpha. \tag{49}$$

We note that the dual and the inner product (metric) are independent concepts. Especially the duality can be introduced without the help of metric.

Customary, tensors which are represented by components with superscripts (subscripts) are designated as contravariant (covariant) tensors. With this terminology, a vector (covector) is a contravariant (covariant) tensor.

The symmetric second order tensor $g = g_{\alpha\beta} \boldsymbol{e}^\alpha \boldsymbol{e}^\beta$ is called a metric tensor. Its components are

$$g_{\alpha\beta} = \begin{cases} -1 & (\alpha = \beta = 0) \\ 1 & (\alpha = \beta \neq 0) \\ 0 & (\text{other cases}) \end{cases} . \tag{50}$$

For a fixed four vector $\underline{z}$, we can find a four covector $\underline{\boldsymbol{a}} = a_\beta \boldsymbol{e}^\beta$ that satisfy

$$\underline{\boldsymbol{a}} \cdot \underline{x} = (\underline{z}, \underline{x}) \tag{51}$$

for any $\underline{x}$. The left and right hand sides can be written as

$$\begin{aligned} \underline{\boldsymbol{a}} \cdot \underline{x} &= a_\beta x^\alpha \boldsymbol{e}^\beta \cdot \boldsymbol{e}_\alpha = a_\beta x^\alpha \delta^\beta_\alpha = a_\beta x^\beta, \\ (\underline{z}, \underline{x}) &= z^\alpha x^\beta (\boldsymbol{e}_\alpha, \boldsymbol{e}_\beta) = z^\alpha g_{\alpha\beta} x^\beta, \end{aligned} \tag{52}$$

respectively. By comparing these, we obtain $a_\beta = z^\alpha g_{\alpha\beta}$. We write this covector $\underline{\boldsymbol{a}}$ determined by $\underline{z}$ as

$$\underline{\boldsymbol{a}} = \underline{z}^\top = z^\alpha g_{\alpha\beta} \boldsymbol{e}^\beta = z_\beta \boldsymbol{e}^\beta, \tag{53}$$

which is called the conjugate of $\underline{z}$. We see that $(\boldsymbol{e}_0)^\top = -\boldsymbol{e}^0$, $(\boldsymbol{e}_i)^\top = \boldsymbol{e}^i$ $(i = 1, 2, 3)$, i.e., $(\boldsymbol{e}_\alpha)^\top = g_{\alpha\beta} \boldsymbol{e}^\beta$ [1]. With $g^{\alpha\beta} = (\boldsymbol{e}^\alpha, \boldsymbol{e}^\beta)$, the conjugate of a covector $\underline{\boldsymbol{a}}$ can be defined similarly with $z^\alpha = g^{\alpha\beta} a_\beta$ as $\underline{z} = \underline{\boldsymbol{a}}^\top$.

[1] An equation $\boldsymbol{e}_\alpha = g_{\alpha\beta} \boldsymbol{e}^\beta$, which we may write carelessly, is not correct.

We have $(\underline{z}^\top)^\top = \underline{z}$, $(\underline{a}^\top)^\top = \underline{a}$, namely, ${}^{\top\top} = 1$.

We introduce the four dimensional completely anti-symmetric tensor of order 4 as

$$\begin{aligned}\mathfrak{E} &= e^0 \wedge e^1 \wedge e^2 \wedge e^3 \\ &= \epsilon_{\alpha\beta\gamma\delta} e^\alpha e^\beta e^\gamma e^\delta.\end{aligned} \tag{54}$$

The components $\epsilon_{\alpha\beta\gamma\delta}$, which are called the Levi-Civita symbol[2] can be written explicitly as

$$\epsilon_{\alpha\beta\gamma\delta} = \begin{cases} 1 & (\alpha\beta\gamma\delta \text{ is an even permutation of } 0123) \\ -1 & (\text{an odd permutation}) \\ 0 & (\text{other cases}) \end{cases} . \tag{55}$$

We note that

$$\begin{aligned}\mathfrak{E}^\top &= (e^0)^\top \wedge (e^1)^\top \wedge (e^2)^\top \wedge (e^3)^\top \\ &= (-e_0) \wedge e_1 \wedge e_2 \wedge e_3 \\ &= \epsilon_{\alpha\beta\gamma\delta} g^{\alpha\mu} g^{\beta\nu} g^{\gamma\sigma} g^{\delta\tau} e_\mu e_\nu e_\sigma e_\tau \\ &= \epsilon^{\mu\nu\sigma\tau} e_\mu e_\nu e_\sigma e_\tau,\end{aligned} \tag{56}$$

where we introduced, $\epsilon^{\mu\nu\sigma\tau} = \epsilon_{\alpha\beta\gamma\delta} g^{\alpha\mu} g^{\beta\nu} g^{\gamma\sigma} g^{\delta\tau}$.

The conjugate of the metric tensor is given by

$$\underline{g}^\top = g_{\alpha\beta}(e^\alpha)^\top (e^\beta)^\top = g_{\alpha\beta} g^{\alpha\mu} g^{\beta\nu} e_\mu e_\nu = g^{\mu\nu} e_\mu e_\nu. \tag{57}$$

6.2 Levi-Civita symbol

Here we will confirm some properties of the completely anti-symmetric tensor of order 4. From the relation between covariant and contravariant components

$$\epsilon^{\alpha\beta\gamma\delta} = g^{\alpha\mu} g^{\beta\nu} g^{\gamma\sigma} g^{\delta\tau} \epsilon_{\mu\nu\sigma\tau}, \tag{58}$$

and $g^{00} = -1$, we see that $\epsilon^{0123} = -\epsilon_{0123}$ and similar relations hold for other components. Here we note $\epsilon^{0123} = -1$.

With respect to contraction, we have

$$\epsilon^{\alpha\beta\gamma\delta}\epsilon_{\alpha\beta\gamma\delta} = -24 \quad (= -4!) \tag{59}$$

$$\epsilon^{\alpha\beta\gamma\delta}\epsilon_{\alpha\beta\gamma\tau} = -6\delta^\delta_\tau \tag{60}$$

$$\epsilon^{\alpha\beta\gamma\delta}\epsilon_{\alpha\beta\sigma\tau} = -2(\delta^\gamma_\sigma\delta^\delta_\tau - \delta^\gamma_\tau\delta^\delta_\sigma) = -4\delta^\gamma_{[\sigma}\delta^\delta_{\tau]} \tag{61}$$

$$\epsilon^{\alpha\beta\gamma\delta}\epsilon_{\alpha\nu\sigma\tau} = -6\delta^\beta_{[\nu}\delta^\gamma_\sigma\delta^\delta_{\tau]}. \tag{62}$$

where $_{[]}$ in the subscript represents the anti-symmetrization with respect to the indices. For example, we have $A_{[\alpha\beta}B_{\gamma]} = (A_{\alpha\beta}B_\gamma + A_{\beta\gamma}B_\alpha + A_{\gamma\alpha}B_\beta - A_{\beta\alpha}B_\gamma - A_{\gamma\beta}B_\alpha - A_{\alpha\gamma}B_\beta)/6$. We note $\underline{A} \wedge \underline{B} = A_{\alpha\beta}B_\gamma e^\alpha \wedge e^\beta \wedge e^\gamma = 6A_{[\alpha\beta}B_{\gamma]} e^\alpha e^\beta e^\gamma$.

---

[2] In the case of three dimension, the parity of a permutation can simply be discriminated by the cyclic or anti-cyclic order. In the case of four dimension, the parity of $0ijk$ follows that of ijk and those of $i0jk$, $ij0k$, $ijk0$ is opposite to that of ijk.

6.3 Hodge dual of anti-symmetric 2nd-order tensors

The four-dimensional Hodge dual $(\underline{*}A)_{\alpha\beta}$ of a second order tensor $A_{\alpha\beta}$ is defined to satisfy

$$(\underline{*}A)_{[\alpha\beta}B_{\gamma\delta]} = \frac{1}{2}(A^{\mu\nu}B_{\mu\nu})\epsilon_{\alpha\beta\gamma\delta}, \tag{63}$$

for an arbitrary tensor $B_{\gamma\delta}$ of order $(d-2)$ (Flanders (1989)). This relation is independent of the basis [3].

Here, we will show that

$$(\underline{*}A)_{\alpha\beta} = \frac{1}{2}\epsilon_{\alpha\beta}{}^{\mu\nu}A_{\mu\nu}. \tag{64}$$

Substituting into the left hand side of Eq. (63) and contracting with $\epsilon^{\alpha\beta\gamma\delta}$, we have

$$\epsilon^{\alpha\beta\gamma\delta}\frac{1}{2}A_{\mu\nu}\epsilon_{[\alpha\beta}{}^{\mu\nu}B_{\gamma\delta]} = 3\epsilon^{\alpha\beta\gamma\delta}\epsilon_{\alpha\beta}{}^{\mu\nu}A_{\mu\nu}B_{\gamma\delta} = 3\epsilon^{\alpha\beta\gamma\delta}\epsilon_{\alpha\beta\mu\nu}A^{\mu\nu}B_{\gamma\delta} = -12A^{\gamma\delta}B_{\gamma\delta}. \tag{65}$$

With Eq. (59), the right hand side of Eq. (63) yields $-12A^{\mu\nu}B_{\mu\nu}$ with the same contraction. We also note

$$\begin{aligned}(\underline{**}A)_{\alpha\beta} &= \frac{1}{4}\epsilon_{\alpha\beta}{}^{\gamma\delta}\epsilon_{\gamma\delta}{}^{\mu\nu}A_{\mu\nu} = \frac{1}{4}\epsilon_{\alpha\beta\gamma\delta}\epsilon^{\gamma\delta\mu\nu}A_{\mu\nu} \\ &= -\frac{1}{2}(\delta^{\alpha}_{\mu}\delta^{\beta}_{\nu} - \delta^{\alpha}_{\nu}\delta^{\beta}_{\mu})A_{\mu\nu} = -\frac{1}{2}(A_{\alpha\beta} - A_{\beta\alpha}) = -A_{\alpha\beta},\end{aligned} \tag{66}$$

i.e., $\underline{**} = -1$, which is different from the three dimensional case; $** = 1$.

7. Differential forms in Minkowski spacetime

7.1 Standard formulation

According to Jackson's textbook (Jackson (1998)), we rearrange the ordinary scalar-vector form of Maxwell's equation in three dimension into a relativistic expression. We use the SI system and pay attention to the dimensions. We start with the source equations

$$\nabla \times \boldsymbol{H} - \frac{\partial}{\partial(c_0 t)}(c_0\boldsymbol{D}) = \boldsymbol{J}, \quad \nabla \cdot (c_0\boldsymbol{D}) = c_0\varrho. \tag{67}$$

Combining field quantities and differential operators as four-dimensional tensors and vectors as

$$(\tilde{G}^{\alpha\beta}) = \begin{bmatrix} 0 & c_0D_x & c_0D_y & c_0D_z \\ -c_0D_x & 0 & H_z & -H_y \\ -c_0D_y & -H_z & 0 & H_x \\ -c_0D_z & H_y & -H_x & 0 \end{bmatrix} \overset{\text{SI}}{\sim} \text{A/m}, \tag{68}$$

[3] With the four-dimensional volume form $\mathfrak{E} = \boldsymbol{e}^0 \wedge \boldsymbol{e}^1 \wedge \boldsymbol{e}^2 \wedge \boldsymbol{e}^3$, the Hodge dual for p form ($p = 1, 2, 3$) can be defined as $(\underline{*}\boldsymbol{A}) \wedge \underline{\boldsymbol{B}} = (\underline{\boldsymbol{A}}, \underline{\boldsymbol{B}})\mathfrak{E}$, $(\underline{*}A) \wedge \underline{B} = (\underline{A}, \underline{B})\mathfrak{E}$, and $(\underline{*}\mathcal{A}) \wedge \underline{\mathcal{B}} = (\underline{\mathcal{A}}, \underline{\mathcal{B}})\mathfrak{E}$. The inner product for p forms is defined as $(\boldsymbol{a}_1 \wedge \cdots \wedge \boldsymbol{a}_p, \boldsymbol{b}_1 \wedge \cdots \wedge \boldsymbol{b}_p) = \det(\boldsymbol{a}_i, \boldsymbol{b}_j)$. (Flanders (1989))

$$(\partial_\beta) = \begin{bmatrix} c_0^{-1}\partial_t \\ \partial_x \\ \partial_y \\ \partial_z \end{bmatrix} \overset{\text{SI}}{\sim} 1/\text{m}, \quad (\tilde{J}^\alpha) = \begin{bmatrix} c_0\varrho \\ J_x \\ J_y \\ J_z \end{bmatrix} \overset{\text{SI}}{\sim} \text{A}/\text{m}^2, \tag{69}$$

we have a relativistic equation

$$\partial_\beta \tilde{G}^{\alpha\beta} = \tilde{G}^{\alpha\beta}{}_{,\beta} = \tilde{J}^\alpha. \tag{70}$$

We append "~", by the reason described later. The suffix 0 represents the time component, and the suffixes $1, 2, 3$ correspond to x, y, z-components. The commas in suffixes "," means the derivative with respect to the following spatial component, e.g., $H_{2,1} = (\partial/\partial x_1)H_2$.

On the other hand, the force equations

$$\nabla \times \boldsymbol{E} + \frac{\partial}{\partial(c_0 t)}(c_0\boldsymbol{B}) = 0, \quad \nabla \cdot (c_0\boldsymbol{B}) = 0, \tag{71}$$

are rearranged with

$$(\tilde{F}^{\alpha\beta}) = \begin{bmatrix} 0 & c_0B_x & c_0B_y & c_0B_z \\ -c_0B_x & 0 & -E_z & E_y \\ -c_0B_y & E_z & 0 & -E_x \\ -c_0B_z & -E_y & E_x & 0 \end{bmatrix} \overset{\text{SI}}{\sim} \text{V}/\text{m}, \tag{72}$$

as

$$\partial_\beta \tilde{F}^{\alpha\beta} = \tilde{F}^{\alpha\beta}{}_{,\beta} = 0. \tag{73}$$

Thus the four electromagnetic field quantities $\boldsymbol{E}$, $\boldsymbol{B}$, $\boldsymbol{D}$, and $\boldsymbol{H}$ are aggregated into two second order, antisymmetric tensors $\tilde{F}^{\alpha\beta}$, $\tilde{G}^{\alpha\beta}$.

In vacuum, the constitutive relations $\boldsymbol{D} = \varepsilon_0\boldsymbol{E}$, $\boldsymbol{H} = \mu_0^{-1}\boldsymbol{B}$ hold, therefore, these tensors are related as

$$\tilde{G}^{\alpha\beta} = Y_0(\underline{*}\tilde{F})^{\alpha\beta}, \quad \text{or} \quad \tilde{F}^{\alpha\beta} = -Z_0(\underline{*}\tilde{G})^{\alpha\beta}, \tag{74}$$

where $Z_0 = 1/Y_0 = \sqrt{\mu_0/\varepsilon_0} \overset{\text{SI}}{\sim} \Omega$ is the vacuum impedance.

The operator $\underline{*}$ is the four-dimensional Hodge's star operator. From Eq. (64), the action for a 2nd-order tensor is written as

$$(\underline{*}A)^{ij} = A^{0k}, \quad (\underline{*}A)^{0i} = -A^{jk}, \tag{75}$$

where i, j, k $(i, j = 1, 2, 3)$ are cyclic. We note that $\underline{**} = -1$, i.e., $\underline{*}^{-1} = -\underline{*}$ holds.

Equation (74) is a relativistic version of constitutive relations of vacuum and carries two roles. First it connect dimensionally different tensors $\tilde{G}$ and $\tilde{F}$ with the vacuum impedance Z_0. Secondly it represents the Hodge's dual relation. The Hodge operator depends both on the handedness of the basis[4] and the metric.

[4] We note $\epsilon_{\alpha\beta\gamma\delta}$ is a pseudo form rather than a form. Therefore, the Hodge operator makes a form into a pseudo form, and a pseudo form into a normal form.

Finally, the Maxwell equations can be simply represented as

$$\partial_\beta \tilde{G}^{\alpha\beta} = \tilde{J}^\alpha, \quad \partial_\beta \tilde{F}^{\alpha\beta} = 0, \quad \tilde{G}^{\alpha\beta} = Y_0(\underline{*}\tilde{F})^{\alpha\beta}. \tag{76}$$

This representation, however, is quite unnatural in the view of two points. First of all, the components of field quantity should be covariant and should have lower indices. Despite of that, here, all quantities are contravariant and have upper indices in order to contract with the spatial differential operator ∂_α with a lower index. Furthermore, it is unnatural that in Eqs. (68) and (72), D and B, which are 2-forms with respect to space, have indices of time and space, and E and H have two spatial indices.

The main reason of this unnaturalness is that we have started with the conventional, scalar-vector form of Maxwell equations rather than from those in differential forms.

7.2 Bianchi identity

In general textbooks, the one of equations in Eq. (76) is further modified by introducing a covariant tensor $F_{\alpha\beta} = \frac{1}{2}\epsilon_{\alpha\beta\gamma\delta}\tilde{F}^{\gamma\delta}$. Solving it as $\tilde{F}^{\alpha\beta} = -\frac{1}{2}\epsilon^{\alpha\beta\gamma\delta}F_{\gamma\delta}$ and substituting into Eq. (73), we have

$$0 = \partial_\beta \epsilon^{\alpha\beta\gamma\delta} F_{\gamma\delta} = \epsilon^{\alpha\beta\gamma\delta}\partial_\beta F_{\gamma\delta}. \tag{77}$$

Considering α as a fixed parameter, we have six non-zero terms that are related as

$$\begin{aligned} 0 &= \partial_\beta(F_{\gamma\delta} - F_{\delta\gamma}) + \partial_\gamma(F_{\delta\beta} - F_{\beta\delta}) + \partial_\delta(F_{\beta\gamma} - F_{\gamma\beta}) \\ &= 2\left(\partial_\beta F_{\gamma\delta} + \partial_\gamma F_{\delta\beta} + \partial_\delta F_{\beta\gamma}\right) \quad (\beta, \gamma, \delta = 0, \ldots, 3). \end{aligned} \tag{78}$$

Although there are many combinations of indices, this represents substantially four equations. To be specific, we introduce the matrix representation of $F_{\alpha\beta}$ as

$$(F_{\alpha\beta}) = \begin{bmatrix} 0 & -E_x & -E_y & -E_z \\ E_x & 0 & c_0B_z & -c_0B_y \\ E_y & -c_0B_z & 0 & c_0B_x \\ E_z & c_0B_y & -c_0B_x & 0 \end{bmatrix}. \tag{79}$$

Comparing this with $\tilde{G}^{\alpha\beta}$ in Eq. (68) and considering the constitutive relations, we find that the signs of components with indices for time "0" are reversed. Therefore, with the metric tensor, we have

$$\tilde{G}^{\alpha\beta} = Y_0 g^{\alpha\gamma} g^{\beta\delta} F_{\gamma\delta} = Y_0 F^{\alpha\beta}. \tag{80}$$

Substitution into Eq. (70) yields $Y_0\partial_\beta F^{\alpha\beta} = \tilde{J}^\alpha$. After all, relativistically, the Maxwell equations are written as

$$\partial_\beta F^{\alpha\beta} = Z_0\tilde{J}^\alpha, \quad \partial_\alpha F_{\beta\gamma} + \partial_\beta F_{\gamma\alpha} + \partial_\gamma F_{\alpha\beta} = 0. \tag{81}$$

Even though this common expression is simpler than that for the non relativistic version, symmetry is somewhat impaired. The covariant and contravariant field tensors are mixed. The reason is that the constitutive relations, which contains the Hodge operator, is eliminated.

7.3 Differential forms

Here we start with the Maxwell equations (31) in three-dimensional differential forms. We introduce a basis $\{e_0, e_1, e_2, e_3\}$, and the corresponding dual basis $\{e^0, e^1, e^2, e^3\}$, i.e., $e^\mu \cdot e_\nu = \delta^\mu_\nu$. With $\underline{G} = e^0 \wedge \boldsymbol{H} + c_0 \boldsymbol{D}$, $\underline{\mathcal{J}} = e^0 \wedge (-\boldsymbol{J}) + c_0 \mathcal{R}$, and $\underline{\boldsymbol{\nabla}} = c_0^{-1}\partial_t e^0 + \boldsymbol{\nabla}$, the source equations

$$\boldsymbol{\nabla} \wedge \boldsymbol{H} - \frac{\partial}{\partial(c_0 t)}(c_0 \boldsymbol{D}) = \boldsymbol{J}, \quad \boldsymbol{\nabla} \wedge (c_0 \boldsymbol{D}) = \mathcal{R}, \tag{82}$$

are unified as

$$\underline{\boldsymbol{\nabla}} \wedge \underline{G} = \underline{\mathcal{J}}, \tag{83}$$

where $\wedge$ represent the anti-symmetric tensor product or the wedge product. In components, Eq. (83) is

$$\partial_{[\gamma} G_{\alpha\beta]} = G_{[\alpha\beta,\gamma]} = J_{\alpha\beta\gamma}/3. \tag{84}$$

The tensor $G_{\alpha\beta}$ can be written as

$$(G_{\alpha\beta}) = (\underline{G} : e_\alpha e_\beta) = \begin{bmatrix} 0 & H_x & H_y & H_z \\ -H_x & 0 & c_0 D_z & -c_0 D_y \\ -H_y & -c_0 D_z & 0 & c_0 D_x \\ -H_z & c_0 D_y & -c_0 D_x & 0 \end{bmatrix}. \tag{85}$$

The covariant tensors (forms) $G_{\alpha\beta}$ and $J_{\alpha\beta\gamma}$ are related to $\tilde{G}^{\alpha\beta}$ and $\tilde{J}^\alpha$ in the previous subsection as

$$G_{\alpha\beta} = \frac{1}{2}\epsilon_{\alpha\beta\gamma\delta}\tilde{G}^{\gamma\delta}, \quad J_{\alpha\beta\gamma} = -\epsilon_{\alpha\beta\gamma\delta}\tilde{J}^\delta, \quad \text{or} \quad \tilde{G}^{\alpha\beta} = -\frac{1}{2}\epsilon^{\alpha\beta\gamma\delta}G_{\gamma\delta}, \quad \tilde{J}^\alpha = \frac{1}{6}\epsilon^{\alpha\beta\gamma\delta}J_{\beta\gamma\delta}. \tag{86}$$

Similarly,

$$\boldsymbol{\nabla} \wedge \boldsymbol{E} + \frac{\partial}{\partial(c_0 t)}(c_0 \boldsymbol{B}) = 0, \quad \boldsymbol{\nabla} \wedge (c_0 \boldsymbol{B}) = 0, \tag{87}$$

can be written as

$$\underline{\boldsymbol{\nabla}} \wedge \underline{F} = 0, \tag{88}$$

with $\underline{F} = e^0 \wedge (-\boldsymbol{E}) + c_0 \boldsymbol{B}$. In components,

$$\partial_{[\gamma} F_{\alpha\beta]} = F_{[\alpha\beta,\gamma]} = 0, \tag{89}$$

where

$$(F_{\alpha\beta}) = (\underline{F} : e_\alpha e_\beta) = \begin{bmatrix} 0 & -E_x & -E_y & -E_z \\ E_x & 0 & c_0 B_z & -c_0 B_y \\ E_y & -c_0 B_z & 0 & c_0 B_x \\ E_z & c_0 B_y & -c_0 B_x & 0 \end{bmatrix}. \tag{90}$$

The covariant tensor (form) $F_{\alpha\beta}$ is related to $\tilde{F}^{\alpha\beta}$ in the previous subsection as

twisted/untwisted	order	quantities
untwisted	1-form	$\underline{V} = \phi \boldsymbol{e}^0 + c_0(-\boldsymbol{A})$
untwisted	2-form	$\underline{F} = \boldsymbol{e}^0 \wedge (-\boldsymbol{E}) + c_0 B$
twisted	2-form	$\underline{G} = \boldsymbol{e}^0 \wedge \boldsymbol{H} + c_0 D$
twisted	2-form	$\underline{I} = \boldsymbol{e}^0 \wedge (-\boldsymbol{M}) + c_0 P$
twisted	3-form	$\underline{\mathcal{J}} = \boldsymbol{e}^0 \wedge (-J) + c_0 \mathcal{R}$

Table 3. Four dimensional electromagnetic field quantities as twisted and untwisted n-forms

$$F_{\alpha\beta} = \frac{1}{2}\epsilon_{\alpha\beta\gamma\delta}\tilde{F}^{\gamma\delta}, \quad \text{or,} \quad \tilde{F}^{\alpha\beta} = -\frac{1}{2}\epsilon^{\alpha\beta\gamma\delta}F_{\gamma\delta}. \tag{91}$$

The Hodge operator acts on a four-dimensional two form as

$$\underline{*}(\boldsymbol{e}^0 \wedge \boldsymbol{X} + Y) = \boldsymbol{e}^0 \wedge (-(*Y)) + (*\boldsymbol{X}) = \boldsymbol{e}^0 \wedge (-Y) + X, \tag{92}$$

where $\boldsymbol{X}$ and Y are a three-dimensional 1-form and a three-dimensional 2-form, and $*$ is the three-dimensional Hodge operator[5]. Now the constitutive relations $D = \varepsilon_0(*\boldsymbol{E})$ and $\boldsymbol{H} = \mu_0^{-1}(*B)$ are four-dimensionally represented as

$$\underline{G} = -Y_0(\underline{*}\underline{F}), \quad \text{or} \quad \underline{F} = Z_0(\underline{*}\underline{G}). \tag{93}$$

With components, these are represented as

$$G_{\alpha\beta} = -Y_0(\underline{*}F)_{\alpha\beta}, \quad \text{or} \quad F_{\alpha\beta} = Z_0(\underline{*}G)_{\alpha\beta}, \tag{94}$$

with the action of Hodge's operator on anti-symmetric tensors of rank 2:

$$(\underline{*}A)_{\alpha\beta} = \frac{1}{2}\epsilon_{\alpha\beta}{}^{\gamma\delta}A_{\gamma\delta} = \frac{1}{2}g_{\alpha\mu}g_{\beta\nu}\epsilon^{\mu\nu\gamma\delta}A_{\gamma\delta}. \tag{95}$$

Now we have the Maxwell equations in the four-dimensional forms with components:

$$\partial_{[\gamma}G_{\alpha\beta]} = J_{\alpha\beta\gamma}/3, \quad \partial_{[\gamma}F_{\alpha\beta]} = 0, \quad F_{\alpha\beta} = Z_0(\underline{*}G)_{\alpha\beta}, \tag{96}$$

and in basis-free representations:

$$\underline{\boldsymbol{\nabla}} \wedge \underline{G} = \underline{\mathcal{J}}, \quad \underline{\boldsymbol{\nabla}} \wedge \underline{F} = 0, \quad \underline{F} = Z_0(\underline{*}\underline{G}), \tag{97}$$

or

$$\underline{\mathrm{d}}\underline{G} = \underline{\mathcal{J}}, \quad \underline{\mathrm{d}}\underline{F} = 0, \quad \underline{F} = Z_0(\underline{*}\underline{G}). \tag{98}$$

with the four-dimensional exterior derivative $\underline{\mathrm{d}}\,\llcorner = \underline{\boldsymbol{\nabla}} \wedge \llcorner$. These are much more elegant and easier to remember compared with Eqs. (76) and (81). A similar type of reformulation has been given by Sommerfeld (Sommerfeld (1952)).

[5] We note the similarity with the calculation rule for complex numbers: $\mathrm{i}(X + \mathrm{i}Y) = -Y + \mathrm{i}X$. If we can formally set as $\boldsymbol{G} = \boldsymbol{H} + \mathrm{i}c_0\boldsymbol{D}$ and $\boldsymbol{F} = -\boldsymbol{E} + \mathrm{i}c_0\boldsymbol{B}$, we have $\boldsymbol{G} = -\mathrm{i}Y_0\boldsymbol{F}$ and $\boldsymbol{H} = \mathrm{i}Z_0\boldsymbol{G}$.

7.4 Potentials and the conservation of charge

We introduce a four-dimensional vector potential $\underline{V} = \phi e^0 + c_0(-A)$, i.e.,

$$(V_\alpha) = (\underline{V} \cdot e_\alpha)$$
$$= (\phi, -c_0 A_x, -c_0 A_y, -c_0 A_z). \tag{99}$$

Then we have $\underline{\nabla} \wedge \underline{V} = -\underline{F}$, or

$$\partial_{[\alpha} V_{\beta]} = -F_{\alpha\beta}/2, \tag{100}$$

which is a relation between the potential and the field strength. Utilizing the potential, the force equation becomes very trivial,

$$0 = \underline{\nabla} \wedge (\underline{\nabla} \wedge \underline{V}) = \underline{\nabla} \wedge \underline{F}, \tag{101}$$

since $\underline{\nabla} \wedge \underline{\nabla} = 0$ or $\underline{\mathrm{dd}} = 0$ holds.

The freedom of gauge transformation with a 0-form Λ can easily be understood; $\underline{V}' = \underline{V} + \underline{\mathrm{d}\Lambda}$ gives no difference in the force quantities, i.e., $\underline{F}' = \underline{F}$. A similar degree of freedom exist for the source fields (Hirst (1997)). With a 1-form $\underline{L}$, we define the transformation $\underline{G}' = \underline{G} + \underline{\mathrm{d}L}$, which yields $\underline{\mathcal{J}}' = \underline{\mathcal{J}}$.

The conservation of charge is also straightforward;

$$0 = \underline{\nabla} \wedge \underline{\nabla} \wedge \underline{G} = \underline{\nabla} \wedge \underline{\mathcal{J}}$$
$$= e^0 \wedge (\partial_t \mathcal{R} + \nabla \wedge J). \tag{102}$$

7.5 Relativistic representation of the Lorentz force

Changes in the energy E and momentum p of a charged particle moving at velocity $\boldsymbol{u}$ in an electromagnetic field are

$$\frac{\mathrm{d}E}{\mathrm{d}t} = q\boldsymbol{E} \cdot \boldsymbol{u},$$
$$\frac{\mathrm{d}\boldsymbol{p}}{\mathrm{d}t} = q\boldsymbol{E} + \boldsymbol{u} \times \boldsymbol{B}. \tag{103}$$

By introducing the four dimensional velocity $u^\alpha = [c_0\gamma, u_x, u_y, u_z]$, and the four dimensional momentum $p_\alpha = [-E/c_0, p_x, p_y, p_z]$, we have the equation of motion

$$\frac{\mathrm{d}p_\alpha}{\mathrm{d}\tau} = qF_{\alpha\beta}u^\beta, \tag{104}$$

where $\mathrm{d}\tau = \mathrm{d}t/\gamma$ the proper time of moving charge, and $\gamma = (1 - u^2/c_0^2)^{-1/2}$ is the Lorentz factor. The change in action ΔS can be written

$$\Delta S = p_\alpha \Delta x^\alpha = -E\Delta t + \boldsymbol{p} \cdot \Delta \boldsymbol{x}. \tag{105}$$

$$
\begin{array}{ccc}
\underline{\text{F quantities}} & & \underline{\text{S quantities}} \\
\underline{V} = \phi e^0 - c_0 A & & \\
\downarrow \underline{\mathsf{d}} & Y_0 & \\
\underline{F} = -e^0 \wedge E + c_0 B & \leftarrow \underline{*} \rightarrow & \underline{G} = e^0 \wedge H + c_0 D \\
\downarrow \underline{\mathsf{d}} & Z_0 & \downarrow \underline{\mathsf{d}} \\
0 & & \underline{\mathcal{J}} = -e^0 \wedge J + c_0 \mathcal{R} \\
 & & \downarrow \underline{\mathsf{d}} \\
 & & 0
\end{array}
$$

Fig. 2. Relations of electromagnetic field forms in four dimension

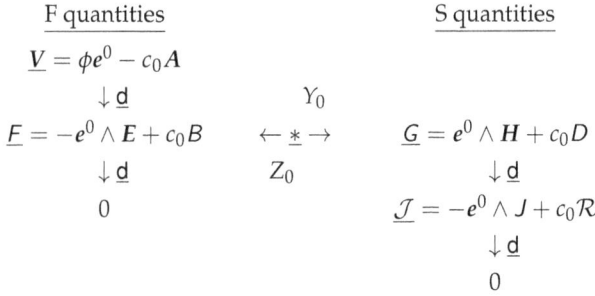

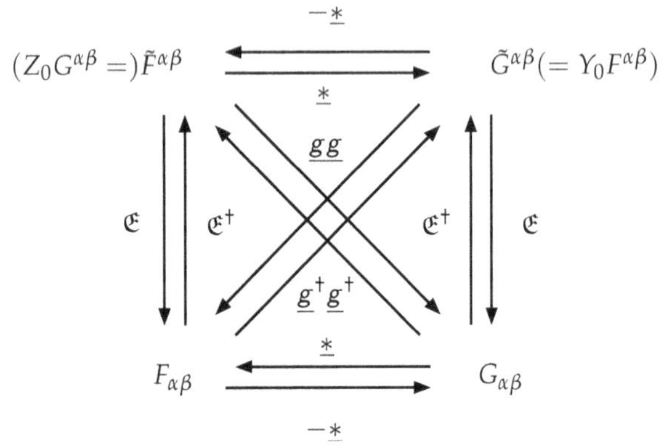

Fig. 3. Various kinds of tensors of order 2 used in the relativistic Maxwell equations

7.6 Summary for relativistic relations

In Fig. 2, the relativistic quantities are arranged as a diagram, the rows of which correspond to the orders of tensors ($n = 1, 2, 3, 4$). In the left column, the quantities related to the electromagnetic forces (F quantities), and in the right column, the quantities related to the electromagnetic sources (S quantities) are listed. The exterior derivative "$\underline{\mathsf{d}}$" connects a pair of quantities by increasing the tensor order by one. These differential relations correspond to the definition of (scalar and vector) potentials, the Maxwell's equations, and the charge conservation (See Fig. 1). Hodge's star operator "$\underline{*}$" connects two 2-forms: $\underline{F}$ and $\underline{G}$. This corresponds to the constitutive relations for vacuum and here appears the vacuum impedance $Z_0 = 1/Y_0$ as the proportional factor.

In Fig. 3, various kinds of tensors of order 2 in the relativistic Maxwell equations and their relations are shown. The left column corresponds to the source fields (D, H), the right column corresponds to the force fields (E, B). Though not explicitly written, due to the difference in dimension, the conversions associate the vacuum impedance (or admittance). "$\mathfrak{E}$" and "$\mathfrak{E}^\top$" represent the conversion by Levi-Civita (or by its conjugate), "$\underline{*}$" represents the conversion by Hodge's operator. Associated with the diagonal arrows, "$\underline{g}^\top \underline{g}^\top$", and "$\underline{g}\underline{g}$" represent raising and lowering of the indices with the metric tensors, respectively. The tensors in the upper row are derived from the scalar-vector formalism and those in the lower row are derived from the differential forms. In order to avoid the use of the Hodge operator, the diagonal pair $F_{\alpha\beta}$, $F^{\alpha\beta}$ $(= Z_0 \tilde{G}^{\alpha\beta})$ are used conventionally and the symmetry is sacrificed.

8. Conclusion

In this book chapter, we have reformulated the electromagnetic theory. First we have confirmed the role of vacuum impedance Z_0 as a fundamental constant. It characterizes the electromagnetism as the gravitational constant G characterizes the theory of gravity. The velocity of light c_0 in vacuum is the constant associated with space-time, which is a framework in which electromagnetism and other theories are constructed. Then, Z_0 is a single parameter characterizing electromagnetism, and $\varepsilon_0 = 1/(Z_0 c_0)$ and $\mu_0 = Z_0/c_0$ are considered derived parameters.

Next, we have introduced anti-symmetric covariant tensors, or differential forms, in order to represent EM field quantities most naturally. It is a significant departure from the conventional scalar-vector formalism. But we have tried not to be too mathematical by carrying over the conventional notations as many as possible for continuous transition. In this formalism, the various field quantities are defined through the volume form, which is the machinery to calculate the volume of parallelepipedon spanned by three tangential vectors. To be precise, it is a pseudo (twisted) form, whose sign depends on the orientation of basis.

Even though the constitutive relation seems as a simple proportional relation, it associates the conversion by the Hodge dual operation and the change in physical dimensions by the vaccum impedance. We have found that this non-trivial relation is the keystone of the EM theory.

The EM theory has the symmetry with respect to the space inversion, therefore, each field quantity has a definite parity, even or odd. We have shown that the parity is determined by the tensorial order and the pseudoness (twisted or untwisted).

The Maxwell equations can be formulated most naturally in the four dimensional space-time. However, the conventional expression with tensor components (with superscripts or subscripts) is somewhat abstract and hard to read out its geometrical or physical meaning. Moreover, sometimes contravariant tensors are introduced in order to avoid the explicit use of the Hodge dual with sacrificing the beauty of equations. It has been shown that the four-dimensional differential forms (anti-symmetric covariant tensors) are the most suitable tools for expressing the structure of the EM theory.

The structured formulation helps us to advance electromagnetic theories to various areas. For example, the recent development of new type of media called metamaterials, for which we have to deal with electric and magnetic interactions simultaneously, confronts us to reexamine theoretical frameworks. It may also be helpful to resolve problems on the electromagnetic momentum within dielectric media.

9. Acknowledgment

I thank Yosuke Nakata for helpful discussions. This research was supported by a Grant-in-Aid for Scientific Research on Innovative Areas (No. 22109004) and the Global COE program "Photonics and Electronics Science and Engineering" at Kyoto University.

10. References

Burke, W. L. (1985). *Applied Differential Geometry*, Cambridge University Press.

Deschamps, G. (1981). Electromagnetics and differential forms, *Proceedings of the IEEE* 69(6): 676 – 696.

Flanders, H. (1989). *Differential Forms with Applications to the Physical Sciences*, Dover Publications.

Frankel, T. (2004). *The Geometry of Physics*, Cambridge University Press, Cambridge.

Hehl, F. W. & Obukhov, Y. N. (2003). *Foundations of Classical Electrodynamics (Progress in Mathematical Physics)*, Birkhäuser, Boston.

Hehl, F. W. & Obukhov, Y. N. (2005). Dimensions and units in electrodynamics, *General Relativity and Gravitation* 37: 733–749.

Hirst, L. L. (1997). The microscopic magnetization: concept and application, *Rev. Mod. Phys.* 69: 607–628.

Jackson, J. D. (1998). *Classical Electrodynamics Third Edition*, Wiley.

Kitano, M. (2009). The vacuum impedance and unit systems, *IEICE Trans. Electronics* EC92-C(1): 3–8.

Klitzing, K. v., Dorda, G. & Pepper, M. (1980). New method for high-accuracy determination of the fine-structure constant based on quantized hall resistance, *Phys. Rev. Lett.* 45: 494–497.

Maxwell, J. C. (1865). A dynamical theory of the electromagnetic field, *Philosophical Transactions of the Royal Society of London* 155: pp. 459–512.

Misner, C. W., Thorne, K. S. & Wheele, J. A. (1973). *Gravitation (Physics Series)*, W. H. Freeman.

Pendry, J. B. & Smith, D. R. (2004). Reversing light: Negative refraction, *Physics Today* 57(6): 37–43.

Sakurai, J. J. (1993). *Modern Quantum Mechanics (Revised Edition)*, Addison Wesley.

Schelkunoff, S. A. (1938). The impedance concept and its application to problems of reflection, refraction, shielding, and power absorption, *Bell System Tech. J.* 17: 17–48.

Sommerfeld, A. (1952). *Electrodynamics — Lectures on Theoretical Physics*, Academic Press, New York.

Weinreich, G. (1998). *Geometrical Vectors (Chicago Lectures in Physics)*, University of Chicago Press.

2

Current-Carrying Wires and Special Relativity

Paul van Kampen
Centre for the Advancement of Science and Mathematics Teaching and Learning &
School of Physical Sciences,
Dublin City University
Ireland

1. Introduction

This chapter introduces the main concepts of electrostatics and magnetostatics: charge and current, Coulomb's Law and the Biot-Savart Law, and electric and magnetic fields. Using linear charge distributions and currents makes it possible to do this without recourse to vector calculus. Special relativity is invoked to demonstrate that electricity and magnetism are, in a sense, two different ways of looking at the same phenomenon: in principle, from a knowledge of either electricity or magnetism and special relativity, the third theory could be derived. The three theories are shown to be mutually consistent in the case of linear currents and charge distributions.

This chapter brings together the results from a dozen or so treatments of the topic in an internally consistent manner. Certain points are emphasized that tend to be given less prominence in standard texts and articles. Where integration is used as a tool to deal with extended charge distributions, non-obvious antiderivatives are obtained from an online integrator; this is rarely encountered in textbooks, and gives the approach a more contemporary feel (admittedly, at the expense of elegance). This enables straightforward derivation of expressions for the electric and magnetic fields of radially symmetric charge and current distributions without using Gauss' or Ampère's Laws. It also allows calculation of the extent of "self-pinching" in a current-carrying wire; this appears to be a new result.

2. Electrostatics

2.1 Charge

When certain objects are rubbed together, they undergo a dramatic change. Whereas before these objects exerted no noticeable forces on their environment, they now do. For example, if you hold one of the objects near a small piece of paper, the piece of paper may jump up towards and attach itself to the object. Put this in perspective: *the entire Earth* is exerting a gravitational pull on the piece of paper, but a comparatively small object is able to exert a force big enough to overcome this pull (Arons, 1996).

If we take the standard example of rubber rods rubbed with cat fur, and glass rods rubbed with silk, we observe that all rubber rods repel each other as do all glass rods, while all rubber rods attract all glass rods. It turns out that all charged objects ever experimented on either behave like a rubber rod, or like a glass rod. This leads us to postulate that there only two types of charge state, which we call positive and negative charge for short.

As it turns out, there are also two types of charge: a positive charge as found on protons, and a negative charge as found on electrons. In this chapter, a wire will be modeled as a line of positively charged ions and negatively charged electrons; these two charge states come about through separation of one type of charge (due to electrons) from previously neutral atoms. However, the atoms themselves were electrically neutral due to equal amounts of the type of charge due to the protons in the nucleus, and the type of charge due to electrons.

Charged objects noticeably exert forces on each other when there is some distance between them. Since the 19th century, we have come to describe this behaviour in terms of electric fields. The idea is that one charged object generates a field that pervades the space around it; this field, in turn, acts on the second object.

2.2 Coulomb's Law

Late in the 18th century, Coulomb used a torsion balance to show that two small charged spheres exert a force on each other that is proportional to the inverse square of the distance between the centres of the spheres, and acts along the line joining the centres (Shamos, 1987a). He also showed that, as a consequence of this inverse square law, all charge on a conductor must reside on the surface. Moreover, by the shell theorem (Wikipedia, 2011) the forces between two perfectly spherical hollow shells are exactly as if all the charge were concentrated at the centre of each sphere. This situation is very closely approximated by two spherical insulators charged by friction, the deviation arising from a very small polarisation effect.

Coulomb also was the first person to quantify charge. For example, having completed one measurement, he halved the charge on a sphere by bringing it in contact with an identical sphere. When returning the sphere to the torsion balance, he measured that the force between the spheres had halved (Arons, 1996). When he repeated this procedure with the other sphere in the balance, the force between the spheres became one-quarter of its original value.

In modern notation, Coulomb thus found the law that bears his name: the electrostatic force $\vec{F}_E$ between two point-like objects a distance r apart, with charge Q and q respectively, is given by

$$\vec{F}_E = \frac{1}{4\pi\epsilon_0}\frac{Qq}{r^2}\hat{r}. \tag{1}$$

In SI units, the constant of proportionality is given as $1/4\pi\epsilon_0$ for convenience in calculations. The constant ϵ_0 is called the permittivity of vacuum.

It is often useful to define the charge per unit length, called the linear charge density (symbol: λ); the charge per unit (surface) area, symbol: σ; and the charge per unit volume, symbol ρ.

We are now in a position to define the electric field $\vec{E}$ mathematically. The electric field is defined as the ratio of the force on an object and its charge. Hence, generally,

$$\vec{E} \equiv \frac{\vec{F}_E}{q}, \tag{2}$$

and for the field due to a point charge Q,

$$\vec{E} = \frac{1}{4\pi\epsilon_0}\frac{Q}{r^2}\hat{r}. \tag{3}$$

Finally, experiments show that Coulomb's Law obeys the superposition principle; that is to say, the force exerted between two point-like charged objects is unaffected by the presence or absence of other point-like charged objects, and the net electrostatic force on a point-like object is found by adding all individual electrostatic forces acting on it. Of course, macroscopic objects generally are affected by other charges, for example through polarization.

2.3 An infinite line charge

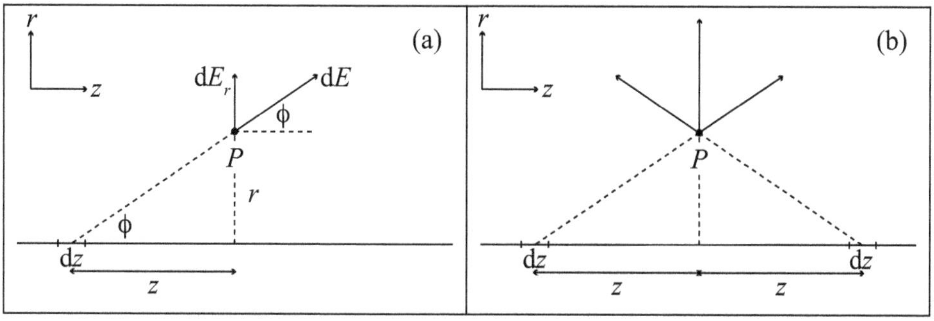

Fig. 1. Linear charges: (a) field due to a small segment of length dl, (b) net field due to two symmetrically placed segments.

Imagine an infinitely long line of uniform linear charge density λ. Take a segment of length dz, a horizontal distance z from point P which has a perpendicular distance r to the line charge. By Coulomb's Law, the magnitude of the electric field at P due this line segment is

$$\mathrm{d}E = \frac{\lambda \mathrm{d}z}{4\pi\epsilon_0(r^2+z^2)}. \tag{4}$$

A second segment of the same length dz a distance z from P (see Fig. 1b) gives rise to an electric field of the same magnitude, but pointing in a different direction. The z components cancel, leaving only the r component:

$$\mathrm{d}E_r = \frac{\lambda \mathrm{d}z \sin\phi}{4\pi\epsilon_0(r^2+z^2)}. \tag{5}$$

To find the net field at P, we add the contributions due to all line segments. This net field is thus an infinite sum, given by the integral

$$E = \frac{\lambda}{4\pi\epsilon_0}\int_{-\infty}^{\infty}\frac{\mathrm{d}z\sin\phi}{r^2+z^2}. \tag{6}$$

The integral in (6) contains two variables, z and ϕ; we must eliminate either. It can be seen from Fig. 1a that

$$\sin\phi = \frac{dE_r}{dE} = \frac{r}{(r^2+z^2)^{1/2}}, \tag{7}$$

which allows us to eliminate ϕ, yielding

$$E = \frac{\lambda r}{4\pi\epsilon_0}\int_{-\infty}^{\infty}\frac{dz}{(r^2+z^2)^{3/2}}. \tag{8}$$

The antiderivative is readily found manually, by online integrator, or from tables; the integration yields

$$\int_{-\infty}^{\infty}\frac{dz}{(r^2+z^2)^{3/2}} = \left.\frac{z}{r^2(r^2+z^2)^{1/2}}\right|_{-\infty}^{\infty} = \frac{2}{r^2}. \tag{9}$$

Hence, the electric field due to an infinity linear charge at a distance r from the line charge is given by

$$E = \frac{\lambda}{2\pi\epsilon_0 r}. \tag{10}$$

2.4 Electric field due to a uniformly charged hollow cylinder

Consider an infinitely long, infinitely thin hollow cylinder of radius R, with uniform surface charge density σ. A cross sectional view is given in Figure 2. What is the electric field at a point P, a distance y_0 from the centre of the cylinder axis? By analogy with the shell theorem,

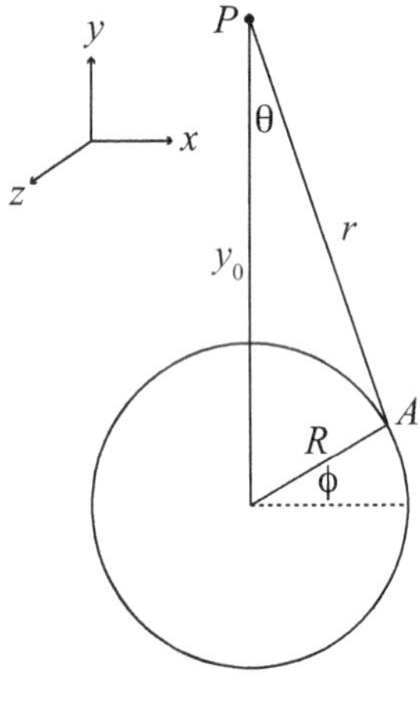

Fig. 2. Uniformly charged hollow cylinder of radius R, with auxiliary variables defined.

one might expect that the answer is the same as if all the charge were placed at the central axis. For an infinite cylinder, this turns out to be true. Think of the hollow cylinder as a collection of infinitely many parallel infinitely long line charges arranged in a circular pattern. If the angular width of each line charge is $d\phi$, then each has linear charge density $\sigma R d\phi$; by (10), each gives rise to an electric field of magnitude

$$dE = \frac{\sigma R d\phi}{2\pi\epsilon_0 r} \tag{11}$$

along the direction AP pointing away from the line charge, as shown in Figure 2.

The net field at any point P follows from superposition. We use a righthanded Cartesian coordinate system where the positive y-axis points up and the positive z-axis points out of the page. When comparing the contributions from the right half of the cylinder to the electric field with those from the left half, it is clear by symmetry that the y-components are equal and add, while the x-components are equal and subtract to yield zero. Hence

$$E = 2\int_{-\pi/2}^{\pi/2} dE_y = \frac{\sigma R}{\pi\epsilon_0}\int_{-\pi/2}^{\pi/2} \frac{\cos\theta}{r} d\phi \tag{12}$$

The integrand in (12) contains 3 variables, r, ϕ, and θ. We may write r and $\cos\theta$ in terms of ϕ and constants:

$$\begin{cases} r = \sqrt{(R\cos\phi)^2 + (R\sin\phi - y_0)^2} = \sqrt{R^2 + y_0^2 - 2Ry_0\sin\phi} \\ \cos\theta = \frac{y_0 - R\sin\phi}{r} \end{cases}; \tag{13}$$

hence

$$E = \frac{\sigma R}{\pi\epsilon_0}\int_{-\pi/2}^{\pi/2} \frac{y_0 - R\sin\phi}{R^2 + y_0^2 - 2y_0 R\sin\phi} d\phi. \tag{14}$$

When entering the integral into the Mathematica online integrator (2011), the antiderivative is given as

$$\left. \frac{-\arctan(\frac{R\cos x/2 - y_0\sin x/2}{y_0\cos x/2 - R\sin x/2})}{2y_0} + \frac{\arctan(\frac{y_0\sin x/2 - R\cos x/2}{y_0\cos x/2 - R\sin x/2})}{2y_0} + \frac{x}{2y_0} \right|_{-\pi/2}^{\pi/2},$$

which is admittedly ugly, but not difficult to use. Since arctan is an odd function, the first two terms are identical, and the antiderivative simplifies to

$$\left. \frac{1}{y_0}\left[\arctan\left(\frac{y_0\sin x/2 - R\cos x/2}{y_0\cos x/2 - R\sin x/2}\right) + \frac{x}{2}\right]\right|_{-\pi/2}^{\pi/2}.$$

Substitution eventually yields that the value of the integral is π/y_0. Hence Equation (14) gives for the electric field E outside the hollow cylinder:

$$E = \frac{\sigma R}{\pi\epsilon_0}\frac{\pi}{y_0}, \tag{15}$$

which, defining $\lambda = \sigma \cdot 2\pi R$, simplifies to

$$E = \frac{\lambda}{2\pi\epsilon_0 y_0}, \tag{16}$$

as expected.

2.5 Electric field due to a uniformly charged cylinder

It follows from (16) that for any cylindrical charge distribution of radius R that is a function of r only, i.e., $\rho = \rho(r)$, the electric field for $r > R$ is given by

$$E = \frac{\lambda}{2\pi\epsilon_0 r}, \tag{17}$$

where the linear charge density λ is equal to the volume charge density ρ integrated over the radial and polar coordinates.

3. Magnetic fields and current-carrying wires

3.1 Current

The flow of charge is called current. To be more precise, define a cross sectional area A through which a charge $\mathrm{d}Q$ flows in a time interval $\mathrm{d}t$. The current I through this area is defined as

$$I \equiv \frac{\mathrm{d}Q}{\mathrm{d}t}. \tag{18}$$

It is often convenient to define a current density J, which is the current per unit cross sectional area A:

$$J \equiv I/A. \tag{19}$$

A steady current flowing through a homogeneous wire can be modeled as a linear charge density λ moving at constant drift speed v_d. In that case, the total charge flowing through a cross sectional area in a time interval Δt is given by $\lambda v_d \Delta t$, and

$$I = \lambda v_d. \tag{20}$$

3.2 Magnetic field due to a linear current

In this chapter, we will only concern ourselves with magnetic effects due to straight current-carrying wires. Oersted found experimentally that a magnet (compass needle) gets deflected when placed near a current-carrying wire (Shamos, 1987b). As in electrostatics, we model this behaviour by invoking a field: the current in the wire creates a magnetic field B that acts on the magnet.

In subsequent decades, experiments showed that moving charged objects are affected by magnetic fields. The magnetostatic force (so called because the source of the magnetic field is steady; it is also often called the Lorentz force) is proportional to the charge q, the speed v, the field B, and the sine of the angle ϕ between v and B; it is also perpendicular to v and B. In vector notation,

$$\vec{F}_m = q\vec{v} \times \vec{B}; \tag{21}$$

in scalar notation,

$$F_m = qvB \sin\phi. \tag{22}$$

As a corollary, two parallel currents exert a magnetostatic force on each other, as the charges in each wire move in the magnetic field of the other wire.

Just as Coulomb was able to abstract from a charged sphere to a point charge, the effect of a current can be abstracted to a steady "point-current" of length dl. (Note that a single moving point charge does not constitute a *steady* point-current.) In fact, there is a close analogy between the electric field due to a line of static charges and the magnetic field due to a line segment of moving charges – i.e., a steady linear current. The Biot-Savart law states that the magnetic field at a point P due to a steady point current is given by

$$\mathrm{d}B = \frac{\mu_0 I}{2\pi} \frac{\mathrm{d}z \sin\phi}{R^2}, \tag{23}$$

where μ_0 is a constant of proportionality called the permeability of vacuum, I is the current, dz is the length of an infinitesimal line segment, ϕ is the angle between the wire and the line connecting the segment to point P, the length of which is R; see Figure 3. Maxwell (1865) showed that μ_0 and ϵ_0 are related; their product is equal to $1/c^2$, where c is the speed of light in vacuum.

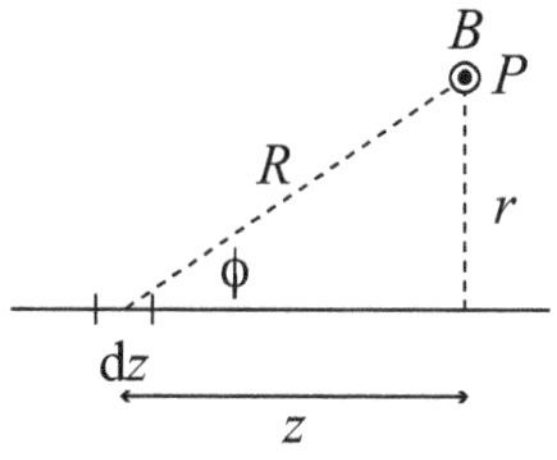

Fig. 3. The Biot-Savart law: magnetic field due to a small segment carrying a current I. The direction of the magnetic field is out of the page.

The magnetostatic force at point P due to an infinitely long straight current-carrying wire is then

$$B = \frac{\mu_0 I}{2\pi} \int_{-\infty}^{\infty} \frac{\mathrm{d}z \sin\phi}{r^2 + z^2}, \tag{24}$$

which has the exact same form as (6).

Because the current distribution must have radial symmetry, all conclusions reached from (6) can be applied here. Thus, the magnetic field due to a steady current I in an infinitely long wire, hollow cylinder, or solid cylinder where the current density only depends on the distance from the centre of the wire, varies with the distance r as

$$B = \frac{\mu_0 I}{2\pi r} \tag{25}$$

outside the wire.

4. Special relativity

4.1 Relativity in Newtonian mechanics

Newton's laws of motion were long assumed to be valid for all inertial reference frames. In Newton's model, an observer in one reference frame measures the position x of an object at various times t. An observer in a second reference frame moves with speed v relative to the first frame, with identical, synchronized clocks and metre sticks. Time intervals and lengths are assumed to be same for both observers.

The second observer sees the first observer move away at speed v. The distance between the two observers at a time t' is given by vt'. Hence, the second observer can use the measurements of the first observer, provided the following changes are made:

$$x' = x - vt \tag{26}$$

$$t' = t \tag{27}$$

Equations (26) and (27) are known as a Galilean transformation. It is easy to see that if Newton's second law holds for one observer, it automatically holds for the other. For an object moving at speed u we find that

$$u' \equiv \frac{\mathrm{d}x'}{\mathrm{d}t'} = \frac{\mathrm{d}x'}{\mathrm{d}t} = \frac{\mathrm{d}x}{\mathrm{d}t} - v = u - v, \tag{28}$$

so we get

$$a' = \frac{\mathrm{d}^2x'}{\mathrm{d}t'^2} = \frac{\mathrm{d}^2x'}{\mathrm{d}t^2} = \frac{\mathrm{d}^2x}{\mathrm{d}t^2} = a. \tag{29}$$

Hence, in both reference frames, the accelerations are the same, and hence the forces are the same, too.

4.2 The wave equation in two inertial reference frames

A problem occurs when we consider light waves. The transformation (28) implies that, in a rest frame travelling at the speed of light c with respect to an emitter, light would be at rest – it is not clear how that could be.

To put this problem on a firmer mathematical footing, we derive the general linear transformation of the wave equation; we then substitute in the Galilean transformation. For an electromagnetic wave, the electric field E satisfies, in one reference frame,

$$\frac{\partial^2 E}{\partial x^2} - \frac{1}{c^2}\frac{\partial^2 E}{\partial t^2} = 0. \tag{30}$$

We can express the derivative with respect to x in terms of variables used in another reference frame, x' and t', by using the chain rule:

$$\frac{\partial E}{\partial x} = \frac{\partial E}{\partial x'}\frac{\partial x'}{\partial x} + \frac{\partial E}{\partial t'}\frac{\partial t'}{\partial x}. \tag{31}$$

The second derivative contains five terms:

$$\frac{\partial^2 E}{\partial x^2} = \frac{\partial^2 E}{\partial x'^2}\left(\frac{\partial x'}{\partial x}\right)^2 + 2\frac{\partial^2 E}{\partial x' \partial t'}\frac{\partial x'}{\partial x}\frac{\partial t'}{\partial x} + \frac{\partial^2 x'}{\partial x^2}\frac{\partial E}{\partial x'} + \frac{\partial^2 E}{\partial t'^2}\left(\frac{\partial t'}{\partial x}\right)^2 + \frac{\partial^2 t'}{\partial x^2}\frac{\partial E}{\partial t'}. \tag{32}$$

For linear transformations, the third and fifth terms are zero. Hence we obtain:

$$\frac{\partial^2 E}{\partial x^2} = \frac{\partial^2 E}{\partial x'^2}\left(\frac{\partial x'}{\partial x}\right)^2 + 2\frac{\partial^2 E}{\partial x' \partial t'}\frac{\partial x'}{\partial x}\frac{\partial t'}{\partial x} + \frac{\partial^2 E}{\partial t'^2}\left(\frac{\partial t'}{\partial x}\right)^2. \tag{33}$$

The second derivative with respect to time is, likewise:

$$\frac{\partial^2 E}{\partial t^2} = \frac{\partial^2 E}{\partial x'^2}\left(\frac{\partial x'}{\partial t}\right)^2 + 2\frac{\partial^2 E}{\partial x' \partial t'}\frac{\partial x'}{\partial t}\frac{\partial t'}{\partial t} + \frac{\partial^2 E}{\partial t'^2}\left(\frac{\partial t'}{\partial t}\right)^2. \tag{34}$$

Substituting all this back into the wave equation, and grouping judiciously, we obtain

$$\frac{\partial^2 E}{\partial x'^2}\left[\left(\frac{\partial x'}{\partial x}\right)^2 - \frac{1}{c^2}\left(\frac{\partial x'}{\partial t}\right)^2\right] - \frac{1}{c^2}\frac{\partial^2 E}{\partial t'^2}\left[\left(\frac{\partial t'}{\partial t}\right)^2 - c^2\left(\frac{\partial t'}{\partial x}\right)^2\right] = 2\frac{\partial^2 E}{\partial x' \partial t'}\left[\frac{1}{c^2}\frac{\partial x'}{\partial t}\frac{\partial t'}{\partial t} - \frac{\partial x'}{\partial x}\frac{\partial t'}{\partial x}\right]. \tag{35}$$

To retain the wave equation (30), it is clear that the right-hand side of this equation must be zero while the terms in square brackets on the left-hand side must be equal. This is not true for the Galilean transformation, since we obtain:

$$\frac{\partial^2 E}{\partial x'^2}\left(1 - \frac{v^2}{c^2}\right) - \frac{1}{c^2}\frac{\partial^2 E}{\partial t'^2} = -\frac{2v}{c^2}\frac{\partial^2 E}{\partial x' \partial t'}. \tag{36}$$

4.3 Principles of special relativity

Einstein's theory of special relativity resolved the problem. In special relativity, velocities measured in two different reference frames can no longer be added as Newton did, because one observer disagrees with the time intervals and lengths measured by the other observer. As a result, the wave equation has the same form to all inertial observers, with the same value for the speed of light, c. Newton's laws of motion are modified in such a way that in all situations they were originally developed for (e.g., uncharged objects moving at speeds much smaller than the speed of light), the differences are so small as to be practically immeasurable. However, when we look at currents it turns out that these very small differences do matter in everyday situations.

In special relativity, all inertial frames are equivalent – meaning that all laws of physics are the same, as they are in Galilean relativity. However, rather than postulating that time and space are the same ("invariant") for all inertial observers, it is postulated that the speed of light c is invariant: it is measured to be the same in all reference frames by all inertial observers. As a consequence, measurements of time and space made in one reference frame that is moving with respect to another are different – even though the measurements may be made in the exact same way as seen from within each reference system. Seen from one reference system, a clock travelling at constant speed appears to be ticking more slowly, and appears contracted in the direction of motion. Also, if there is more than one clock at different locations, the clocks can only be synchronized according to one observer, but not simultaneously to another observer in a different reference frame.

These ideas can be investigated with an imaginary device – a light clock. Because both observers agree that light travels at speed c in both reference frames, this allows us to compare measurements in the two reference frames. Both observers agree that their own light clock consists of two mirrors mounted on a ruler a distance l_0 apart, and that it takes a light pulse a time interval Δt_0 for a round trip. Both observers agree that l_0 and Δt_0 are related by

$$l_0 = c \cdot \frac{1}{2}\Delta t_0, \tag{37}$$

and both agree that this is true irrespective of the orientation of the light clock.

However, when comparing each other's measurements, the observers are in for some surprises. As motion in one direction is independent from motion in an orthogonal direction, it makes sense to distinguish between lengths parallel and perpendicular to the relative motion of the two reference frames. A very useful sequence of looking at the light clock was given by Mermin (1989):

1. length perpendicular to motion
2. time intervals
3. length parallel to motion
4. synchronization of clocks

4.4 Lengths perpendicular to motion are unaffected

In the first thought experiments, each observer has a light clock. They are parallel to each other, and perpendicular to their relative motion (see Figure 4a). We can imagine that a piece of chalk is attached to each end of each clock, so that when the two clocks overlap, each makes a mark on the other.

We arrive at a result by reductio ad absurdum. Say that observer 1 sees clock 2 contract (but his own does not, of course – the observed contraction would be due purely to relative motion). Both observers would agree on the marks made by the pieces of chalk on clock 2 – they are inside the ends of clock 1. They would also both agree that the ends of clock 1 do not mark clock 2. Special relativity demands that the laws of physics are the same for both observers: so observer 2 must see clock 1 shrink by the same factor, clock 2 retains the same length; and hence the chalk marks on clock 2 are inside the ends, while there are no marks on clock 1. Thus, we arrive at a contradiction. Assuming one observer sees the other's clocks expand lead to the same conundrum. The only possible conclusion: both observers agree that both clocks have length l_0 in both frames.

4.5 Time intervals: moving clocks run more slowly

In the same set-up, observer 1 sees the light pulse in his clock move vertically, while the light pulse in clock 2 moves diagonally (see Figure 4a). Observer 1 uses his measurements only, plus the information that clock 2 has length l_0 and that the light pulse of clock 2 moves at speed c, also as measured by observer 1. Observer 1 measures that a pulse in clock 2 goes from the bottom mirror to the top and back again in an interval Δt, which must be greater than Δt_0, as the light bouncing between the mirrors travels further at the same speed. Thus, as seen by observer 1, clock 2 takes longer to complete a tick, and runs slow; clock 1 has already started a second cycle when clock 2 completes its first.

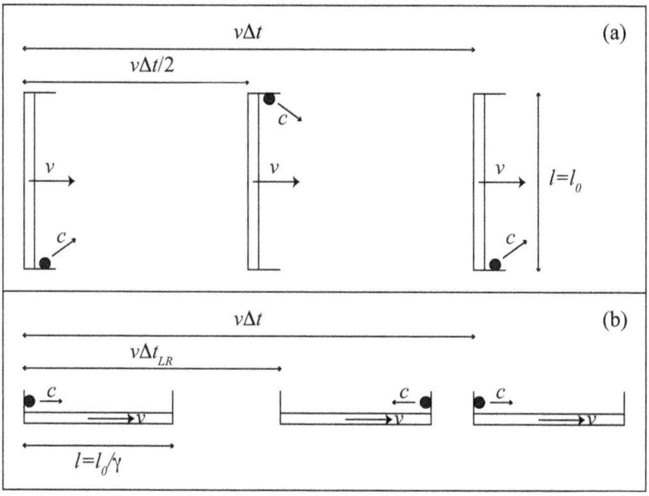

Fig. 4. A light clock in frame 2 as measured by observer 1, when the light clock is (a) perpendicular and (b) parallel to the relative motion of the two frames at speed v. The dot indicates a photon travelling at speed c in the direction indicated by the arrow. Other variables are defined in the text.

This reasoning can be quantified using Pythagoras' Theorem. Observer 1 sees that

$$\left(c \cdot \frac{1}{2}\Delta t\right)^2 - \left(v \cdot \frac{1}{2}\Delta t\right)^2 = l_0^2. \tag{38}$$

The time interval Δt can be related to the time on clock 1, Δt, because $l_0 = c \cdot \frac{1}{2}\Delta t_0$; hence

$$\Delta t^2 \left(1 - \frac{v^2}{c^2}\right) = \Delta t_0^2. \tag{39}$$

Now, defining

$$\gamma \equiv \frac{1}{\sqrt{1 - \frac{v^2}{c^2}}}, \tag{40}$$

substituting, taking the square root and dividing by γ, we conclude that

$$\Delta t = \gamma \Delta t_0. \tag{41}$$

Since all processes in frame 2 are in sync with clock 2, observer 1 sees all processes in frame 2 run slower than those in frame 1 by a factor γ. Conversely, to observer 2, everything is normal in frame 2; but observer 2 sees all processes in frame 1 run slow by the same factor γ.

4.6 Lengths parallel to relative motion are contracted

Now both observers turn their clocks through 90 degrees, so the light travels parallel to their relative motion, as shown in Figure 4b. Within their own reference frames, the clocks still run at the same rate; hence each observer sees the other's clock run slow by a factor γ as before.

However, to observer 1, after the light pulse leaves the left mirror of clock 2, that whole clock travels to the right. The pulse thus travels by a distance $l + v \cdot \Delta t_{LR}$ at speed c during a time interval Δt_{LR} before it hits the right mirror, where l is the length of the clock as measured by observer 1. Hence

$$l + v \cdot \Delta t_{LR} = c \cdot \Delta t_{LR}. \tag{42}$$

Similarly, after the pulse reflects it travels a distance $l - v \cdot \Delta t_{RL}$ before it hits the left mirror again. Observer 1 finds for the total time Δt

$$\Delta t = \Delta t_{LR} + \Delta t_{RL} = \frac{l}{c - v} + \frac{l}{c + v} = \frac{2l}{c} \cdot \frac{1}{1 - \frac{v^2}{c^2}} = \gamma^2 \frac{2l}{c}. \tag{43}$$

The time interval Δt can be linked to the time interval in frame 1, Δt_0, by (41), which, in turn, is linked to the length in frame 1, Δl_0, by (37). Straightforward substitution yields

$$l = \frac{l_0}{\gamma}. \tag{44}$$

Thus, as measured by observer 1, all lengths in frame 2 parallel to the motion are shorter than in frame 1 by a factor γ (but both perpendicular lengths are the same). As seen by observer 2, everything is normal in frame 2, but all parallel lengths in frame 1 are contracted by the same factor γ. When the two observers investigate each other's metre sticks, they both agree on how many atoms there are in the each stick, but disagree on the spacing between them.

4.7 Synchronization of clocks is only possible in one frame at a time

As it stands, it is hard to see how the observations in both frames can be reconciled. How can *both* observers see the other clocks run slowly, and the other's lengths contracted? The answer lies in synchronization. Without going into much detail, we outline some key points here.

Measuring the length of an object requires, in principle, the determination of two locations (the ends of the object) at the same time. However, when two clocks are synchronized in frame 1 according to observer 1, they are not according to observer 2. As the frames move with respect to each other, observer 2 concludes that observer 1 moved his ruler while determining the position of each end of the object. In the end, each observer can explain all measurements in a consistent fashion. For an accessible yet rigorous in-depth discussion see Mermin (1989). The end result is the transformation laws

$$x' = \gamma(x - vt) \tag{45}$$

$$t' = \gamma(t - vx/c^2) \tag{46}$$

4.8 Transformation of forces and invariance of the wave equation in special relativity

Substituting the transformations of special relativity into the wave equation (35) shows that the wave equation has the same form in both frames: the two factors in square brackets are equal to 1, and the right hand side is equal to zero.

However, Newton's Second Law does not transform in special relativity. In the situations under discussion in Section 5.1, all forces are perpendicular to the relative speed v. In that case, a force of magnitude F_0 in the rest frame is measured by an observer in a moving frame to have magnitude F given by

$$F = F_0/\gamma. \tag{47}$$

An operational definition for a transverse force is given by Martins (1982). For the sake of completeness we note that a parallel force transforms as $F = F_0/\gamma^3$.

5. Electric fields, magnetic fields, and special relativity

The considerations of the three previous sections can be brought together quite neatly. We model a current-carrying wire as a rigid lattice of ions, and a fluid of electrons that are free to move through the lattice. In the reference frame of the ions, then, the electrons move with a certain drift speed, v_d. But, by the same token, in the frame of the electrons, the ions move with a speed v_d.

We will consider four cases:

1. An infinitely thin current-carrying wire;
2. A current-carrying wire of finite width;
3. An charged object moving parallel to a current-carrying wire at speed v_d;
4. Two parallel current-carrying wires.

5.1 Length contraction in a current carrying wire

Experimental evidence shows that a stationary charge is not affected by the presence of a current-carrying wire. This absence of a net electrostatic force implies that the ion and electron charge densities in a current-carrying wire must have the same magnitude. This statement is more problematic than it may seem at first glance.

Consider the case of zero current. Call the linear charge density of the ions λ_0. By charge neutrality, the linear charge density of the electrons must be equal to $-\lambda_0$. Now let the electrons move at drift speed v_d relative to the ions, causing a current I. Experimentally, both linear charge densities remain unchanged, since a stationary charged object placed near the wire does not experience a net force. So, as seen in the ion frame, the linear electron charge density is given by:

$$\lambda_- = -\lambda_0. \tag{48}$$

In the electron frame, the linear charge density of the electrons must be

$$\lambda'_- = -\lambda_0/\gamma, \tag{49}$$

so that

$$\lambda_- = \gamma\lambda'_- = \gamma \cdot (-\lambda_0/\gamma) = -\lambda_0. \tag{50}$$

Moreover, in the electron frame, the ions are moving, and hence their linear charge density is

$$\lambda'_+ = \gamma\lambda_0. \tag{51}$$

The net charge density in the electron frame, λ', is then given by

$$\lambda' = \lambda'_+ + \lambda'_- = \gamma\lambda_0 - \lambda_0/\gamma = \lambda_0\gamma\left(1 - 1/\gamma^2\right) = \lambda_0\gamma v_d^2/c^2. \tag{52}$$

Thus, in the electron frame, the wire is charged. We cannot, however, simply assume that Coulomb's Law (6) holds; that law was obtained from experiments on stationary charges, while the ions are moving in the electron frame. In fact, the magnitude of the electric field dE due to a point charge λdz moving at speed v_d is given by (French, 1968; Purcell, 1984)

$$\mathrm{d}E = \frac{\lambda \mathrm{d}z\,(1 - v_d^2/c^2)}{4\pi\epsilon_0(r^2 + z^2)\left(1 - \frac{v_d^2}{c^2}\frac{r^2}{r^2+z^2}\right)}, \tag{53}$$

using the notation of Figure 1a. However, when we integrate the radial component of this electric field, we *do* obtain the same result; switching to primed coordinates to denote the electron frame,

$$E' = \frac{\lambda'}{2\pi\epsilon_0 r'} = \frac{\lambda_0\gamma v_d^2}{2\pi\epsilon_0 r}. \tag{54}$$

We have used the fact that lengths perpendicular to motion do not contract; hence $r = r'$.

5.2 Current and charge distribution within a wire

Now consider a wire of finite radius, R. We can model this as an infinite number of parallel infinitely thin wires placed in a circle. Assume that each wire starts out as discussed above.

As seen in the ion frame, there are many electron currents in the same direction; each current will set up a magnetostatic field, the net effect of which will be an attraction towards the centre. However, once the electrons start to migrate towards the centre, a net negative charge is created in the centre of the wire; equilibrium is established when the two cancel (Gabuzda, 1993; Matzek & Russell, 1968).

As seen in the electron frame, there is a current of positive ions, but since the ion frame is assumed perfectly rigid, no redistribution of charge occurs as a result. However, due to length contraction there is also a net positive charge density, which will attract the electrons towards the centre of the wire (Gabuzda, 1993). This must occur in such a way that the net electric field is zero; this, in turn, can only happen if the net volume charge density is zero. Consequently, the linear electron density is distorted: within a radius a, a uniform electron volume charge density is established that is equal to the ion volume charge density; between a and R, the electron density is zero.

The magnitude of this effect can be calculated easily. The uniform ion volume charge density is given by

$$\rho'_+ = \frac{\lambda'_+}{\pi R^2} = \frac{\gamma\lambda_0}{\pi R^2}; \tag{55}$$

this must be equal to (minus) the uniform electron volume density over a radius a. Hence we obtain

$$\rho'_- = \frac{\lambda'_-}{\pi a^2} = -\frac{\lambda_0}{\gamma\pi a^2}. \tag{56}$$

Combining the two yields

$$a = R/\gamma. \tag{57}$$

Thus, the wire is electrically neutral between 0 and R/γ, and has a positive volume charge density given by (55) between R/γ and R. Because in practice the outer shell is very thin, it can be approximated as a surface density:

$$\sigma' = \rho'_+ \cdot 2\pi R = 2\gamma\lambda_0/R. \tag{58}$$

As we have seen, lengths perpendicular to the motion do not change. Hence in the ion frame the electrons comprise a uniform line of electrons moving at speed v_d; in other words, there is a "self-pinched" uniform current density given by

$$J = \frac{I}{\pi a^2} = \frac{\gamma^2 I}{\pi R^2} \tag{59}$$

between 0 and R/γ, and zero current density between R/γ and R.

Figure 5 shows some relevant current and volume charge densities in both reference frames. Note that, by the considerations of Section 2.4, for $r > R$ we may treat the wire as if all current and charge were located on the central axis of the wire.

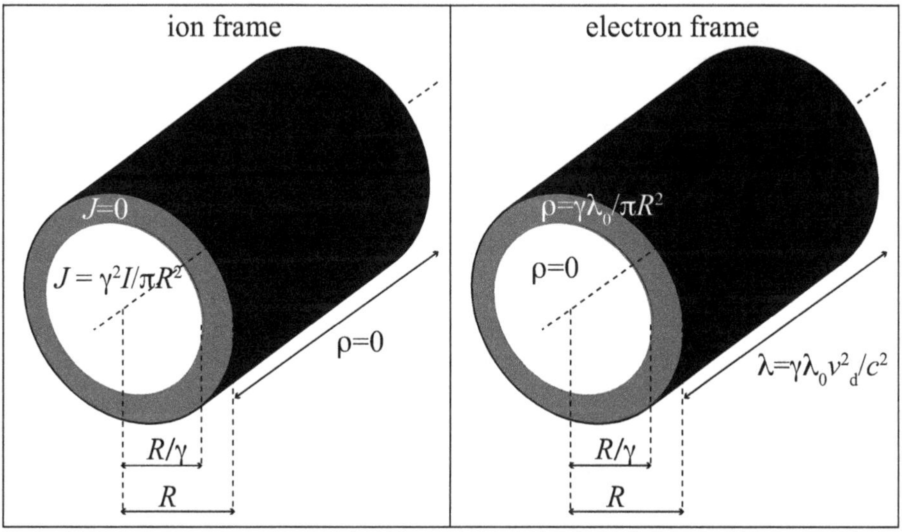

Fig. 5. Current and charge densities in a current-carrying wire. $\gamma = 1/\sqrt{1 - v_d^2/c^2}$, where v_d is the relative speed of the ions and electrons, and λ_0 is the linear ion density as seen in the ion frame.

5.3 A charged object near a current carrying wire

We have established that in the ion frame, a current-carrying wire does not exert an electrostatic force; but in the electron frame, it does. There is nothing wrong with this, but we must make sure that the effect on charges near the wire is the same in both frames; otherwise, the principle of relativity would be violated.

First, consider a point-like object of charge q stationary in the ion frame. There is no electrostatic force on the object, since there is no net charge; nor is there a magnetostatic force, because the speed of the object is zero. In the electron frame, the electrostatic force F_e' is given by

$$F_e' = qE' = \frac{q\lambda_0 \gamma v_d^2}{2\pi\epsilon_0 c^2 r'}. \tag{60}$$

There is also a magnetostatic force, F_m', as in the electron frame the object is moving with speed v_d parallel to a current of positive ions; hence

$$F_m' = qv_d B' = \frac{qv_d \mu_0 I}{2\pi r'}. \tag{61}$$

The two forces are readily shown to be equal, as $I = \lambda_+' v_d = \gamma\lambda_0 v_d$ and $\mu_0 = 1/\epsilon_0 c^2$. The forces cancel, because a current of positive ions attracts a positively charged object while the positive charge density repels it.

As a second case, a point-like object of charge q, moving parallel to a current carrying wire at speed v_d in the ion frame, experiences a purely magnetostatic force due to the electrons in the wire, given by:

$$F = qv_d B = \frac{qv_d \mu_0 I}{2\pi r} = \frac{qv_d^2 \lambda_0}{2\pi\epsilon_0 c^2 r}, \tag{62}$$

since $I = \lambda_- v_d = \lambda_0 v_d$. In the electron frame, the speed of q is zero so the force is purely electrostatic:

$$F' = qE' = \frac{q\lambda'}{2\pi\epsilon_0 r} = \frac{\gamma q v_d^2 \lambda_0}{2\pi\epsilon_0 c^2 r}. \tag{63}$$

The electron frame is the rest frame for q; hence (47) becomes

$$F = F'/\gamma, \tag{64}$$

which is obviously satisfied. Hence, what appears as a purely magnetostatic force in the ion frame appears as a purely electrostatic force in the electron frame.

In the general case, where a point-like object of charge q is moving at any speed v (say, in the ion frame), the ions and electrons are contracted by different factors, but always in such a way that the resulting net electrostatic force is balanced by the net magnetostatic force between the charge q and both on and electron currents. This case is discussed in detail by Gabuzda (1987).

5.4 Two parallel wires

As a final case, consider two parallel wires, each carrying a current I. When considering the effect of wire 1 on wire 2, we must consider the electrons and ions in wire 2 separately, as they have no common rest frame (Redžić, 2010). We cannot reify one segment of length l in one frame and transform it as a whole, even though coincidentally the same formulae can be obtained (van Kampen, 2008; 2010).

One convenient way of looking at the problem is by considering a segment of wire 2 of length l, as measured in the ion frame. This consists of a segment of ions of length l/γ as seen in the electron frame, and a segment of electrons of length γl as seen in the electron frame. The transformation leaves the total charge unaltered: it is $\lambda_0 l$.

In the ion frame, we can find the total force on the ion segment by dividing it up into point-like segments of charge density λ_0, and integrating over the entire length l. An identical procedure holds for the electrons. In the electron frame, we integrate ion segments of charge density $\gamma\lambda_0$ over a length l/γ, and electron segments of charge density λ_0/γ over a length γl. The net result is that all expressions found in the previous paragraph hold, if we replace q with $\lambda_0 l$:

$$F_{e+} = F_{m+} = F_{e-} = F'_{m-} = 0 \tag{65}$$

$$F'_{e+} = F'_{e-} = F'_{m+} = \frac{\gamma\lambda_0^2 v_d^2 l}{2\pi\epsilon_0 c^2 r} \tag{66}$$

$$F_{m-} = \frac{\lambda_0^2 v_d^2 l}{2\pi\epsilon_0 c^2 r} \tag{67}$$

6. Conclusion

In this chapter, we have outlined how electrostatics, magnetostatics and special relativity give consistent results for a few cases involving infinitely long current carrying wires. We have used an online integrator to obtain expression for the electrostatic field due to a hollow uniformly charged cylinder, and derived expressions for a solid uniformly charged cylinder and current-carrying wires from it. We have also derived expressions for self-pinching in a current-carrying wire, by a factor γ, and the creation of a surface charge density.

7. Acknowledgement

The author gratefully acknowledges fruitful discussions with Enda McGlynn and Mossy Kelly.

8. References

Arons, A.B. (1996). *Teaching Introductory Physics*, 167–187, John Wiley and Sons, ISBN 0-471-13707-3, New York.

French, A.P. (1968). *Special Relativity*, 221–225, 251–254, 262–264, Nelson, ISBN 0-393-09793-5, London.

Gabuzda, D.C. (1987). Magnetic force due to a current-carrying wire: a paradox and its resolution. Am. J. Phys. 55, 420–422.

Gabuzda, D.C. (1993). The charge densities in a current-carrying wire. Am. J. Phys. 61, 360–362.

Martins, R. de A. (1982). Force measurement and force transformation in special relativity. Am. J. Phys. 50, 1008–1011.

Mathematica online integrator, http://integrals.wolfram.com/; last accessed 30 August 2011.

Matzek, M.A. & Russell, B.R. (1968). On the Transverse Electric Field within a Conductor Carrying a Steady Current. Am. J. Phys. 36, 905–907.

Maxwell, J.C. (1865). A Dynamical Theory of the Electromagnetic Field, Philosophical Transactions of the Royal Society of London 155, 459–512.

Mermin, N.D. (1989). *Space and Time in Special Relativity*, Waveland Press, ISBN 0-881-33420-0, Long Grove.

Purcell, E.M. (1984). *Electricity and Magnetism*, Second Edition, McGraw-Hill, ISBN 0-070-04908-4, New York.

Redžić, D.V. (2010) Comment on 'Lorentz contraction and current-carrying wires'. Eur. J. Phys. 31, L25–L27.

Shamos, M.H. (1987a). *Great Experiments in Physics*, 59–75, Dover, ISBN 0-486-25346-5, New York.

Shamos, M.H. (1987b). *Great Experiments in Physics*, 121–128, Dover, ISBN 0-486-25346-5, New York.

van Kampen, P. (2008). Lorentz contraction and current-carrying wires. Eur. J. Phys. 29, 879–883.

van Kampen, P. (2010) Reply to 'Comment on "Lorentz contraction and current-carrying wires" '. Eur. J. Phys. 31, L25–L27.

Wikipedia, http://en.wikipedia.org//wiki/Shell_theorem/; last accessed 30 August 2011.

3

Foundations of Electromagnetism, Equivalence Principles and Cosmic Interactions

Wei-Tou Ni[1,2]
[1]Center for Gravitation and Cosmology
Department of Physics, National Tsing Hua University, Hsinchu,
[2]Shanghai United Center for Astrophysics
Shanghai Normal University, Shanghai,
[1]Taiwan, ROC
[2]China

1. Introduction

Standard electromagnetism is based on Maxwell equations and Lorentz force law. It can be derived by a least action with the following Lagrangian density for a system of charged particles in Gaussian units (e.g., Jackson, 1999),

$$L_{EMS}=L_{EM}+L_{EM\text{-}P}+L_P=-(1/(16\pi))[(1/2)\eta^{ik}\eta^{jl}-(1/2)\eta^{il}\eta^{kj}]F_{ij}F_{kl}-A_kj^k-\Sigma_I m_I[(ds_I)/(dt)]\delta(\mathbf{x}-\mathbf{x}_I), \quad (1)$$

where $F_{ij} \equiv A_{j,i} - A_{i,j}$ is the electromagnetic field strength tensor with A_i the electromagnetic 4-potential and comma denoting partial derivation, η^{ij} is the Minkowskii metric with signature (+, -, -, -), m_I the mass of the Ith charged particle, s_I its 4-line element, and j^k the charge 4-current density. Here, we use Einstein summation convention, i.e., summation over repeated indices. There are three terms in the Lagrangian density L_{EMS} – (i) L_{EM} for the electromagnetic field, (ii) $L_{EM\text{-}P}$ for the interaction of electromagnetic field and charged particles and (iii) L_P for charged particles.

The electromagnetic field Lagrangian density can be written in terms of the electric field **E** [$\equiv (E_1, E_2, E_3) \equiv (F_{01}, F_{02}, F_{03})$] and the magnetic induction **B** [$\equiv (B_1, B_2, B_3) \equiv (F_{32}, F_{13}, F_{21})$] as

$$L_{EM} = (1/8\pi)[\mathbf{E}^2-\mathbf{B}^2]. \quad (2)$$

This classical Lagrangian density is based on the photon having zero mass. To include the effects of nonvanishing photon mass m_{photon}, a mass term L_{Proca},

$$L_{Proca} = (m_{photon}{}^2c^2/8\pi\hbar^2)(A_kA^k), \quad (3)$$

needs to be added (Proca, 1936a, 1936b, 1936c, 1937, 1938). We use η^{ij} and its inverse η_{ij} to raise and lower indices. With this term, the Coulomb law is modified to have the electric potential A_0,

$$A_0 = q(e^{-\mu r}/r), \quad (4)$$

where q is the charge of the source particle, r is the distance to the source particle, and μ ($\equiv m_{photon}c/\hbar$) gives the inverse range of the interaction.

Experimental test on Coulomb's law (Williams, Faller & Hill, 1971) gives a constraint of the photon mass as

$$m_{photon} \leq 10^{-14}\ \text{eV}\ (= 2 \times 10^{-47}\ \text{g}), \tag{5}$$

on the interaction range μ^{-1} as

$$\mu^{-1} \geq 2 \times 10^{7}\ \text{m}. \tag{6}$$

Photon mass affects the structure and the attenuation of magnetic field and therefore can be constrained by measuring the magnetic field of Earth, Sun or an astronomical body (Schrödinger, 1943; Bass & Schrödinger, 1955). From the magnetic field measurement of Jupiter during Pioneer 10 flyby, constraints are set as (Davis, Goldhaber & Nieto, 1975)

$$m_{photon} \leq 4 \times 10^{-16}\ \text{eV}\ (= 7 \times 10^{-49}\ \text{g});\ \mu^{-1} \geq 5 \times 10^{8}\ \text{m}. \tag{7}$$

Using the plasma and magnetic field data of the solar wind, constraints on the photon mass are set recently as (Ryutov, 2007)

$$m_{photon} \leq 10^{-18}\ \text{eV}\ (= 2 \times 10^{-51}\ \text{g});\ \mu^{-1} \geq 2 \times 10^{11}\ \text{m}. \tag{8}$$

Large-scale magnetic fields in vacuum would be direct evidence for a limit on their exponential attenuation with distance, and hence a limit on photon mass. Using observations on galactic sized fields, Chibisov limit is obtained (Chibisov, 1976)

$$m_{photon} \leq 2 \times 10^{-27}\ \text{eV}\ (= 4 \times 10^{-60}\ \text{g});\ \mu^{-1} \geq 10^{20}\ \text{m}. \tag{9}$$

For a more detailed discussion of this work and for a comprehensive review on the photon mass, please see Goldhaber and Nieto (2010).

As larger scale magnetic field discovered and measured, the constraints on photon mass and on the interaction range may become more stringent. If cosmic scale magnetic field is discovered, the constraint on the interaction range may become bigger or comparable to Hubble distance (of the order of radius of curvature of our observable universe). If this happens, the concept of photon mass may lose significance amid gravity coupling or curvature coupling of photons.

Now we turn to quantum corrections to classical electrodynamics. In classical electrodynamics, the Maxwell equations are linear in the electric field **E** and magnetic field **B**, and we have the principle of superposition of electromagnetic field in vacuum. However, in the electrodynamics of continuous matter, media are usually nonlinear and the principle of superposition of electromagnetic field is not valid. In quantum electrodynamics, due to loop diagrams like the one in Fig. 1, photon can scatter off photon in vacuum. This is the origin of invalidity of the principle of superposition and makes vacuum a nonlinear medium also. The leading order of this effect in slowly varying electric and magnetic field is derived in Heisenberg and Euler (1936) and can be incorporated in the Heisenberg-Euler Lagrangian density

$$L_{Heisenberg\text{-}Euler} = [2\alpha^2\hbar^2/45(4\pi)^2 m^4 c^6][(\mathbf{E}^2\text{-}\mathbf{B}^2)^2 + 7(\mathbf{E}\cdot\mathbf{B})^2], \tag{10}$$

where α is the fine structure constant and m the electron mass. In terms of critical field strength B_c defined as

$$B_c \equiv E_c \equiv m^2c^3/e\hbar = 4.4\text{x}10^{13}\ \text{G} = 4.4\text{x}10^9\ \text{T} = 4.4\text{x}10^{13}\ \text{statvolt/cm} = 1.3\text{x}10^{18}\ \text{V/m}, \quad (11)$$

this Lagrangian density can be written as

$$L_{Heisenberg\text{-}Euler} = (1/8\pi)\ B_c^{-2}\ [\eta_1(\mathbf{E}^2\text{-}\mathbf{B}^2)^2 + 4\eta_2(\mathbf{E}\cdot\mathbf{B})^2], \quad (12)$$

$$\eta_1 = \alpha/(45\pi) = 5.1\text{x}10^{-5} \text{ and } \eta_2 = 7\alpha/(180\pi) = 9.0\ \text{x}10^{-5}. \quad (13)$$

For time varying and space varying effects of external fields, and higher order corrections in quantum electrodynamics, please see Dittrich and Reuter (1985) and Kim (2011a, 2011b) and references therein.

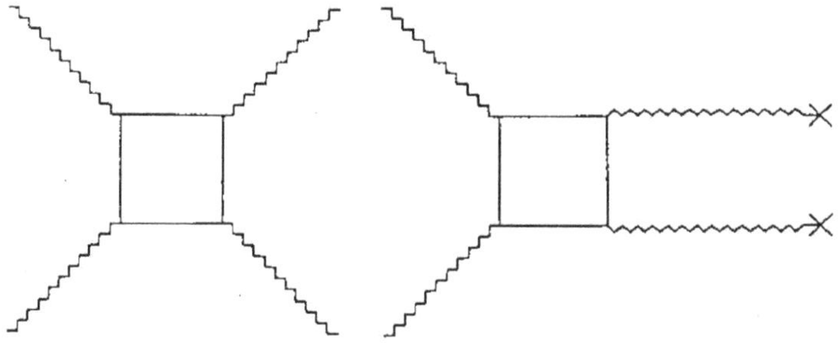

Fig. 1. On the left is the basic diagram for light-light scattering and for nonlinear electrodynamics; on the right is the basic diagram for the nonlinear light (electromagnetic-wave) propagation in strong electric and/or magnetic field.

Before Heisenberg & Euler (1936), Born and Infeld (Born, 1934; Born & Infeld, 1934) proposed the following Lagrangian density for the electromagnetic field

$$L_{Born\text{-}Infeld} = -(b^2/4\pi)\ [1 - (\mathbf{E}^2\text{-}\mathbf{B}^2)/b^2 - (\mathbf{E}\cdot\mathbf{B})^2/b^4]^{1/2}, \quad (14)$$

where b is a constant which gives the maximum electric field strength. For field strength small compared with b, (14) can be expanded into

$$L_{Born\text{-}Infeld} = (1/8\pi)\ [(\mathbf{E}^2\text{-}\mathbf{B}^2) + (\mathbf{E}^2\text{-}\mathbf{B}^2)^2/b^2 + (\mathbf{E}\cdot\mathbf{B})^2/b^2 + O(b^{-4})]. \quad (15)$$

The lowest order of Born-Infeld electrodynamics agrees with the classical electrodynamics. The next order corrections are of the form of Eq. (12) with

$$\eta_1 = \eta_2 = B_c^2/b^2. \quad (16)$$

In the Born-Infeld electrodynamics, b is the maximum electric field. Electric fields at the edge of heavy nuclei are of the order of 10^{21} V/m. If we take b to be 10^{21} V/m, then, $\eta_1 = \eta_2 = 5.9 \times 10^{-6}$.

For formulating a phenomenological framework for testing corrections to Maxwell-Lorentz classical electrodynamics, we notice that ($\mathbf{E}^2$-$\mathbf{B}^2$) and ($\mathbf{E}\cdot\mathbf{B}$) are the only Lorentz invariants second order in the field strength, and ($\mathbf{E}^2$-$\mathbf{B}^2$)^2, ($\mathbf{E}\cdot\mathbf{B}$)^2 and ($\mathbf{E}^2$-$\mathbf{B}^2$) ($\mathbf{E}\cdot\mathbf{B}$) are the only Lorentz invariants fourth order in the field strength. However, ($\mathbf{E}\cdot\mathbf{B}$) is a total divergence and, by itself in the Lagrangian density, does not contribute to the equation of motion (field equation). Multiplying (**E**·**B**) by a pseudoscalar field Φ, the term $\Phi(\mathbf{E}\cdot\mathbf{B})$ is the Lagrangian density for the pseudoscalar-photon (axion-photon) interaction. When this term is included together with the fourth-order invariants, we have the following phenomenological Lagrangian density for our Parametrized Post-Maxwell (PPM) Lagrangian density including various corrections and modifications to be tested by experiments and observations,

$$L_{PPM} = (1/8\pi)\{(\mathbf{E}^2-\mathbf{B}^2)+\xi\Phi(\mathbf{E}\cdot\mathbf{B})+B_c^{-2}[\eta_1(\mathbf{E}^2-\mathbf{B}^2)^2+4\eta_2(\mathbf{E}\cdot\mathbf{B})^2+2\eta_3(\mathbf{E}^2-\mathbf{B}^2)(\mathbf{E}\cdot\mathbf{B})]\}. \quad (17)$$

This PPM Lagrangian density contains 4 parameters ξ, η_1, η_2 & η_3, and is an extension of the two-parameteer (η_1 and η_2) post-Maxwellian Lagrangian density of Denisov, Krivchenkov and Kravtsov (2004). The manifestly Lorentz covariant form of Eq. (17) is

$$L_{PPM} = (1/(32\pi))\{-2F^{kl}F_{kl} -\xi\Phi F^{*kl}F_{kl}+B_c^{-2}[\eta_1(F^{kl}F_{kl})^2+\eta_2(F^{*kl}F_{kl})^2+\eta_3(F^{kl}F_{kl})(F^{*ij}F_{ij})]\}, \quad (18)$$

where

$$F^{*ij} \equiv (1/2)e^{ijkl}\, F_{kl}, \quad (19)$$

with e^{ijkl} defined as

$$e^{ijkl} \equiv 1 \text{ if } (ijkl) \text{ is an even permutation of (0123); -1 if odd; 0 otherwise.} \quad (20)$$

In section 2, we derive the PPM nonlinear electrodynamic equations, and in section 3, we use them to derive the light propagation equation in PPM nonlinear electrodynamics. In section 4, we discuss ultra-high precision laser interferometry experiments to measure the parameters of PPM electrodynamics. In section 5, we treat electromagnetism in curved spacetime using Einstein Equivalence Principle, and discuss redshift as an application with examples from astrophysics and navigation. In section 6, we discuss empirical tests of electromagnetism in gravity and the χ-g framework and find pseudoscalar-photon interaction uniquely standing out. In section 7, we discuss the pseudoscalar-photon interaction and its relation to other approaches. In section 8, we use Cosmic Microwave Background (CMB) observations to constrain the cosmic polarization rotation and discuss radio galaxy observations. In section 9, we present a summary and an outlook briefly.

2. Equations for nonlinear electrodynamics

In analogue with the nonlinear electrodynamics of continuous media, we can define the electric displacement **D** and the magnetic field **H** as follows:

$$\mathbf{D}\equiv4\pi(\partial L_{PPM}/\partial\mathbf{E})=[1+2\eta_1(\mathbf{E}^2-\mathbf{B}^2)B_c^{-2}+2\eta_3(\mathbf{E}\cdot\mathbf{B})B_c^{-2}]\mathbf{E}+[\Phi+4\eta_2(\mathbf{E}\cdot\mathbf{B})B_c^{-2}+\eta_3(\mathbf{E}^2-\mathbf{B}^2)B_c^{-2}]\mathbf{B}, \quad (21)$$

$$\mathbf{H}\equiv-4\pi(\partial L_{PPM}/\partial\mathbf{B})=[1+2\eta_1(\mathbf{E}^2-\mathbf{B}^2)B_c^{-2}+2\eta_3(\mathbf{E}\cdot\mathbf{B})B_c^{-2}]\mathbf{B}-[\Phi+4\eta_2(\mathbf{E}\cdot\mathbf{B})B_c^{-2}+\eta_3(\mathbf{E}^2-\mathbf{B}^2)B_c^{-2}]\mathbf{E}. \quad (22)$$

From **D** & **H**, we can define a second-rank G_{ij} tensor, just like from **E** & **B** to define F_{ij} tensor. With these definitions and following the standard procedure in electrodynamics [see, e.g., Jackson (1999), p. 599], the nonlinear equations of the electromagnetic field are

$$\text{curl}\,\mathbf{H} = (1/c)\,\partial\mathbf{D}/\partial t + 4\pi\,\mathbf{J}, \quad (23)$$

$$\text{div}\,\mathbf{D} = 4\pi\,\rho, \quad (24)$$

$$\text{curl } \mathbf{E} = -(1/c)\, \partial \mathbf{B}/\partial t, \tag{25}$$

$$\text{div } \mathbf{B} = 0. \tag{26}$$

We notice that it has the same form as in macroscopic electrodynamics. The Lorentz force law remains the same as in classical electrodynamics:

$$d[(1-\mathbf{v}_I^2/c^2)^{-1/2} m_I \mathbf{v}_I]/dt = q_I[\mathbf{E} + (1/c)\mathbf{v}_I \times \mathbf{B}] \tag{27}$$

for the I-th particle with charge q_I and velocity $\mathbf{v}_I$ in the system. The source of Φ in this system is $(\mathbf{E}\cdot\mathbf{B})$ and the field equation for Φ is

$$\partial^i L_\Phi / \partial(\partial^i \Phi) - \partial L_\Phi/\partial \Phi = \mathbf{E}\cdot\mathbf{B}, \tag{28}$$

where L_Φ is the Lagrangian density of the pseudoscalar field Φ.

3. Electromagnetic wave propagation in PPM electrodynamics

Here we follow the previous method (Ni et al., 1991; Ni, 1998), and separate the electric field and the magnetic induction field into the wave part (small compared to external part) and external part as follows:

$$\mathbf{E} = \mathbf{E}^{\text{wave}} + \mathbf{E}^{\text{ext}}, \tag{29}$$

$$\mathbf{B} = \mathbf{B}^{\text{wave}} + \mathbf{B}^{\text{ext}}. \tag{30}$$

We use the following expressions to calculate the displacement field $\mathbf{D}^{\text{wave}}$ [= $(D^{\text{wave}}{}_a)$ = $(D^{\text{wave}}{}_1, D^{\text{wave}}{}_2, D^{\text{wave}}{}_3)$] and the magnetic field $\mathbf{H}^{\text{wave}}$ [= $(H^{\text{wave}}{}_a)$ = $(H^{\text{wave}}{}_1, H^{\text{wave}}{}_2, H^{\text{wave}}{}_3)$] of the electromagnetic waves:

$$D^{\text{wave}}{}_a = D_a - D^{\text{ext}}{}_a = (4\pi)[(\partial L_{\text{PPM}}/\partial E_a)_{\mathbf{E\&B}} - (\partial L_{\text{PPM}}/\partial E_a)_{\text{ext}}], \tag{31}$$

$$H^{\text{wave}}{}_a = H_a - H^{\text{ext}}{}_a = -(4\pi)[(\partial L_{\text{PPM}}/\partial B_a)_{\mathbf{E\&B}} - (\partial L_{\text{PPM}}/\partial B_a)_{\text{ext}}], \tag{32}$$

where $(\ldots)_{\mathbf{E\&B}}$ means that the quantity inside paranthesis is evaluated at the total field values $\mathbf{E}$ & $\mathbf{B}$ and $(\ldots)_{\text{ext}}$ means that the quantity inside paranthesis is evaluated at the external field values $\mathbf{E}^{\text{ext}}$ & $\mathbf{B}^{\text{ext}}$.

Since both the total field and the external field satisfy Eqs. (23)-(26), the wave part also satisfy the same form of Eqs. (23)-(26) with the source terms subtracted:

$$\text{curl } \mathbf{H}^{\text{wave}} = (1/c)\, \partial \mathbf{D}^{\text{wave}}/\partial t, \tag{33}$$

$$\text{div } \mathbf{D}^{\text{wave}} = 0, \tag{34}$$

$$\text{curl } \mathbf{E}^{\text{wave}} = -(1/c)\, \partial \mathbf{B}^{\text{wave}}/\partial t, \tag{35}$$

$$\text{div } \mathbf{B}^{\text{wave}} = 0. \tag{36}$$

After calculating $D^{\text{wave}}{}_a$ and $H^{\text{wave}}{}_a$ from Eqs. (31) & (32), we express them in the following form:

$$D^{\text{wave}}_{a} = \Sigma_{\beta=1}^{3}\, \varepsilon_{\alpha\beta}\, E^{\text{wave}}_{\beta} + \Sigma_{\beta=1}^{3}\, \lambda_{\alpha\beta}\, B^{\text{wave}}_{\beta}, \tag{37}$$

$$H^{\text{wave}}_{a} = \Sigma_{\beta=1}^{3}\, (\mu^{-1})_{\alpha\beta}\, B^{\text{wave}}_{\beta} - \Sigma_{\beta=1}^{3}\, \lambda_{\beta a}\, E^{\text{wave}}_{\beta}, \tag{38}$$

where

$$\varepsilon_{\alpha\beta}=\delta_{\alpha\beta}[1+2\eta_1(\mathbf{E}^2-\mathbf{B}^2)B_c^{-2}+2\eta_3(\mathbf{E}\cdot\mathbf{B})B_c^{-2}]+4\eta_1E_\alpha E_\beta B_c^{-2}+4\eta_2B_\alpha B_\beta B_c^{-2}+2\eta_3(E_\alpha B_\beta+E_\beta B_\alpha)B_c^{-2}, \tag{39}$$

$$(\mu^{-1})_{\alpha\beta}=\delta_{\alpha\beta}[1+2\eta_1(\mathbf{E}^2-\mathbf{B}^2)B_c^{-2}+2\eta_3(\mathbf{E}\cdot\mathbf{B})B_c^{-2}]-4\eta_1B_\alpha B_\beta B_c^{-2}-4\eta_2E_\alpha E_\beta B_c^{-2}+2\eta_3(E_\alpha B_\beta+E_\beta B_\alpha)B_c^{-2}, \tag{40}$$

$$\lambda_{\alpha\beta}=\delta_{\alpha\beta}[\xi\varphi+4\eta_2(\mathbf{E}\cdot\mathbf{B})\, B_c^{-2}+\eta_3(\mathbf{E}^2-\mathbf{B}^2)B_c^{-2}]-4\eta_1E_\alpha B_\beta B_c^{-2}+4\eta_2B_\alpha E_\beta B_c^{-2}+2\eta_3(E_\alpha E_\beta+B_\alpha B_\beta)B_c^{-2}, \tag{41}$$

and we have dropped the upper indices 'ext' for simplicity. Note that the coefficients of B^{wave}_{β} in Eq. (37) is the negative transpose of the coefficients of E^{wave}_{β} in Eq. (38) and vice versa. This is a property derivable from the existence of Lagrangian. It is a reciprocity relation; or simply, action equals reaction.

Using eikonal approximation, we look for plane-wave solutions. Choose the z-axis in the propagation direction. Solving the dispersion relation for ω, we obtain

$$\omega_{\pm} = k\,\{1 + (1/4)\,[(J_1+J_2) \pm [(J_1-J_2)^2 + 4J^2]^{1/2}]\}, \tag{42}$$

where

$$J_1 \equiv (\mu^{-1})_{22} - \varepsilon_{11} - 2\lambda_{12}, \tag{43}$$

$$J_2 \equiv (\mu^{-1})_{11} - \varepsilon_{22} + 2\lambda_{21}, \tag{44}$$

$$J \equiv -\,\varepsilon_{12} - (\mu^{-1})_{12} + \lambda_{11} - \lambda_{22}. \tag{45}$$

Since the index of refraction n is

$$n = k/\omega, \tag{46}$$

we find

$$n_{\pm} = 1 - (1/4)\,\{(J_1+J_2) \pm [(J_1-J_2)^2 + 4J^2]^{1/2}\}. \tag{47}$$

From this formula, we notice that "no birefringence" is equivalent to $J_1=J_2$ and $J=0$. A sufficient condition for this to happen is $\eta_1 = \eta_2$, $\eta_3 = 0$, and no constraint on ξ. We will show in the following that this is also a necessary condition. The Born-Infeld electrodynamics satisfies this condition and has no birefringence in the theory.

For $\mathbf{E}^{\text{ext}} = 0$, we now derive the refractive indices in the transverse external magnetic field $\mathbf{B}^{\text{ext}}$ for the linearly polarized lights whose polarizations (electric fields) are parallel and orthogonal to the magnetic field. First, we use Eqs. (39)-(41) & Eqs. (43)-(46) to obtain

$$\varepsilon_{\alpha\beta}=\delta_{\alpha\beta}[1-2\eta_1\mathbf{B}^2B_c^{-2}]+4\eta_2B_\alpha B_\beta B_c^{-2}, \tag{48}$$

$$(\mu^{-1})_{\alpha\beta}=\delta_{\alpha\beta}[1-2\eta_1\,\mathbf{B}^2B_c^{-2}]-4\eta_1B_\alpha B_\beta B_c^{-2}, \tag{49}$$

$$\lambda_{\alpha\beta}=\delta_{\alpha\beta}[\varphi-\eta_3\mathbf{B}^2B_c^{-2}]+2\eta_3B_\alpha B_\beta B_c^{-2}, \tag{50}$$

$$J_1 =-4\eta_1B_2^{\,2}B_c^{-2} - 4\eta_2B_1^{\,2}B_c^{-2} - 4\eta_3B_1B_2B_c^{-2}, \tag{51}$$

$$J_2 = -4\eta_1 B_1^2 B_c^{-2} - 4\eta_2 B_2^2 B_c^{-2} + 4\eta_3 B_1 B_2 B_c^{-2}, \tag{52}$$

$$J = 4\eta_1 B_1 B_2 B_c^{-2} - 4\eta_2 B_1 B_2 B_c^{-2} + 2\eta_3 (B_1^2 - B_2^2) B_c^{-2}. \tag{53}$$

Using Eq. (47), we obtain the indices of refraction for this case:

$$n_\pm = 1 + \{(\eta_1+\eta_2) \pm [(\eta_1-\eta_2)^2 + \eta_3^2]^{1/2}\} (B_1^2+B_2^2) B_c^{-2}. \tag{54}$$

The condition of no birefringence in Eq. (54) means that $[(\eta_1-\eta_2)^2 + \eta_3^2]$ vanishes, i.e.,

$$\eta_1 = \eta_2,\ \eta_3 = 0, \text{ and no constraint on } \xi \tag{55}$$

This shows that Eq. (55) is a necessary condition for no birefringence. For $\mathbf{E}^{ext} = 0$, the refractive indices in the transverse external magnetic field $\mathbf{B}^{ext}$ for the linearly polarized lights whose polarizations are parallel and orthogonal to the magnetic field, are as follows:

$$n_\| = 1 + \{(\eta_1+\eta_2) + [(\eta_1-\eta_2)^2 + \eta_3^2]^{1/2}\} (\mathbf{B}^{ext})^2 B_c^{-2} \quad (\mathbf{E}^{wave} \,\|\, \mathbf{B}^{ext}), \tag{56}$$

$$n_\perp = 1 + \{(\eta_1+\eta_2) - [(\eta_1-\eta_2)^2 + \eta_3^2]^{1/2}\} (\mathbf{B}^{ext})^2 B_c^{-2} \quad (\mathbf{E}^{wave} \perp \mathbf{B}^{ext}). \tag{57}$$

For $\mathbf{B}^{ext} = 0$, we derive in the following the refractive indices in the transverse external electric field $\mathbf{E}^{ext}$ for the linearly polarized lights whose polarizations (electric fields) are parallel and orthogonal to the electric field. First, we use (39)-(41) & (43)-(46) to obtain

$$\varepsilon_{\alpha\beta} = \delta_{\alpha\beta}[1+2\eta_1 \mathbf{E}^2 B_c^{-2}] + 4\eta_1 E_\alpha E_\beta B_c^{-2}, \tag{58}$$

$$(\mu^{-1})_{\alpha\beta} = \delta_{\alpha\beta}[1+2\eta_1 \mathbf{E}^2 B_c^{-2}] - 4\eta_2 E_\alpha E_\beta B_c^{-2}, \tag{59}$$

$$\lambda_{\alpha\beta} = \delta_{\alpha\beta}[\varphi+\eta_3 \mathbf{E}^2 B_c^{-2}] + 2\eta_3 E_\alpha E_\beta B_c^{-2}, \tag{60}$$

$$J_1 = -4\eta_1 E_1^2 B_c^{-2} - 4\eta_2 E_2^2 B_c^{-2} - 4\eta_3 E_1 E_2 B_c^{-2}, \tag{61}$$

$$J_2 = -4\eta_1 E_2^2 B_c^{-2} - 4\eta_2 E_1^2 B_c^{-2} + 4\eta_3 E_1 E_2 B_c^{-2}, \tag{62}$$

$$J = -4\eta_1 E_1 E_2 B_c^{-2} + 4\eta_2 E_1 E_2 B_c^{-2} + 2\eta_3 (E_1^2 - E_2^2) B_c^{-2}. \tag{63}$$

Using (47), we obtain the indices of refraction for this case:

$$n_\pm = 1 + \{(\eta_1+\eta_2) \pm [(\eta_1-\eta_2)^2 + \eta_3^2]^{1/2}\} (E_1^2+E_2^2) B_c^{-2}. \tag{64}$$

The condition of no birefringence in (64) is the same as (55), i.e., that $[(\eta_1-\eta_2)^2 + \eta_3^2]$ vanishes. For $\mathbf{B}^{ext} = 0$, the refractive indices in the transverse external magnetic field $\mathbf{E}^{ext}$ for the linearly polarized lights whose polarizations are parallel and orthogonal to the magnetic field, are as follows:

$$n_\| = 1 + \{(\eta_1+\eta_2) + [(\eta_1-\eta_2)^2 + \eta_3^2]^{1/2}\} (\mathbf{E}^{ext})^2 B_c^{-2} \quad (\mathbf{E}^{wave} \,\|\, \mathbf{E}^{ext}), \tag{65}$$

$$n_\perp = 1 + \{(\eta_1+\eta_2) - [(\eta_1-\eta_2)^2 + \eta_3^2]^{1/2}\} (\mathbf{E}^{ext})^2 B_c^{-2} \quad (\mathbf{E}^{wave} \perp \mathbf{E}^{ext}). \tag{66}$$

The magnetic field near pulsars can reach 10^{12} G, while the magnetic field near magnetars can reach 10^{15} G. The astrophysical processes in these locations need nonlinear electrodynamics to model. In the following section, we turn to experiments to measure the parameters of the PPM electrodynamics.

4. Measuring the parameters of the PPM electrodynamics

There are four parameters η_1, η_2, η_3, and ξ in PPM electrodynamics to be measured by experiments. For the QED (Quantum Electrodynamics) corrections to classical electrodynamics, $\eta_1 = \alpha/(45\pi) = 5.1\times10^{-5}$, $\eta_2 = 7\alpha/(180\pi) = 9.0\times10^{-5}$, $\eta_3 = 0$, and $\xi = 0$. There are three vacuum birefringence experiments on going in the world to measure this QED vacuum birefringence - the BMV experiment (Battesti et al., 2008), the PVLAS experiment (Zavattini et al.. 2008) and the Q & A experiment (Chen et al., 2007; Mei et al., 2010). The birefringence Δn in the QED vacuum birefringence in a magnetic field $\mathbf{B}^{ext}$ is

$$\Delta n = n_{\parallel} - n_{\perp} = 4.0 \times 10^{-24} (\mathbf{B}^{ext}/1\mathrm{T})^2. \quad (67)$$

For 2.3 T field of the Q & A rotating permanent magnet, Δn is 2.1×10^{-23}. This is about the same order of magnitude change in fractional optical path-length that ground interferometers for gravitational-wave detection aim at. Quite a lot of techniques developed in the gravitational-wave detection community are readily applicable for vacuum birefringence detection (Ni et al., 1991).

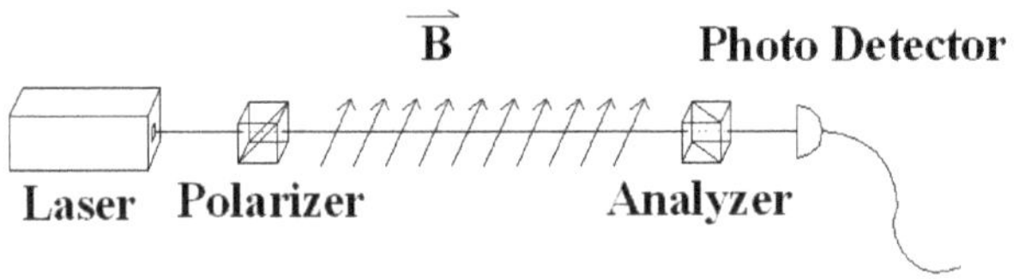

Fig. 2. Principle of vacuum dichroism and birefringence measurement.

The basic principle of these experimental measurements is shown as Fig. 2. The laser light goes through a polarizer and becomes polarized. This polarized light goes through a region of magnetic field. Its polarization status is subsequently analyzed by the analyzer-detector subsystem to extract the polarization effect imprinted in the region of the magnetic field. Since the polarization effect of vacuum birefringence in the magnetic field that can be produced on earth is extremely small, one has to multiply the optical pass through the magnetic field by using reflections or Fabry-Perot cavities. An already performed experiment, the BFRT experiment (Cameron et al., 1993) used multiple reflections; PVLAS, Q & A, BMV experiments all use Fabry-Perot cavities. For polarization experiment, Fabry-Perot cavity has the advantage of normal incidence of laser light which suppressed the part of polarization due to slant angle of reflections. With Fabry-Perot cavity, one needs to control the laser frequency and/or the cavity length so that the cavity is in resonance. With a finesse of 30,000, the resonant width (FWHM) is 17.7 pm for light with 1064 nm wavelength; when rms cavity length control is 10 % of this width, the precision would be 2.1 pm. Hence, one needs a feedback mechanism to lock the cavity to the laser or vice versa. For this, a commonly used scheme is Pound-Drever-Hall method (Drever et al., 1983). Vibration introduces noises in the Fabry-Perot cavity mirrors and hence, in the light intensity and light polarization transmitted through the Fabry-Perot cavity. Since the analyzer-detector subsystem detects light intensity to deduce the polarization effect, both intensity noise and polarization noise will contribute to the measurement results. Gravitational-wave community has a long-standing R & D on this. We benefit from their research advancements.

Now we illustrate with our Q & A experiment. Since 1991 we have worked on precision interferometry --- laser stabilization schemes, laser metrology and Fabry-Perot interferometers. With these experiences, we started in 1994 to build a 3.5 m prototype interferometer for measuring vacuum birefringence and improving the sensitivity of axion search as part of our continuing effort in precision interferometry. In 2002, we finished Phase I of constructing the 3.5 m prototype interferometer and made some Cotton-Mouton coefficient and Verdet coefficient measurements with a 1T electromagnet (Wu et al., 2002). The two vacuum tanks shown on the left photo of Fig. 3 house the two 5 cm-diameter Pabry-Perot mirrors with their suspensions; the 1T electromagnet had been in place of permanent magnet in the middle of the photo.

Fig. 3. Photo on the left-hand side shows the Q & A apparatus for Phase II experiment; photo on the right-hand side shows the Q & A apparatus for Phase III experiment.

Starting 2002, we had been in Phase II of Q & A experiment until 2008. The results of Phase II on dichroism and Cotton-Mouton effect measurement had been reported (Chen et al., 2007; Mei et al. 2009). At the end of Phase II, our sensitivity was still short from detection of QED vacuum birefringence by 3 orders of magnitude; so was the PVLAS experiment and had been the BFRT experiment. In 2009, we started Phase III of the Q & A experiment to extend the 3.5 m interferometer to 7 m with various upgrades. Photo on the left of Fig. 3 shows the apparatus for Phase II; photo on the right side of Fig. 3 shows the apparatus for Phase III, with the big (front) tank moved further to the front (out of the photo). The laser has been changed to 532 nm wavelength and is located next and beyond the front tank. We have installed a new 1.8 m 2.3 T permanent magnet (in the middle to bottom of right side photo) capable of rotation up to 13 cycles per second to enhance the physical effects. We are working with 532 nm Nd:YAG laser as light source with cavity finesse around 100,000, and aim at 10 nrad$(Hz)^{-1/2}$ optical sensitivity. With all these achieved and the upgrading of vacuum, for a period of 50 days (with duty cycle around 78 % as performed before) the vacuum birefringence measurement would be improved in precision by 3-4 orders of magnitude, and QED birefringence would be measured to 28 % (Mei et al., 2010). To enhance the physical effects further, another 1.8 m magnet will be added in the future.

All three ongoing experiments - PVLAS, Q &A, and BMV - are measuring the birefringence Δn, and hence, $\eta_1-\eta_2$ in case η_3 is assumed to be zero. To measure η_1 and η_2 separately, one-arm common path polarization measurement interferometer is not enough. We need a two-arm interferometer with the paths in two arms in magnetic fields with different strengths (or one with no magnetic field).

To measure η_3 in addition, one needs to use both external electric and external magnetic field. One possibility is to let light goes through strong microwave cavity and interferes. Suppose light propagation direction is the same as the microwave propagation direction which is perpendicular to the microwave fields. Let's choose z-axis to be in the propagation direction, x-axis in the $\mathbf{E}^{ext}$ direction and y-axis in the $\mathbf{B}^{ext}$ direction, i.e., $\mathbf{k} = (0, 0, k)$, $\mathbf{E}^{ext} = (E, 0, 0)$ and $\mathbf{B}^{ext} = (0, B, 0)$. We calculate the indices of refraction using Eqs. (39)-(47) without first assuming $E = B$ and obtain the following

$$\varepsilon_{\alpha\beta}: \varepsilon_{11}=1+2\eta_1(E^2-B^2)B_c^{-2}+4\eta_1E^2B_c^{-2};\ \varepsilon_{22}=1+2\eta_1(E^2-B^2)B_c^{-2}+4\eta_2B^2B_c^{-2};$$

$$\varepsilon_{33}=1+2\eta_1(E^2-B^2)B_c^{-2};\ \varepsilon_{12}=\varepsilon_{21}=2\eta_3EBB_c^{-2};\ \varepsilon_{13}=\varepsilon_{23}=\varepsilon_{31}=\varepsilon_{32}=0, \quad (68)$$

$$(\mu^{-1})_{\alpha\beta}: (\mu^{-1})_{11}=1+2\eta_1(E^2-B^2)B_c^{-2}-4\eta_2E^2B_c^{-2};\ (\mu^{-1})_{22}=1+2\eta_1(E^2-B^2)B_c^{-2}-4\eta_1B^2B_c^{-2};$$

$$(\mu^{-1})_{33}=1+2\eta_1(E^2-B^2)B_c^{-2};\ (\mu^{-1})_{12}=(\mu^{-1})_{21}=2\eta_3EBB_c^{-2};\ (\mu^{-1})_{13}=(\mu^{-1})_{23}=(\mu^{-1})_{13}=(\mu^{-1})_{23}=0, \quad (69)$$

$$\lambda_{\alpha\beta}: \lambda_{11}=\xi\varphi+\eta_3(E^2-B^2)B_c^{-2}+2\eta_3E_2B_c^{-2};\ \lambda_{22}=\xi\varphi+\eta_3(E^2-B^2)B_c^{-2}+2\eta_3B^2B_c^{-2};$$

$$\lambda_{33}=\xi\varphi+\eta_3(E^2-B^2)B_c^{-2};\ \lambda_{12}=-4\eta_1EBB_c^{-2};\ \lambda_{21}=4\eta_2EBB_c^{-2};\ \lambda_{13}=\lambda_{23}=\lambda_{31}=\lambda_{32}=0, \quad (70)$$

$$J_1 \equiv -4\eta_1(E^2+B^2)B_c^{-2}+4\eta_1EBB_c^{-2}, \quad (71)$$

$$J_2 \equiv -4\eta_2(E^2+B^2)B_c^{-2}+4\eta_2EBB_c^{-2}, \quad (72)$$

$$J \equiv 2\eta_3(E^2-B^2)B_c^{-2}, \quad (73)$$

$$n_\pm = 1 + (\eta_1+\eta_2)(E^2+B^2-EB)B_c^{-2} \pm [(\eta_1-\eta_2)^2(E^2+B^2-EB)^2+\eta_3^2(E^2-B^2)]^{1/2}\ B_c^{-2}. \quad (74)$$

As a consistent check, there is no birefringence in Eq. (74) for $\eta_1 = \eta_2$, $\eta_3 = 0$.

Now, we consider two special cases for Eq. (74): (i) $E=B$ as in the strong microwave cavity, the indices of refraction for light is

$$n_\pm = 1 + (\eta_1+\eta_2)B^2B_c^{-2} \pm (\eta_1-\eta_2)B^2B_c^{-2}, \quad (75)$$

with birefringence Δn given by

$$\Delta n = 2(\eta_1-\eta_2)B^2B_c^{-2}; \quad (76)$$

(ii) $E=0$, $B\neq 0$, the indices of refraction for light is

$$n_\pm = 1 + (\eta_1+\eta_2)B^2B_c^{-2} \pm [(\eta_1-\eta_2)^2+\eta_3^2]^{1/2}B^2B_c^{-2}, \quad (77)$$

with birefringence Δn given by

$$\Delta n = 2[(\eta_1-\eta_2)^2+\eta_3^2]^{1/2}B^2B_c^{-2}. \quad (78)$$

Equation (77) agrees with (54) derived earlier.

To measure η_1, η_2 and η_3, we could do the following three experiments to determine them: (i) to measure the birefringence $\Delta n = 2(\eta_1-\eta_2)B^2B_c^{-2}$ of light with the external field provided by a strong microwave cavity or wave guide to determine $\eta_1-\eta_2$; (ii) to measure the birefringence $\Delta n = 2[(\eta_1-\eta_2)^2+\eta_3^2]^{1/2}B^2B_c^{-2}$ of light with the external magnetic field provided by a strong magnet to determine η_3 with $\eta_1-\eta_2$ determined by (i); (iii) to measure η_1 and η_2 separately using two-arm interferometer with the paths in two arms in magnetic fields with different strengths (or one with no magnetic field).

As to the term $\xi\Phi$ and parameter ξ, it does not give any change in the index of refraction. However, as we will see in section 7 and section 8, it gives a polarization rotation and the effect can be measured though observations with astrophysical and cosmological propagation of electromagnetic waves.

5. Electromagnetism in curved spacetime and the Einstein equivalence principle

In the earth laboratory, where variation of gravity is small, we can use standard Maxwell equations together with Lorentz force law for ordinary measurements and experiments. However, in precision experiments on earth, in space, in the astrophysical situation or in the cosmological setting, the gravity plays an important role and is non-negligible. In the remaining part of this chapter, we address to the issue of electromagnetism in gravity and more empirical tests of electromagnetism and special relativity. The standard way of including gravitational effects in electromagnetism is to use the comma-goes-to-semicolon rule, i. e., the principle of equivalence (the minimal coupling rule). This is the essence of Einstein Equivalence Principle (EEP) which states that everywhere in the 4-dimensional spacetime, locally, the physics is that of special relativity. This guarantees that the 4-dimensional geometry can be described by a metric g_{ij} which can be transformed into the Minkowski metric locally. In curved spacetime, η_{ij} is replaced by g_{ij} with partial derivative (comma) replaced by the covariant derivative in the g_{ij} metric (semi-colon) in the Lagrangian density for a system of charged particles. When this is done the Lagrangian density becomes

$$L_I = - (1/(16\pi))\chi_{GR}{}^{ijkl}\, F_{ij}\, F_{kl} - A_k\, j^k\, (-g)^{(1/2)} - \Sigma_I\, m_I\, (ds_I)/(dt)\; \delta(\boldsymbol{x}-\boldsymbol{x}_I), \tag{79}$$

where the GR (General Relativity) constitutive tensor $\chi_{GR}{}^{ijkl}$ is given by

$$\chi_{GR}{}^{ijkl} = (-g)^{1/2}\,[(1/2)\, g^{ik}\, g^{jl} - (1/2)\, g^{il}\, g^{kj}], \tag{80}$$

and g is the determinant of g_{ij}. In general relativity or metric theories of gravity where EEP holds, the line element near a world point (event) P is given by

$$ds^2 = g_{ij}\, dx^i\, dx^j = g_{AB}\, dx^A\, dx^B = [\eta_{AB} + \mathrm{O}((\Delta x^C)^2)]\, dx^A\, dx^B, \tag{81}$$

where $\{x^i\}$ is an arbitrary coordinate system, $\{x^A\}$ is a locally inertial frame, and g_{ij} & g_{AB} are the metric tensor in their respective frames. According to the definition of locally inertial frame, we have $g_{AB} = \eta_{AB} + \mathrm{O}((\Delta x^C)^2)$. Therefore, in the locally inertial system near P, special relativity holds up to the curvature ambiguity, and the definition of rods and clocks is the same as in the special relativity including local quantum mechanics and electromagnetism.

Nevertheless, for long range propagation and large-scale phenomenon, curvature effects are important. For long range electromagnetic propagation, wavelength/frequency shift is important. From distant quasars, the redshift factor z exceeds 6, i.e., the wavelength changes by more than 6-fold. The gravitational redshift is given by

$$\Delta\tau_A/\Delta\tau_B = g_{00}(B)/g_{00}(A), \tag{82}$$

where $\Delta\tau_A$ and $\Delta\tau_B$ are the proper periods of a light signal emitted by a source A and received by B respectively. This formula applies equally well to the solar system, to galaxies and to cosmos. Its realm of practical application is in clock and frequency comparisons. In the weak gravitational field such as near earth or in the solar system, we have

$$g_{00} = 1\text{-}2U/c^2, \tag{83}$$

in the first approximation, where U is the Newtonian potential. On the surface of earth, $U/c^2 \approx 0.7 \times 10^{-9}$ and the redshift is a fraction of it. This redshift is measured in the laboratory and in space borne missions. It is regularly corrected for the satellite navigation systems such as GPS, GLONASS, Galileo and Beidou. Another effect of electromagnetic propagation in gravity is its deflection with important application to gravitational lensing effects in astrophysics.

6. Empirical tests of electromagnetism in gravity and the χ-g framework

In section 1, we have discussed the constraints on Proca part of Lagrangian density, i.e., photon mass. In this section, we discuss the empirical foundation of the Maxwell (main) part of electromagnetism. First we need a framework to interpret experimental tests. A natural framework is to extend the GR constitutive tensor $\chi_{GR}{}^{ijkl}$ [equation (80)] into a general form, and look for experimental and observational evidences to test it to see how much it is constrained to the GR form. The general framework we adopt is the χ-g framework (Ni, 1983a, 1984a, 1984b, 2010).

The χ-g framework can be summarized in the following interaction Lagrangian density

$$L_I = -(1/(16\pi))\chi^{ijkl} F_{ij} F_{kl} - A_k j^k (-g)^{(1/2)} - \Sigma_I m_I (ds_I)/(dt)\, \delta(\mathbf{x}\text{-}\mathbf{x}_I), \tag{84}$$

where $\chi^{ijkl} = \chi^{klij} = -\chi^{jikl}$ is a tensor density of the gravitational fields (e.g., g_{ij}, φ, etc.) or fields to be investigated and $F_{ij} \equiv A_{j,i} - A_{i,j}$ etc. have the usual meaning in classical electromagnetism. The gravitation constitutive tensor density χ^{ijkl} dictates the behaviour of electromagnetism in a gravitational field and has 21 independent components in general. For general relativity or a metric theory (when EEP holds), χ^{ijkl} is determined completely by the metric g_{ij} and equals $(-g)^{1/2}\,[(1/2)\, g^{ik} g^{jl} - (1/2)\, g^{il} g^{jk}]$.

In the following, we use experiments and observations to constrain the 21 degrees of freedom of χ^{ijkl} to see how close we can reach general relativity. This procedure also serves to reinforce the empirical foundations of classical electromagnetism as EEP locally is based on special relativity including classical electromagnetism.

In the χ-g framework, for a weak gravitational field,

$$\chi^{ijkl} = \chi^{(0)ijkl} + \chi^{(1)ijkl}, \tag{85}$$

where

$$\chi^{(0)ijkl} = (1/2)\eta^{ik}\eta^{jl} - (1/2)\eta^{il}\eta^{kj}, \quad (86)$$

with η^{ij} the Minkowski metric and $|\chi^{(1)ijkl}| << 1$ for all i, j, k, and l. The small special relativity violation (constant part), if any, is put into the $\chi^{(1)ijkl}$'s. In this field the dispersion relation for ω for a plane-wave propagating in the z-direction is

$$\omega_{\pm} = k\{1+(1/4)[(K_1+K_2) \pm [(K_1-K_2)^2 + 4 K^2]^{1/2}]\}, \quad (87)$$

where

$$K_1 = \chi^{(1)1010} - 2\chi^{(1)1013} + \chi^{(1)1313}, \quad (88)$$

$$K_2 = \chi^{(1)2020} - 2\chi^{(1)2023} + \chi^{(1)2323}, \quad (89)$$

$$K = \chi^{(1)1020} - \chi^{(1)1023} - \chi^{(1)1320} + \chi^{(1)1323}. \quad (90)$$

Photons with two different polarizations propagate with different speeds $V_{\pm} = \omega_{\pm}/k$ and would split in 4-dimensional spacetime. The conditions for no splitting (no retardation) is $\omega_+ = \omega_-$, i.e.,

$$K_1 = K_2, \quad K = 0. \quad (91)$$

Eq. (91) gives two constraints on the $\chi^{(1)ijkl}$'s (Ni, 1983a, 1984a, 1984b).

Constraints from no birefringence. The condition for no birefringence (no splitting, no retardation) for electromagnetic wave propagation in all directions in the weak field limit gives ten independent constraint equations on the constitutive tensor χ^{ijkl}'s. With these ten constraints, the constitutive tensor χ^{ijkl} can be written in the following form

$$\chi^{ijkl} = (-H)^{1/2}[(1/2)H^{ik} H^{jl} - (1/2)H^{il} H^{kj}]\psi + \varphi e^{ijkl}, \quad (92)$$

where $H = \det(H_{ij})$ and H_{ij} is a metric which generates the light cone for electromagnetic propagation (Ni, 1983a, 1984a,b). Note that (92) has an axion degree of freedom, φe^{ijkl}, and a 'dilaton' degree of freedom, ψ. Lämmerzahl and Hehl (2004) have shown that this non-birefringence guarantees, without approximation, Riemannian light cone, i.e., Eq. (92) holds without the assumption of weak field also. To fully recover EEP, we need (i) good constraints from no birefringence, (ii) good constraints on no extra physical metric, (iii) good constraints on no ψ ('dilaton'), and (iv) good constraints on no φ (axion) or no pseudoscalar-photon interaction.

Eq. (92) is verified empirically to high accuracy from pulsar observations and from polarization measurements of extragalactic radio sources. With the null-birefringence observations of pulsar pulses and micropulses before 1980, the relations (92) for testing EEP are empirically verified to 10^{-14} - 10^{-16} (Ni, 1983a, 1984a, 1984b). With the present pulsar observations, these limits would be improved; a detailed such analysis is given by Huang (2002). Analyzing the data from polarization measurements of extragalactic radio sources, Haugan and Kauffmann (1995) inferred that the resolution for null-birefringence is 0.02 cycle at 5 GHz. This corresponds to a time resolution of 4×10^{-12} s and gives much better constraints. With a detailed analysis and more extragalactic radio observations, (92) would be tested down to 10^{-28}-10^{-29} at cosmological distances. In 2002, Kostelecky and Mews (2002) used polarization measurements of light from cosmologically distant astrophysical sources to yield stringent constraints down to 2×10^{-32}. For a review, see Ni (2010). In the remaining part of this subsection, we assume (92) to be correct.

Constraints on one physical metric and no 'dilaton' (ψ). Let us now look into the empirical constraints for H^{ij} and ψ. In Eq. (84), ds is the line element determined from the metric g_{ij}. From Eq. (92), the gravitational coupling to electromagnetism is determined by the metric H_{ij} and two (pseudo)scalar fields φ 'axion' and ψ 'dilaton'. If H_{ij} is not proportional to g_{ij}, then the hyperfine levels of the lithium atom, the beryllium atom, the mercury atom and other atoms will have additional shifts. But this is not observed to high accuracy in Hughes-Drever experiments (Hughes et al., 1960; Beltran-Lopez et al., 1961; Drever, 1961; Ellena et al., 1987; Chupp et al., 1989). Therefore H_{ij} is proportional to g_{ij} to certain accuracy. Since a change of H^{ik} to λH^{ij} does not affect χ^{ijkl} in Eq. (92), we can define $H_{11} = g_{11}$ to remove this scale freedom (Ni, 1983a, 1984a). For a review, see Ni (2010).

Eötvös-Dicke experiments (Eötvös, 1890; Eötvös et al., 1922; Roll et al., 1964; Braginsky and Panov, 1971; Schlamminger et al., 2008 and references therein) are performed on unpolarized test bodies. In essence, these experiments show that unpolarized electric and magnetic energies follow the same trajectories as other forms of energy to certain accuracy. The constraints on Eq. (92) are

$$| 1-\psi | / U < 10^{-10}, \tag{93}$$

and

$$| H_{00} - g_{00} | / U < 10^{-6}, \tag{94}$$

where U ($\sim 10^{-8}$) is the solar gravitational potential at the earth.

In 1976, Vessot et al. (1980) used an atomic hydrogen maser clock in a space probe to test and confirm the metric gravitational redshift to an accuracy of 1.4×10^{-4}, i. e.,

$$| H_{00} - g_{00} | / U \leq 1.4 \times 10^{-4}, \tag{95}$$

where U is the change of earth gravitational field that the maser clock experienced.

With constraints from (i) no birefringence, (ii) no extra physical metric, (iii) no ψ ('dilaton'), we arrive at the theory (84) with χ^{ijkl} given by

$$\chi^{ijkl} = (-g)^{1/2} [(1/2) g^{ik} g^{jl} - (1/2) g^{il} g^{kj} + \varphi \varepsilon^{ijkl}], \tag{96}$$

i.e., an axion theory (Ni, 1983a, 1984a; Hehl and Obukhov 2008). Here ε^{ijkl} is defined to be $(-g)^{-1/2} e^{ijkl}$. The current constraints on φ from astrophysical observations and CMB polarization observations will be discussed in section 8. Thus, from experiments and observations, only one degree of freedom of χ^{ijkl} is not much constrained.

Now let's turn into more formal aspects of equivalence principles. We proved that for a system whose Lagrangian density is given by Eq. (84), the Galileo Equivalence Principle (UFF [Universality of Free Fall; WEP I [Weak Equivalence Principle I]) holds if and only if Eq. (96) holds (Ni, 1974, 1977).

If $\varphi \neq 0$ in (96), the gravitational coupling to electromagnetism is not minimal and EEP is violated. Hence WEP I does not imply EEP and Schiff's conjecture (which states that WEP I implies EEP) is incorrect (Ni, 1973, 1974, 1977). However, WEP I does constrain the 21 degrees of freedom of χ to only one degree of freedom (φ), and Schiff's conjecture is largely right in spirit.

The theory with $\varphi \neq 0$ is a pseudoscalar theory with important astrophysical and cosmological consequences (section 8). This is an example that investigations in fundamental physical laws lead to implications in cosmology. Investigations of CP problems in high energy physics leads to a theory with a similar piece of Lagrangian with φ the axion field for QCD [Quantum Chromodynamics] (Peccei and Quinn, 1977; Weinberg, 1978; Wilczek, 1978).

In the nonmetric theory with χ^{ijkl} ($\varphi \neq 0$) given by Eq. (96) (Ni 1973, 1974, 1977), there are anomalous torques on electromagnetic-energy-polarized bodies so that different test bodies will change their rotation state differently, like magnets in magnetic fields. Since the motion of a macroscopic test body is determined not only by its trajectory but also by its rotation state, the motion of polarized test bodies will not be the same. We, therefore, have proposed the following stronger weak equivalence principle (WEP II) to be tested by experiments, which states that in a gravitational field, both the translational and rotational motion of a test body with a given initial motion state is independent of its internal structure and composition (universality of free-fall motion) (Ni 1974, Ni 1977). To put in another way, the behavior of motion including rotation is that in a local inertial frame for test-bodies. If WEP II is violated, then EEP is violated. Therefore from above, in the χ-g framework, the imposition of WEP II guarantees that EEP is valid.

WEP II state that the motion of all six degrees of freedom (3 translational and 3 rotational) must be the same for all test bodies as in a local inertial frame. There are two different scenarios that WEP II would be violated: (i) the translational motion is affected by the rotational state; (ii) the rotational state changes with angular momentum (rotational direction/speed) or species. Recent experimental results of Gravity Probe B experiment with rotating quartz balls in earth orbit (Everitt et al., 2011) not just verifies frame-dragging effect, but also verifies both aspects of WEP II for unpolarized-bodies to an ultimate precision (Ni, 2011).

In this section, we have shown that the empirical foundation of classical electromagnetism is solid except in the aspect of a pseudoscalar-photon interaction. This exception has important consequences in cosmology. In the following two sections, we address this issue.

7. Pseudoscalar-photon interaction

In this section, we discuss the modified electromagnetism in gravity with the pseudoscalar-photon interaction which was reached in the last section, i.e., the theory with the constitutive tensor density (96). Its Lagrangian density is as follows

$$L_I = -(1/(16\pi))(-g)^{1/2}[(1/2)g^{ik}g^{jl} - (1/2)g^{il}g^{kj} + \varphi\, \varepsilon^{ijkl}]F_{ij}F_{kl} - A_k j^k(-g)^{(1/2)} - \Sigma_I m_I (ds_I)/(dt)\delta(\mathbf{x}-\mathbf{x}_I). \quad (97)$$

In the constitutive tensor density and the Lagrangian density, φ is a scalar or pseudoscalar function of relevant variables. If we assume that the φ-term is local CPT invariant, then φ

should be a pseudoscalar (function) since ε^{ijkl} is a pseudotensor. The pseudoscalar(scalar)-photon interaction part (or the nonmetric part) of the Lagrangian density of this theory is

$$L^{(\varphi\gamma\gamma)} = L^{(NM)} = - (1/16\pi)\ \varphi\ e^{ijkl}F_{ij}F_{kl} = - (1/4\pi)\ \varphi_{,i}\ e^{ijkl}A_jA_{k,l}\ (\text{mod div}), \qquad (98)$$

where 'mod div' means that the two Lagrangian densities are related by integration by parts in the action integral. This term gives pseudoscalar-photon-photon interaction in the quantum regime and can be denoted by $L^{(\varphi\gamma\gamma)}$. This term is also the ξ-term in the PPM Lagrangian density L_{PPM} with the $\varphi \equiv (1/4)\xi\Phi$ correspondence. The Maxwell equations (Ni 1973, 1977) from Eq. (97) become

$$F^{ik}{}_{;k} + \varepsilon^{ikml}\ F_{km}\varphi_{,l} = -4\pi j^i, \qquad (99)$$

where the derivation $;$ is with respect to the Christoffel connection of the metric. The Lorentz force law is the same as in metric theories of gravity or general relativity. Gauge invariance and charge conservation are guaranteed. For discussions on the tests of charge conservation, and on the limits of differences in active and passive charges, please see Lämmerzahl et al. (2005, 2007). The modified Maxwell equations (99) are also conformally invariant.

The rightest term in equation (99) is reminiscent of Chern-Simons (1974) term $e^{\alpha\beta\gamma} A_\alpha F_{\beta\gamma}$. There are two differences: (i) Chern-Simons term is in 3 dimensional space; (ii) Chern-Simons term in the integral is a total divergence (Table 1). However, it is interesting to notice that the cosmological time may be defined through the Chern-Simons invariant (Smolin and Soo, 1995).

Term	Dimension	Reference	Meaning
$e^{\alpha\beta\gamma} A_\alpha F_{\beta\gamma}$	3	Chern-Simons (1974)	Intergrand for topological invarinat
$e^{ijkl}\ \varphi\ F_{ij}\ F_{kl}$	4	Ni (1973, 1974, 1977)	Pseudoscalar-photon coupling
$e^{ijkl}\ \varphi\ F^{QCD}{}_{ij}\ F^{QCD}{}_{kl}$	4	Peccei-Quinn (1977) Weinberg (1978) Wilczek (1978)	Pseudoscalar-gluon coupling
$e^{ijkl}\ V_i\ A_j\ F_{kl}$	4	Carroll-Field-Jackiw (1990)	External constant vector coupling

Table 1. Various terms in the Lagrangian and their meaning.

A term similar to the one in equation (98), axion-gluon interaction term, occurs in QCD in an effort to solve the strong CP problem (Peccei & Quinn, 1977; Weinberg, 1978; Wilczek, 1978). Carroll, Field and Jackiw (1990) proposed a modification of electrodynamics with an additional $e^{ijkl}\ V_i\ A_j\ F_{kl}$ term with V_i a constant vector (See also Jackiw, 2007). This term is a special case of the term $e^{ijkl}\ \varphi\ F_{ij}\ F_{kl}$ (mod div) with $\varphi_{,i} = - ½V_i$.

Various terms in the Lagrangians discussed in this subsection are listed in Table 1. Empirical tests of the pseudoscalar-photon interaction (98) will be discussed in next section.

8. Cosmic polarization rotation

For the electromagnetism in gravity with an effective pseudoscalar-photon interaction discussed in the last section, the electromagnetic wave propagation equation is governed by equation (99). In a local inertial (Lorentz) frame of the g-metric, it is reduced to

$$F^{ik}{}_{,k} + e^{ikml} F_{km} \varphi_{,l} = 0. \tag{100}$$

Analyzing the wave into Fourier components, imposing the radiation gauge condition, and solving the dispersion eigenvalue problem, we obtain $k = \omega + (n^\mu\varphi_{,\mu} + \varphi_{,0})$ for right circularly polarized wave and $k = \omega - (n^\mu\varphi_{,\mu} + \varphi_{,0})$ for left circularly polarized wave in the eikonal approximation (Ni 1973). Here n^μ is the unit 3-vector in the propagation direction. The group velocity is

$$v_g = \partial\omega/\partial k = 1, \tag{101}$$

which is independent of polarization. There is no birefringence. For the right circularly polarized electromagnetic wave, the propagation from a point $P_1 = \{x_{(1)}{}^i\} = \{x_{(1)}{}^0; x_{(1)}{}^\mu\} = \{x_{(1)}{}^0, x_{(1)}{}^1, x_{(1)}{}^2, x_{(1)}{}^3\}$ to another point $P_2 = \{x_{(2)}{}^i\} = \{x_{(2)}{}^0; x_{(2)}{}^\mu\} = \{x_{(2)}{}^0, x_{(2)}{}^1, x_{(2)}{}^2, x_{(2)}{}^3\}$ adds a phase of $a = \varphi(P_2) - \varphi(P_1)$ to the wave; for left circularly polarized light, the added phase will be opposite in sign (Ni 1973). Linearly polarized electromagnetic wave is a superposition of circularly polarized waves. Its polarization vector will then rotate by an angle a. Locally, the polarization rotation angle can be approximated by

$$a = \varphi(P_2)-\varphi(P_1) = \Sigma_{i=0}{}^3 [\varphi_{,i} \times (x_{(2)}{}^i - x_{(1)}{}^i)] = \Sigma_{i=0}{}^3 [\varphi_{,i}\Delta x^i] = \varphi_{,0}\Delta x^0 + [\Sigma_{\mu=1}{}^3\varphi_{,\mu}\Delta x^\mu]$$

$$= - (1/2)\, \Sigma_{i=0}{}^3 [V_i\Delta x^i] = - (1/2)\, V_0\Delta x^0 - (1/2)\, [\Sigma_{\mu=1}{}^3 V_\mu\Delta x^\mu] \tag{102}$$

The rotation angle in (102) consists of 2 parts -- $\varphi_{,0}\Delta x^0$ and $[\Sigma_{\mu=1}{}^3\varphi_{,\mu}\Delta x^\mu]$. For light in a local inertial frame, $|\Delta x^\mu| = |\Delta x^0|$. In Fig. 4, space part of the rotation angle is shown. The amplitude of the space part depends on the direction of the propagation with the tip of magnitude on upper/lower sphere of diameter $|\Delta x^\mu| \times |\varphi_{,\mu}|$. The time part is equal to $\Delta x^0 \varphi_{,0}$. ($\nabla\varphi \equiv [\varphi_{,\mu}]$) When we integrate along light (wave) trajectory in a global situation, the total polarization rotation (relative to no φ-interaction) is again $\Delta\varphi = \varphi_2 - \varphi_1$ for φ is a scalar field where φ_1 and φ_2 are the values of the scalar field at the beginning and end of the wave. When the propagation distance is over a large part of our observed universe, we call this phenomenon cosmic polarization rotation (Ni, 2008, 2009a, 2010).

In the CMB polarization observations, there are variations and fluctuations. The variations and fluctuations due to scalar-modified propagation can be expressed as $\delta\varphi(2) - \delta\varphi(1)$, where 2 denotes a point at the last scattering surface in the decoupling epoch and 1 observation point. $\delta\varphi(2)$ is the variation/fluctuation at the last scattering surface. $\delta\varphi(1)$ at the present observation point is zero or fixed. Therefore the covariance of fluctuation $<[\delta\varphi(2) - \delta\varphi(1)]^2>$ gives the covariance of $\delta\varphi^2(2)$ at the last scattering surface. Since our Universe is isotropic to $\sim 10^{-5}$, this covariance is $\sim (\varsigma \times 10^{-5})^2$ where the parameter ς depends on various cosmological models. (Ni, 2008, 2009a, 2010)

Now we must say something about nomenclature.

Birefringence, also called double refraction, refers to the two different directions of propagation that a given incident ray can take in a medium, depending on the direction of polarization. The index of refraction depends on the direction of polarization.

Dichroic materials have the property that their absorption constant varies with polarization. When polarized light goes through dichroic material, its polarization is rotated due to difference in absorption in two principal directions of the material for the two polarization components. This phenomenon or property of the medium is called dichroism.

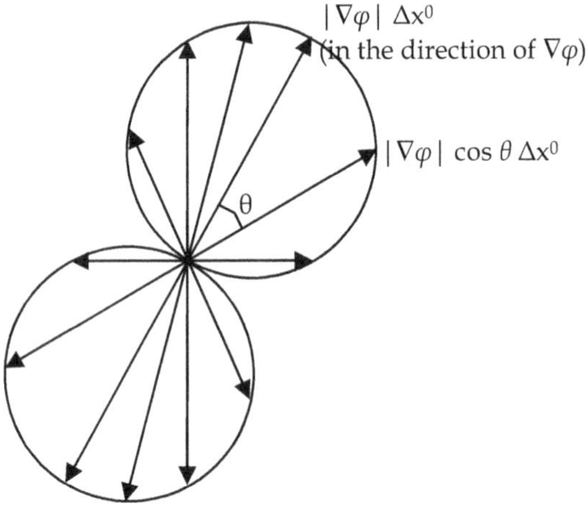

Fig. 4. Space contribution to the local polarization rotation angle -- $[\Sigma_{\mu=1}^{3}\varphi_{,\mu}\Delta x^{\mu}] = |\nabla\varphi|\cos\theta\,\Delta x^0$. The time contribution is $\varphi_{,0}\,\Delta x^0$. The total contribution is $(|\nabla\varphi|\cos\theta + \varphi_{,0})\,\Delta x^0$. ($\Delta x^0 > 0$).

In a medium with optical activity, the direction of a linearly polarized beam will rotate as it propagates through the medium. A medium subjected to magnetic field becomes optically active and the associated polarization rotation is called Faraday rotation.

Cosmic polarization rotation is neither dichroism nor birefringence. It is more like optical activity, with the rotation angle independent of wavelength. Conforming to the common usage in optics, one should not call it cosmic birefringence -- *a misnomer*.

Now we review and compile the constraints of various analyses from CMB polarization observations.

In 2002, DASI microwave interferometer observed the polarization of the cosmic background (Kovac et al., 2002). E-mode polarization is detected with 4.9 σ. The TE correlation of the temperature and E-mode polarization is detected at 95% confidence. This correlation is expected from the Raleigh scattering of radiation. However, with the (pseudo)scalar-photon interaction under discussion, the polarization anisotropy is shifted differently in different directions relative to the temperature anisotropy due to propagation; the correlation will then be downgraded. In 2003, from the first-year data (WMAP1), WMAP found that the polarization and temperature are correlated to more than 10 σ (Bennett *et al* 2003). This gives a constraint of about 10^{-1} for $\Delta\varphi$ (Ni, 2005a, 2005b).

Further results and analyses of CMB polarization observations came out after 2006. In Table 2, we update our previous compilations (Ni 2008, 2010). Although these results look different at 1 σ level, they are all consistent with null detection and with one another at 2 σ level.

Analysis	Constraint [mrad]	Source data
Ni (2005a, b)	±100	WMAP1 (Bennett *et al* 2003)
Feng, Li, Xia, Chen & Zhang (2006)	-105 ± 70	WMAP3 (Spergel *et al* 2007) & BOOMERANG (B03) (Montroy *et al* 2006)
Liu, Lee & Ng (2006)	±24	BOOMERANG (B03) (Montroy *et al* 2006)
Kostelecky & Mews (2007)	209 ± 122	BOOMERANG (B03) (Montroy *et al* 2006)
Cabella, Natoli & Silk (2007)	-43 ± 52	WMAP3 (Spergel *et al* 2007)
Xia, Li, Wang & Zhang (2008)	-108 ± 67	WMAP3 (Spergel *et al* 2007) & BOOMERANG (B03) (Montroy *et al* 2006)
Komatsu *et al* (2009)	-30 ± 37	WMAP5 (Komatsu *et al* 2009)
Xia, Li, Zhao & Zhang (2008)	-45 ± 33	WMAP5 (Komatsu *et al* 2009) & BOOMERANG (B03) (Montroy *et al* 2006)
Kostelecky & Mews (2008)	40 ± 94	WMAP5 (Komatsu *et al* 2009)
Kahniashvili, Durrer & Maravin (2008)	± 44	WMAP5 (Komatsu *et al* 2009)
Wu *et al* (2009)	9.6 ± 14.3 ± 8.7	QuaD (Pryke *et al* 2009)
Brown *et al.* (2009)	11.2 ± 8.7 ± 8.7	QuaD (Brown *et al* 2009)
Komatsu *et al.* (2011)	-19 ± 22 ± 26	WMAP7 (Komatsu *et al* 2011)

Table 2. Constraints on cosmic polarization rotation from CMB polarization observations. [See Ni (2010) for detailed references.]

Both magnetic field and potential new physics affect the propagation of CMB propagation and generate BB power spectra from EE spectra of CMB. The Faraday rotation due to magnetic field is wavelength dependent while the cosmic polarization rotation due to effective pseudoscalar-photon interaction is wavelength-independent. This property can be used to separate the two effects. With the tensor mode generated by these two effects measured and subtracted, the remaining tensor mode perturbations could be analyzed for signals due to primordial (inflationary) gravitational waves (GWs). In Ni (2009a,b), we have discussed the direct detectability of these primordial GWs using space GW detectors.

Observations of radio and optical/UV polarization of radio galaxies are also sensitive to measure/test the cosmic polarization rotation, and give comparable constraints of tens of mrad. These observations have the capability of determining the polarization rotation in various directions. For a recent work, see di Serego Alighieri et al. (2010).

9. Outlook

We have looked at the foundations of electromagnetism in this chapter. For doing this, we have used two approaches. The first one is to formulate a Parametrized Post-Maxwellian framework to include QED corrections and a pseudoscalar photon interaction. We discuss various vacuum birefringence experiments - ongoing and proposed -- to measure these parameters. The second approach is to look at electromagnetism in gravity and various experiments and observations to determine its empirical foundation. We found that the foundation is solid with the only exception of a potentially possible pseudoscalar-photon interaction. We discussed its experimental constraints and look forward to more future experiments.

10. Acknowledgment

I would like to thank the National Science Council (Grants No. NSC100-2119-M-007-008 and No. NSC100-2738-M-007-004) for supporting this work in part.

11. References

Bass, L. & Schrödinger, E. (1955). Must the photon mass be zero?, *Proceedings of the Royal Society of London, Ser. A*, Vol.232, pp. 1–6

Battesti, R. et al. (BMV Collaboration). (2008). The BMV Experiment: a Novel Apparatus to Study the Propagation of Light in a Transverse Magnetic Field. *The European Physical Journal D*, Vol.46, No.2, pp. 323-333

Beltran-Lopez, V.; Robinson, H. G. & Hughes, V. W. (1961). Improved upper limit for the anisotropy of inertial mass from a nuclear magnetic resonance experiment. *Bulletin of American Physical Society*, Vol. 6, p. 424

Bennett, C. L. et al. (2003). First Year Wilkinson Microwave Anisotropy Probe (WMAP) Observations: Preliminary Maps and Basic Results. *Astrophysical Journal Supplements*, Vol.148, pp. 1-27; and references therein

Born, M. (1934). On the Quantum Theory of the Electromagnetic Field. *Proceedings of the Royal Society of London Series A-Mathematical and Physical Sciences*, Vol.143, pp. 410-437

Born, M. & Infeld, L. (1934). Foundations of the new field theory. *Proceedings of the Royal Society of London Series A-Mathematical and Physical Sciences*, Vol.144, pp. 425-451

Braginsky, V. B. & Panov, V. I. (1971). Verification of the equivalence of inertial and gravitational mass. *Zh. Eksp. Teor. Fiz.* Vol.61, pp. 873-879 [*Sov. Phys. JETP*, (1972). Vol.34, pp. 464-466]

Cameron, R. et al. (BFRT Collaboration). (1993). Search for nearly massless, weakly coupled particles by optical techniques. *Physical Review D*, Vol.47, pp. 3707-3725

Carroll, S. M.; Field G. B. & Jackiw, R. (1990). Limits on a Lorentz- and parity-violating modification of electrodynamics. *Physical Review D*, Vol.41, pp. 1231-1240

Chen, S.-J.; Mei, H.-H. & Ni, W.-T. (Q & A Collaboration). (2007). Q & A Experiment to Search for Vacuum Dichroism, Pseudoscalar Photon Interaction and Millicharged Fermions. *Modern Physics Letters A*, Vol.22, pp. 2815-2831 [arXiv:hep-ex/0611050]

Chern, S.-S. & Simons, J. (1974). Characteristic forms and geometric invariants. *The Annals of Mathematics*, 2nd Ser. Vol.99, p. 48

Chibisov, G. V. (1976). Astrophysical upper limits on the photon rest mass. *Uspekhi Fizicheskikh Nauk*, Vol.119, pp. 551–555 [*Soviet Physics Uspekhi*, Vol.19, pp. 624–626 (1976)]

Chupp, T. E.; Hoara, R. J.; Loveman, R. A.; Oteiza, E. R.; Richardson, J. M. & Wagshul, M. E. (1989). Results of a new test of local Lorentz invariance: A search for mass anisotropy in ^{21}Ne. *Physical Review Letters*, Vol.63, pp. 1541-1545

Davis, L., Jr.; Goldhaber, A. S. & Nieto, M. M. (1975). Limit on the photon mass deduced from Pioneer-10 observations of Jupiter's magnetic field. *Physical Review Letters*, Vol.35, pp. 1402–1405

Denisov,V. I.; Krivchenkov, I. V. & Kravtsov, N. V. (2004). Experiment for Measuring the post-Maxwellian Parameters of Nonlinear Electrodynamics of Vacuum with Laser-Interferometer Techniques. *Physical Review D*, Vol.69, 066008

di Serego Alighieri, S.; Finelli, F. & Galaverni, M. (2010). Limits on cosmological birefringence from the UV polarization of distant radio galaxies. *Astrophysical Journal*, Vol.715, p. 33

Dittrich, W. & Reuter, M. (1985). *Effective Lagrangians in Quantum Electrodynamics, Lecture Notes in Physics*, Vol. 220 , Springer, Berlin

Drever, R. W. P. (1961). A search for anisotropy of inertial mass using a free precession technique. *Phililosophical Magazine* Vol.6, pp. 683-687

Drever, R. W. P. et al. (1983). Laser phase and frequency stabilization using an optical resonator. *Applied Physics B*, Vol.31, pp. 97-105

Ellena, J. F.; Ni, W.-T. & Ueng, T.-S. (1987). Precision NMR measurement of Li^7 and limits on spatial anisotropy. *IEEE Transactions on Instrumentation and Measurement* Vol.IM-36, pp. 175-180

Eötvös, R. V.; Pekar, D. & Fekete, E. (1922). Articles on the laws of proportionality of inertia and gravity. *Ann. Phys.* (Leipzig), Vol.68, pp. 11-66

Eötvös, R. V. (1890). Über die Anziehung der Erde auf verschiedene Substanzen. *Math. Naturwiss. Ber. Ungarn* (Budapest), Vol.8, pp. 65-68

Everitt, C. W. F. et al. (2011). Gravity Probe B: Final Results of a Space Experiment to Test General Relativity. *Physical Review Letters*, Vol.106, 221101

Goldhaber, A. S. & Nieto, M. M. (2010). Photon and Graviton Mass Limits. *Review of Modern Physics*, Vol.82, No.1, (January-March 2010), pp. 939-979

Haugan, M. P. & Kauffmann, T. F. (1995). New test of the Einstein equivalence principle and the isotropy of space. *Physical Review D*, Vol.52, pp. 3168-3175

Hehl, F. W. & Obukhov, Yu. N. (2008). Equivalence principle and electromagnetic field: no birefringence, no dilaton, and no axion, *General Relativity and Gravitation*, Vol.40, pp. 1239-1248

Heisenberg, W. & Euler, E. (1936). Folgerungen aus der Diracschen Theorie des Positrons. *Zeitschrift für Physik*, Vol.98, pp. 714-732

Huang, H.-W. (2002). Pulsar Timing and Equivalence Principle Tests. *Master thesis, National Tsing Hua University*

Hughes, V. W.; Robinson, H. G. & Beltran-Lopez, V. (1960). Upper Limit for the Anisotropy of Inertial Mass from Nuclear Resonance Experiments. *Physical Review Letters*, Vol.4, pp. 342-344

Jackiw, R. (2007). Lorentz Violation in a Diffeomorphism-Invariant Theory, *CPT'07 Proceedings*, arXiv: 0709.2348

Jackson, J. D. (1999). *Classical Electrodynamics*, John Wiley & Sons, Hoboken

Kim, S. P. (2011a). QED Effective Actions in Space-Dependent Gauge and Electromagnetic Duality, *arXiv:1105.4382v2 [hep-th]*

Kim, S. P. (2011b). QED Effective Action in Magnetic Field Backgrounds and Electromagnetic Duality, *Physical Review D*, Vol.84, 065004

Kostelecky, V. A. & Mewes, M. (2008). Astrophysical Tests of Lorentz and CPT Violation with Photons. *Astrophysical Journal*, Vol. 689, pp. L1-L4

Kovac, J. M.; Leitch, E. M.; Pryke, C.; Carlstrom, J. E.; Halverson, N. W. & Holzapfel, W. L. (2002). Detection of Polarization in the Cosmic Microwave Background using DASI. *Nature*, Vol.420, pp. 772-787

Lämmerzahl, C. & Hehl, F. W. (2004). Riemannian Light Cone from Vanishing Birefringence in Premetric Vacuum Electrodynamics. *Physical Review D*, Vol.70, 105022

Lämmerzahl, C. ; Macias, A. & Müller, H. (2005). Lorentz invariance violation and charge (non)conservation: A general theoretical frame for extensions of the Maxwell equations. *Physical Review A*, Vol. 71, 025007

Lämmerzahl, C. ; Macias, A. & Müller, H. (2007). Limits to Differences in Active and Passive Charges. *Physical Review A*, Vol.75, 052104

Mei, H.-H.; Ni, W.-T.; Chen, S.-J & Pan, S.-s. (Q & A Collaboration). (2009). Measurement of the Cotton-Mouton Effect in Nitrogen, Oxygen, Carbon Dioxide, Argon, and Krypton with the Q & A Apparatus. *Chemical Physics Letters*, Vol.471, pp. 216-221

Mei, H.-H.; Ni, W.-T.; Chen, S.-J & Pan, S.-s. (Q & A Collaboration). (2010). Axion Search with Q & A Experiment. *Modern Physics Letters A*, Vol. 25, Nos. 11 & 12, pp. 983–993

Ni, W.-T. (1973). A Nonmetric Theory of Gravity. preprint, Montana State University, Bozeman, Montana, USA, http://astrod.wikispaces.com/file/view/A+Non-metric+Theory+of+Gravity.pdf

Ni, W.-T. (1974). Weak Equivalence Principles and Gravitational Coupling. *Bulletin of American Physical Society*, Vol.19, p. 655

Ni, W.-T. (1977). Equivalence Principles and Electromagnetism. *Physical Review Letters*, Vol.38, pp. 301-304

Ni, W.-T. (1983a). Equivalence Principles, Their Empirical Foundations, and the Role of Precision Experiments to Test Them. *Proceedings of the 1983 International School and Symposium on Precision Measurement and Gravity Experiment*, Taipei, Republic of China, January 24-February 2, 1983, W.-T. Ni, (Ed.), (Published by National Tsing Hua University, Hsinchu, Taiwan, Republic of China), 491-517

Ni, W.-T. (1983b). Implications of Hughes-Drever Experiments. *Proceedings of the 1983 International School and Symposium on Precision Measurement and Gravity Experiment*, Taipei, Republic of China, January 24-February 2, 1983, W.-T. Ni, (Ed.), (Published by National Tsing Hua University, Hsinchu, Taiwan, Republic of China), 519-529

Ni, W.-T. (1983c). Spin, Torsion and Polarized Test-Body Experiments. *Proceedings of the 1983 International School and Symposium on Precision Measurement and Gravity Experiment*, Taipei, Republic of China, January 24-February 2, 1983, W.-T. Ni, (Ed.), (Published by National Tsing Hua University, Hsinchu, Taiwan, Republic of China), 531-540

Ni, W.-T. (1984a). Equivalence Principles and Precision Experiments. *Precision Measurement and Fundamental Constants II*, B. N. Taylor and W. D. Phillips, (Ed.), Natl. Bur. Stand. (U S) Spec. Publ. 617, pp 647-651

Ni, W.-T. (1984b). Timing Observations of the Pulsar Propagations in the Galactic Gravitational Field as Precision Tests of the Einstein Equivalence Principle. *Proceedings of the Second Asian-Pacific Regional Meeting of the International Astronomical Union*, B. Hidayat and M. W. Feast (Ed.), (Published by Tira Pustaka, Jakarta, Indonesia), 441-448

Ni, W.-T.; Tsubono, K.; Mio, N.; Narihara, K.; Chen, S.-C.; King, S.-K. & Pan, S.-s. (1991). Test of Quantum Electrodynamics using Ultra-High Sensitive Interferometers. *Modern Physics Letters A*, Vol.6, pp. 3671-3678

Ni, W.-T. (1998). Magnetic Birefringence of Vacuum—Q & A Experiment,. *Frontier Test of QED and Physics of the Vacuum*, Eds. E. Zavattini, D. Bakalov, C. Rizzo, Heron Press, Sofia, pp. 83-97

Ni, W.-T. (2005a). Probing the Microscopic Origin of Gravity via Precision Polarization and Spin Experiments. *Chinese Physical Letters*, Vol.22, pp. 33-35 [arXiv:gr-qc/0407113]

Ni, W.-T. (2005b). Empirical Foundations of the Relativistic Gravity. *International Journal of Modern Physics D*, Vol.14, pp. 901-921 [arXiv:gr-qc/0504116]

Ni, W.-T. (2008). From equivalence principles to cosmology: cosmic polarization rotation, CMB observation, neutrino number asymmetry, Lorentz invariance and CPT. *Progress of Theoretical Physics Supplement*, Vol.172, pp. 49-60 [arXiv:0712.4082]

Ni, W.-T. (2009a). Cosmic polarization rotation, cosmological models, and the detectability of primordial gravitational waves. *International Journal of Modern Physics A* Vol. 18&19, p. 3493 [arXiv:0903.0756]

Ni, W.-T. (2009b). Direct Detection of the Primordial Inflationary Gravitational Waves. *International Journal of Modern Physics D*, Vol.18, pp. 2195-2199 [arXiv:0905.2508]

Ni, W.-T. (2010). Searches for the Role of Spin and Polarization in Gravity. *Reports on Progress in Physics*, Vol.73, 056901

Ni, W.-T. (2011). Rotation, the Equivalence Principle, and the Gravity Probe B Experiment. *Physical Review Letters*, Vol.107, 051103

Peccei, R. D. & Quinn, H. R. (1977). CP Conservation in the presence of pseudoparticles *Physical Review Letters*, Vol.38, pp. 1440-1443

Proca, A. (1936a). Sur la théorie du positron. *Comptes Rendus de l'Académie des Sciences*, Vol.202, pp. 1366–1368

Proca, A. (1936b). Sur la théorie ondulatoire des electrons positifs et negatives. *Journal de Physique et Le Radium*, Vol.7, pp. 347–353

Proca, A. (1936c). Sur les photons et les particules charge pure. *Comptes Rendus de l'Académie des Sciences*, Vol.203, pp. 709–711

Proca, A. (1937). Particles libres: Photons et particules 'charge pure'. *Journal de Physique et Le Radium*, Vol.8, pp. 23–28

Proca, A. (1938). Théorie non relativiste des particles a spin entire. *Journal de Physique et Le Radium*. Vol.9, pp. 61–66

Roll, P. G.; Krotkov, R. & Dicke, R. H. (1964). The equivalence of inertial and passive gravitational mass *Ann. Phys.* (N Y) 26 442-517

Ryutov, D. D. (2007). Using Plasma Physics to Weigh the Photon. *Plasma Physics and Controlled Fusion*, Vol.49, pp. B429–B438

Schlamminger, S.; Choi, K.-Y.; Wagner, T. A.; Gundlach, J. H. & Adelberger, E. G. (2008). Test of the equivalence principle using a rotating torsion balance. *Physical Review Letters* Vol.100, 041101

Schrödinger, E. (1943). The Earth's and the Sun's Permanent Magnetic Fields in the Unitary Field Theory. *Proceedings of the Royal Irish Academy, Section A*, Vol.49, pp. 135–148

Smolin, L. & Soo, C. (1995). The Chern-Simons Invariant as the Natural Time Variable for Classical and Quantum Cosmology. *Nuclear Physics B,* Vol.449, pp. 289-314

Vessot, R. F. C. et al. (1980). Test of relativistic gravitation with a space-borne hydrogen maser. *Physical Revew Letters,* Vol.45, pp. 2081-2084

Weinberg, S. (1978). A New Light Boson? *Physical Review Letters,* Vol.40, p. 233

Wilczek, F. (1978). Problem of strong P and T invariance in the presence of instantons. *Physical Review Letters,* Vol.40, p. 279

Williams, E. R.; Faller, J. E. & Hill, H. A. (1971). New Experimental Test of Coulomb's Law: A Laboratory Upper Limit on the Photon Rest Mass. *Physical Review Letters,* Vol.26, pp. 721–724

Wu, J.-S.; Chen, S.-J. & Ni, W.-T. (Q & A Collaboration). (2004). Building a 3.5 m Prototype Interferometer for the Q & A Vacuum Birefringence Experiment and High-Precision Ellipsometry. *Classical and Quantum Gravity*, Vol.21, pp. S1259-S1263

Zavattini, E. et al. (PVLAS Collaboration). (2008). New PVLAS Results and Limits on Magnetically Induced Optical Rotation and Ellipticity in Vacuum. *Physical Review D,* Vol.77, 032006

Part 2

Electromagnetism, Thermodynamics and Quantum Physics

4

Thermodynamics of Electric and Magnetic Systems

Radu Paul Lungu
Department of Physics, University of Bucharest, Măgurele-Bucharest, Romania

1. Introduction

The systems which contain electric or magnetic media have besides the electric (respectively magnetic) properties also the common properties of a thermodynamic systems (that is thermal, volumic, chemical); moreover, there are correlations between the electric (magnetic) properties and the thermal or volumic properties. Because there are a great variety of situations with the corresponding properties, we shall present briefly only the most important characteristics of the simplest electric or magnetic systems. For these systems there are supplementary difficulties (comparing to the simple neutral fluid) because the inherent non-homogeneity of these systems and also because the special coupling between the electric (respectively magnetic) degree of freedom and the volumic degree of freedom. These difficulties have let to the use of different methods of study in the literature, being necessary to modify some standard thermodynamic quantities (introduced in the standard textbooks for the simple thermodynamic systems).

In order to have a relation with the presentations of other works, we shall discuss the electric (respectively the magnetic) systems with many methods and we shall note some improper use of the different concepts which had been introduced initially in the standard thermodynamics. Although the most of the electric or magnetic systems are solid, in order to maintain a short and also an intelligible exposition we shall present explicitly only the case when the system is of the fluid type, and when it contain a single chemical species, therefore neglecting the anisotropy effects and the complications introduced by the theory of elasticity.

We remark that for the thermodynamics of quasi-static processes must be considered only equilibrium states, so that we will deal only with electrostatic or magneto-static fields. Although there are interference effects between the electric and magnetic phenomena, these are very small; therefore, in order to simplify the exposition, we shall study separately the electric and the magnetic systems, emphasizing the formal similitude between these type of systems.

There are many textbooks which present the basic problems of thermodynamics, some of the most important of them used the classical point of new [1-12], and also other used the neo-gibbsian point of view [13-15]; in the following we shall use the last point of view (i.e. we shall use the neo-gibbsian thermodynamics) [16].

2. Electric systems

2.1 General electrodynamic results

Accordingly to the electrodynamics, the electrostatic field created by a distribution of static electric charges, in an electric medium, is characterized by the vectorial fields *the intensity of the electric field* $\mathcal{E}(\mathbf{r})$ and *the intensity of electric induction* $\mathcal{D}(\mathbf{r})$, which satisfy the electrostatic Maxwell equations [17-20]:

$$\text{rot } \mathcal{E}(\mathbf{r}) = 0 , \tag{1a}$$

$$\text{div } \mathcal{D}(\mathbf{r}) = \rho(\mathbf{r}) , \tag{1b}$$

where $\rho(\mathbf{r})$ is the volumic density of the electric free charge (there are excluded the polarization charges).

From Eq. (1a) it follows that $\mathcal{E}(\mathbf{r})$ is an irrotational field, *i.e.* it derives from an electrostatic potential $\Phi(\mathbf{r})$:

$$\mathcal{E}(\mathbf{r}) = -\text{ grad } \Phi(\mathbf{r}) . \tag{2}$$

Also, on the surface of an conductor having the electric charge, the surface density of electric charge σ is related to the normal component (this is directed towards the conductor) of the electric induction with the relation:

$$\mathcal{D}_n = -\sigma . \tag{3}$$

Under the influence of the electrostatic field, the dielectric polarizes (it appears polarization charges), and it is characterized by the *electric dipolar moment* $\mathcal{P}$, respectively by the *polarization* (the volumic density of dipolar electric moment) $\boldsymbol{P}(\mathbf{r})$:

$$\boldsymbol{P}(\mathbf{r}) \equiv \lim_{\delta V \to 0} \frac{\delta \mathcal{P}(\mathbf{r})}{\delta V(\mathbf{r})} \quad \Longleftrightarrow \quad \mathcal{P} = \int_V \mathrm{d}^3\mathbf{r}\, \boldsymbol{P}(\mathbf{r}) . \tag{4}$$

Using the polarization it results the relation between the characteristic vectors of the electrostatic field:

$$\mathcal{D}(\mathbf{r}) = \bar{\varepsilon}_0\, \mathcal{E}(\mathbf{r}) + \boldsymbol{P}(\mathbf{r}) , \tag{5}$$

where $\bar{\varepsilon}_0$ is the electric permitivity of the vacuum (it is an universal constant depending on the system of units).

The general relation between the intensity of the electric field $\mathcal{E}(\mathbf{r})$ and the polarization $\boldsymbol{P}(\mathbf{r})$ is

$$\boldsymbol{P}(\mathbf{r}) = \bar{\varepsilon}_0\, \hat{\chi}_{\mathrm{e}}(\mathcal{E}, \mathbf{r}) : \mathcal{E}(\mathbf{r}) + \boldsymbol{P}_0(\mathbf{r}) , \tag{6a}$$

where $\boldsymbol{P}_0$ is the spontaneous polarization, and $\hat{\chi}_{\mathrm{e}}$ is the electric susceptibility tensor (generally it is dependent on the electric field).

For simplicity, we shall consider only the particular case when there is no spontaneous polarization (i.e. the absence of ferroelectric phenomena) $\boldsymbol{P}_0 = 0$, and the dielectric is linear and isotropic (then $\hat{\chi}_{\mathrm{e}}$ is reducible to a scalar which is independent on the electric field); in this last case, Eq. (6a) becomes:

$$\boldsymbol{P} = \bar{\varepsilon}_0\, \chi_{\mathrm{e}}\, \mathcal{E} , \tag{6b}$$

and Eq. (5) allows a parallelism and proportionality relation between the field vectors:

$$\mathcal{D} = \bar{\varepsilon}_0 (1 + \chi_{\mathrm{e}})\, \mathcal{E} . \tag{7}$$

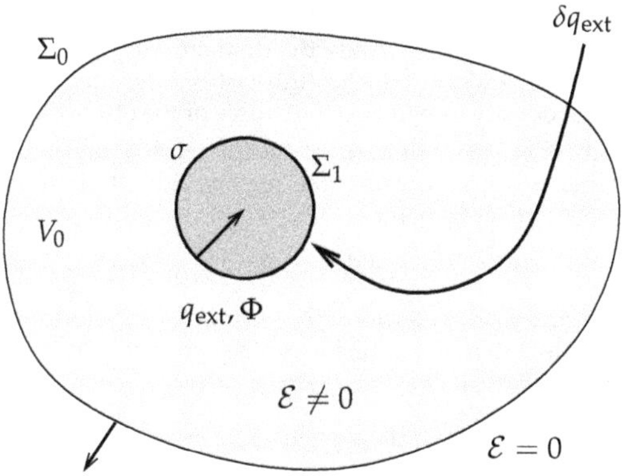

Fig. 1. The system chosen for the evaluation of the electric work.

In the common cases the susceptibility (for the specified types of dielectrics) depends on the temperature and of the particle density (or of the pressure) in the form:

$$\chi_e = \mathfrak{n}\, \overline{\chi}_e(T, \mathfrak{P}) \,, \tag{8}$$

where $\mathfrak{n} \equiv N/V$ is the particle density, $\overline{\chi}_e$ is the specific (per particle) susceptibility, and $\mathfrak{P}$ is the pressure[1].

In the narrow sense, the concrete expression of the electric susceptibility per particle is an empirical information of the thermal state equation type; we shall consider explicitly two simple cases:

– the *ideal* dielectric

$$\overline{\chi}_e(T) = \frac{K}{T} \,, \tag{9a}$$

– the *non-ideal* dielectric

$$\overline{\chi}_e(T, \mathfrak{P}) = \frac{K}{T - \Theta(\mathfrak{P})} \,, \tag{9b}$$

where K is a constant depending on the dielectric (called *Curie constant*), and $\Theta(\mathfrak{P})$ is a function of the pressure, having the dimension of a temperature.

Using the general relations of the electrostatics, we can deduce the expression of the infinitesimal *electric work*, as the energy given to the thermo-isolated dielectric when the electric field varies:

$$\delta \mathcal{L}_e = \int_{V_0} d^3\mathbf{r}\, \boldsymbol{\mathcal{E}} \cdot \delta \boldsymbol{\mathcal{D}} \,, \tag{10}$$

under the condition that the system is located in the domain with the volume $V_0 =$ constant, so that outside to this domain the electric field *vanishes*.

Proof:

[1] We can define the dipolar electric moment per particle as $\boldsymbol{\mathcal{P}} = N\, \mathfrak{p}$, resulting the susceptibility per particle with an analogous relation to (6b) $\mathfrak{p} = \overline{\varepsilon}_0\, \overline{\chi}_e\, \boldsymbol{\mathcal{E}}$.

The electrostatic field from the dielectric is created by charges located on conductors; we consider the situation illustrated in Fig. 1, where inside the domain with the volume V_0 and fixed external surface Σ_0, there are dielectrics and a conductor (this last has the surface Σ_1) with the electric charge q_{ex} (respectively the charge density σ) which is the source of the electrostatic field[2].

In the conditions defined above, the infinitesimal electric work is the energy given for the transport of the small electric charge δq_{ex} from outside (the region without electric field) until the conductor surface Σ_1, which has the electrostatic potential Φ:

$$\delta \mathcal{L}_{\text{e}} = \Phi \, \delta q_{\text{ex}} \, .$$

In order to express the electric work in terms of the vectors for the electrostatic field, we observe that the charge q_{ex} located on the internal conductor can be written with the normal component of the electric induction, accordingly to Eq. (3):

$$q_{\text{ex}} = \oint_{\Sigma_1} \sigma \, \mathrm{d}\mathcal{A} = -\oint_{\Sigma_1} \mathcal{D}_n \, \mathrm{d}\mathcal{A} \, ,$$

and from previous relation it results that a variation of the charge δq_{ex} implies a variation of the electric induction $\delta\mathcal{D}$ (the surface Σ_1 is fixed); because the electrostatic potential Φ is constant on the surface Σ_1, the expression of the electric work can be written as:

$$\delta \mathcal{L}_{\text{e}} = -\oint_{\Sigma_1} \mathrm{d}\mathcal{A} \, \boldsymbol{n}_1 \cdot \Phi \, \delta \boldsymbol{\mathcal{D}} \, .$$

The integral on the surface Σ_1 can be transformed in a volume integral, using the Gauss' theorem:

$$\int_{\mathcal{D}} \mathrm{d}V \ \operatorname{div} \boldsymbol{a} = \oint_{\partial\mathcal{D}} \mathrm{d}\mathcal{A} \, \boldsymbol{n} \cdot \boldsymbol{a} \, ;$$

then we obtain

$$\delta \mathcal{L}_{\text{e}} = -\int_{V} \mathrm{d}V \ \operatorname{div} \left(\Phi \, \delta \boldsymbol{\mathcal{D}}\right) + \oint_{\Sigma_0} \mathrm{d}\mathcal{A} \, \boldsymbol{n}_0 \cdot \Phi \, \delta \boldsymbol{\mathcal{D}} \, .$$

From the defining conditions, on the external surface Σ_0 the electrostatic potential Φ and the electric induction $\mathcal{D}$ vannish, so that the surface integral $\oint_{\Sigma_0} \dots$ has no contribution.

For the volumic integral we can perform the following transformations of the integrand:

$$\begin{aligned} \operatorname{div} \left(\Phi \, \delta \boldsymbol{\mathcal{D}}\right) &= \operatorname{grad} \Phi \cdot \delta \boldsymbol{\mathcal{D}} + \Phi \ \operatorname{div} \left(\delta \boldsymbol{\mathcal{D}}\right) = -\boldsymbol{\mathcal{E}} \cdot \delta \boldsymbol{\mathcal{D}} + \Phi \, \delta\left(\operatorname{div} \boldsymbol{\mathcal{D}}\right) \\ &= -\boldsymbol{\mathcal{E}} \cdot \delta \boldsymbol{\mathcal{D}} \, , \end{aligned}$$

because Eq. (2) and the absence of another free charges inside the dielectric:

$$\operatorname{div} \boldsymbol{\mathcal{D}} = \rho_{\text{int}} = 0 \, .$$

Finally, by combining the previous results, we get Eq. (10). □

We note that the expression (10) for the electric work, implies a domain for integration with a fixed volume (V_0), and in addition the electric field must vanish outside this domain.

Therefore, there are two methods to deal the thermodynamics of dielectrics on the basis of the electric work (and also the necessary conditions for the validity of the corresponding expression): *the open system method* (when the domain of integration is fixed but it has a fictitious frontier and it contains only a part of the dielectric), and *the closed system method* (when the domain of integration has physical a frontier, possibly located at the infinity, but the dielectric is located in a part of this domain); we observe that the second method is more physical, but in the same time it is more complex, because we must consider a compound system and only a part of this total system is of special interest (this is the dielectric). In addition we shall see that the second method implies the change of the common definitions

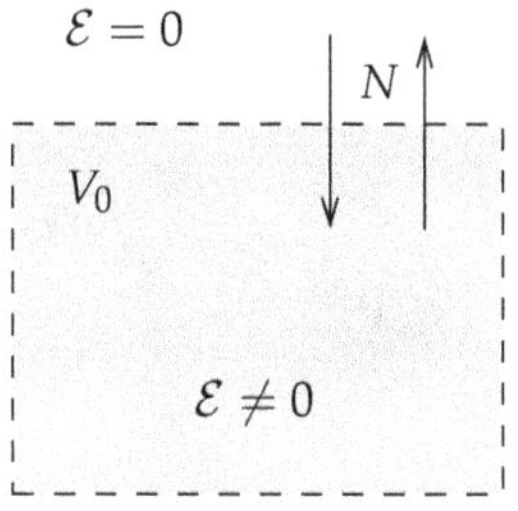

Fig. 2. The open system model.

for the state parameters associated to the electric and volumic degrees of freedom (both the extensive and the intensive).

1. The open system method: we consider a fixed domain (having the volume V = const.) which contains an electrostatic field inside, but outside to this domain the electrostatic field vanishes; the interesting system is the dielectric located inside the above specified domain, as an open thermodynamic system (the dielectric fills completely the domain, but there is a part of this dielectric outside the domain, at vanishing electrostatic field, because the frontier is totally permeable).

We note the following characteristic features of this situation:

– the thermodynamic system (the portion of the dielectric located in electrostatic field) has a fixed volume ($V = V_0$ = constant), but in the same time, it is an *open* thermodynamic system ($N \neq$ constant);

– the electro-striction effect (this is the variation of the volume produced by the variation of the electric field) in this case leads to the variation of the particle number, or in another words, by variation of the particle density $\mathfrak{n} \equiv N/V_0 \neq$ constant (because the volume is fixed, but the frontier is chemical permeable);

– in the simplest case, when we consider a homogeneous electrostatic field[3], inside the domain with the volume V_0, infinitesimal electric work can be written in the form

$$\delta \mathcal{L}_e = \mathcal{E}\ \delta(V_0\, \mathcal{D})\ , \tag{11}$$

and this implies the following definition for the electric state parameters (the extensive and the intensive ones):

$$\begin{cases} X_e = V_0\, \mathcal{D} = \mathfrak{D} \\ P_e = \mathcal{E} \end{cases} \tag{12}$$

[in this case V = const., that is the volumic degree of freedom for this system is frozen; but we emphasize that the expression $\delta \mathcal{L}_e = \mathcal{E}\ \delta(V\, \mathcal{D})$ when the volume of the system V can varies is *incorrect*].

[2] In order to have the general situation, we do not suppose particular properties for the dielectrics, so that we consider the non-homogeneous case.

[3] This situation is realized by considering a plane electric condenser having very closed plates, so that it is possible to consider approximately that the electric field vanishes outside the condenser; the space inside and outside the condenser is filled with a fluid dielectric. Accordingly to the previous definitions, the thermodynamic interesting system is only the part of the dielectric which is located inside the plates of the condenser, and the frontier of this system is fictitious.

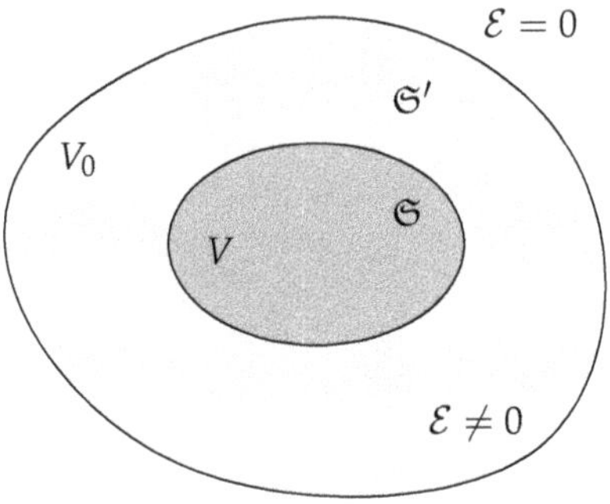

Fig. 3. The model for the closed system.

2. The closed system method: we consider the dielectric surrounded by an non-electric fluid (that is, the dielectric is located on a part of the domain with electric field).
Then it is necessary to define the compound system corresponding to the whole space where exists electric field

$$\mathfrak{S}^{(\tau)} = \mathfrak{S} \bigcup \mathfrak{S}' ,$$

where $\mathfrak{S}$ is the dielectric system with the volume V, and $\mathfrak{S}'$ is an auxiliary non-electric system having the volume $V' = V_0 - V$, as it is illustrated in Fig. 3.

We must remark that the auxiliary system (having negligible electric properties) is necessary in order to obtain the condition $\mathcal{E} \to 0$ towards the frontier of the domain which have the volume V_0, and also it produces a pressure on the dielectric; thus, the volume of the dielectric is not fixed and we can distinguish directly electro-striction effects.

Because the polarization $\boldsymbol{P}$ is non-vanishing only in the domain V, occupied by the system $\mathfrak{S}$, we transform the expression (10) using Eq. (5), in order to extract the electrization work on the subsystem $\mathfrak{S}$

$$\delta\mathcal{L}_{\text{e}}^{(\tau)} = \int_{V_0} \mathrm{d}^3\mathbf{r}\, \boldsymbol{\mathcal{E}} \cdot \delta\boldsymbol{\mathcal{D}} = \int_{V_0} \mathrm{d}^3\mathbf{r}\, \boldsymbol{\mathcal{E}} \cdot \bar{\varepsilon}_0\, \delta\boldsymbol{\mathcal{E}} + \int_{V_0} \mathrm{d}^3\mathbf{r}\, \boldsymbol{\mathcal{E}} \cdot \delta\boldsymbol{P}$$
$$\equiv \delta W_{\mathcal{E}}^{(\tau)} + \delta\mathcal{L}_{\text{p}} ,$$

where $\delta W_{\mathcal{E}}^{(\tau)}$ is the variation of the energy for the field inside the total volume V_0, and $\delta\mathcal{L}_{\text{p}}$ is the work for polarize the dielectric.

The first term allows the separation of the contributions corresponding the two subsystems when the energy of the electrostatic field changes:

$$\delta W_{\mathcal{E}}^{(\tau)} = \int_{V_0} \mathrm{d}^3\mathbf{r}\, \delta\Big(\frac{\bar{\varepsilon}_0\, \mathcal{E}^2}{2}\Big) = \delta\left\{\int_{V} \mathrm{d}^3\mathbf{r}\, \frac{\bar{\varepsilon}_0\, \mathcal{E}^2}{2}\right\} + \delta\left\{\int_{V'} \mathrm{d}^3\mathbf{r}\, \frac{\bar{\varepsilon}_0\, \mathcal{E}^2}{2}\right\}$$
$$= \delta W_{\mathcal{E}} + \delta W'_{\mathcal{E}} .$$

The second term can be interpreted as electric polarization work and it implies only the dielectric; in order to include the possible electro-strictive effects, we shall write this term in the form

$$\delta\mathcal{L}_{\text{p}} = \int_{V_0} \mathrm{d}^3\mathbf{r}\, \boldsymbol{\mathcal{E}} \cdot \delta\boldsymbol{P} = \int_{V_f} \mathrm{d}^3\mathbf{r}\, \boldsymbol{\mathcal{E}} \cdot \boldsymbol{P}_f - \int_{V_i} \mathrm{d}^3\mathbf{r}\, \boldsymbol{\mathcal{E}} \cdot \boldsymbol{P}_i = \delta\left\{\int_{V} \mathrm{d}^3\mathbf{r}\, \boldsymbol{\mathcal{E}} \cdot \boldsymbol{P}\right\}\Bigg|_{\mathcal{E}=\text{const.}}$$

that is, the polarization work implies the variation of the polarization $\delta \boldsymbol{P}$ and of the volume δV of the dielectric with the condition of constant electric field: $\boldsymbol{\mathcal{E}} =$ constant (during the process).

On the basis of the previous results we can separate the contribution of the electric work on the dielectric ($\eth\mathcal{L}_e$) from those on the auxiliary non-electric system ($\delta W'_{\mathcal{E}}$):

$$\eth\mathcal{L}_e^{(\tau)} = \eth\mathcal{L}_e + \delta W'_{\mathcal{E}} \,, \tag{13a}$$

$$\eth\mathcal{L}_e = \eth\mathcal{L}_p + \delta W_{\mathcal{E}} \,. \tag{13b}$$

For the electric work on the dielectric (the subsystem $\mathfrak{S}$) we observe two interpretations in the case when $\mathfrak{S}$ is homogeneous[4]:

1. we take into account only the polarization work $\eth\mathcal{L}_p$ and we neglect systematically the energy of the electric field inside the dielectric $\delta W_{\mathcal{E}}$; then the polarization work in the uniform electric field can be expressed with the dipolar electric moment

$$\eth\mathcal{L}_p = \boldsymbol{\mathcal{E}} \cdot \delta \int_V \mathrm{d}^3\mathbf{r}\, \boldsymbol{P} = \boldsymbol{\mathcal{E}} \cdot \delta \boldsymbol{\mathcal{P}} \,; \tag{14}$$

2. we estimate the contributions of the both terms from Eq. (13b), taking into account the implications due to the homogeneity of the system:

$$\begin{aligned} \text{đ}\mathcal{L}_p &= \mathcal{E}\ \mathrm{d}\mathcal{P} = \mathcal{E}\ \mathrm{d}(VP)\,, \\ \mathrm{d}W_{\mathcal{E}} &= \mathrm{d}\Big(\frac{\bar{\varepsilon}_0\,\mathcal{E}^2}{2}\,V\Big) = -\,\frac{\bar{\varepsilon}_0\,\mathcal{E}^2}{2}\,\mathrm{d}V + \mathcal{E}\ \mathrm{d}(V\,\bar{\varepsilon}_0\,\mathcal{E})\,, \end{aligned}$$

so that we obtain the total electric work performed by the dielectric

$$\text{đ}\mathcal{L}_e = \text{đ}\mathcal{L}_p + \mathrm{d}W_{\mathcal{E}} = -\,\frac{\bar{\varepsilon}_0\,\mathcal{E}^2}{2}\,\mathrm{d}V + \mathcal{E}\ \mathrm{d}(V\,\mathcal{D})\,, \tag{15}$$

and the last expression can be interpreted as an work performed on two degrees of freedom (volumic and electric).

We observe that for isotropic dielectrics the vectors $\boldsymbol{\mathcal{E}}$, $\boldsymbol{\mathcal{D}}$ and $\boldsymbol{P}$ are collinear; therefore, we shall omit the vectorial notation, for simplicity.

2.2 Thermodynamic potentials

We shall discuss, for simplicity, only the case when the electric system is homogeneous and of fluid type, being surrounded by a non-electric environment. Then, the fundamental thermodynamic differential form is:

$$\mathrm{d}\mathcal{U} = \text{đ}\mathcal{Q} + \text{đ}\mathcal{L}_V + \text{đ}\mathcal{L}_e + \text{đ}\mathcal{L}_N \,. \tag{16}$$

For the thermodynamic study of the electric system there are many methods, depending the choice of the fundamental variables (corresponding to the choice of the concrete expression for the electric work $\text{đ}\mathcal{L}_e$).

[4] The condition of homogeneity implies an uniform electric field $\boldsymbol{\mathcal{E}}(\mathbf{r}) =$ const. in the subsystem $\mathfrak{S}$, and this property is realized only when the dielectric is an ellipsoid in an uniform external field.

2.2.1 Pseudo-potentials method

We replace the expression (13b) – (14) for the electric work, and also the expression for the other forms of work and for the heat; then, the fundamental thermodynamic differential form has the explicit expression:

$$\mathrm{d}\mathcal{U} = T\,\mathrm{d}\mathcal{S} - \mathfrak{P}\,\mathrm{d}V + \mathcal{E}\,\mathrm{d}\mathcal{P} + \mathrm{d}W_{\mathcal{E}} + \mu\,\mathrm{d}N\,. \tag{17}$$

We observe that the preceding differential form contains a term $\mathrm{d}W_{\mathcal{E}}$ which is a exact total differential (from the mathematical point of view) and it represents the variation of the energy of the electric field located in the space occupied by the dielectric; we put this quantity in the left side of the above equality we obtain:

$$\mathrm{d}\widetilde{\mathcal{U}} = T\,\mathrm{d}\mathcal{S} - \mathfrak{P}\,\mathrm{d}V + \mathcal{E}\,\mathrm{d}\mathcal{P} + \mu\,\mathrm{d}N\,, \tag{18}$$

where $\widetilde{\mathcal{U}} \equiv \mathcal{U} - W_{\mathcal{E}}$ is called the *internal pseudo-energy* of the dielectric[5].

We present some observations concerning the differential form (18):

• $\widetilde{\mathcal{U}}(\mathcal{S}, V, \mathcal{P}, N)$ is equivalent to the fundamental thermodynamic equation of the system, since it contains the whole thermodynamic information about the system (that is, its derivatives are the state equations)[6]; and on the other side, the pseudo-energy has no specified convexity properties (because it was obtained by subtracting a part of the energy from the total internal energy of the system);

• $\widetilde{\mathcal{U}}(\mathcal{S}, V, \mathcal{P}, N)$ is a homogeneous function of degree 1 (because it is obtained as a difference of two homogeneous functions of degree 1);

• by considering the differential form (18) as similar to the fundamental thermodynamic differential form, it follows that the electric state parameters are[7]

$$\begin{cases} \widetilde{X}_e = \mathcal{P} = PV\,, \\ \widetilde{P}_e = \mathcal{E}\,; \end{cases}$$

• if we perform the Legendre transformations of the function $\widetilde{\mathcal{U}}(\mathcal{S}, V, \mathcal{P}, N)$, then we obtain objects of the thermodynamic potential types (that is, the derivatives of these quantities give the state equations of the dielectric); however, these objects are not true thermodynamic potentials (firstly since they have not the needed properties of convexity-concavity), so that they are usually called *thermodynamic pseudo-potentials*.

In the following we shall present briefly only the most used pseudo-potentials: the *electric pseudo-free energy* and the *electric Gibbs pseudo-potential*.

---

[5] Because we consider in the expression $W_{\mathcal{E}} = V\,\bar{\varepsilon}_0\mathcal{E}^2/2$ the intensity of the electric field in the presence of the dielectric, this energy is due both to the vacuum and to the dielectric; thus, $\widetilde{\mathcal{U}}$ *is not the internal energy of the dielectric* (without the energy of the electric fields in vacuum), but it is an artificial quantity.

[6] We shall show later that the derivatives of the pseudo-energy (more exactly, the derivatives of the pseudo-potentials deduced from the pseudo-energy) are the correct state equations.

[7] We observe that the use of the internal pseudo-energy implies the modification of the extensive electric state parameter ($V\boldsymbol{\mathcal{D}} \rightarrow V\boldsymbol{P}$), but the intensive electric state parameter ($\boldsymbol{\mathcal{E}}$) had been unmodified.

a.1. The electric pseudo-free energy is the Legendre transformation on the thermal and electric degrees of freedom[8]

$$\widetilde{\mathcal{F}}^* \equiv \widetilde{\mathcal{U}} - T\,\mathcal{S} - \mathcal{E}\,\mathcal{P}\,, \tag{19}$$

having the differential form[9]:

$$\mathrm{d}\widetilde{\mathcal{F}}^* = -\,\mathcal{S}\,\mathrm{d}T - \mathfrak{P}\,\mathrm{d}V - \mathcal{P}\,\mathrm{d}\mathcal{E} + \mu\,\mathrm{d}N\,, \tag{20}$$

so that it allows the deduce the state equations in the representation $(T, V, \mathcal{E}, N)$.
We consider the simplest case, when the dielectric behaves as a neutral fluid, having the free energy $\mathcal{F}_0(T, V, N)$, in the absence of the electric field and when the electric susceptibility is $\chi_e(T, V/N)$; then, we can consider as known the electric state equation (the expression of the dipolar electric moment):

$$\left(\frac{\partial\,\widetilde{\mathcal{F}}^*}{\partial\,\mathcal{E}}\right)_{T,V,N} = -\,\mathcal{P}(T, V, \mathcal{E}, N) = -\,\bar{\varepsilon}_0\,\chi_e(T, V/N)\,\mathcal{E}\,V\,.$$

By partial integration, with respect to the electric field, and taking into account that at vanishing electrical field the electric pseudo-free energy reduces to the proper free energy (the Helmholtz potential):

$$\widetilde{\mathcal{F}}^*\Big|_{\mathcal{E}=0} = \mathcal{U}_0 - T\,\mathcal{S}_0 = \mathcal{F}_0(T, V, N)\,,$$

we obtain:

$$\widetilde{\mathcal{F}}^*(T, V, \mathcal{E}, N) = \mathcal{F}_0(T, V, N) + \mathcal{F}^*_{\mathrm{el}}(T, V, \mathcal{E}, N)\,, \tag{21a}$$

$$\mathcal{F}^*_{\mathrm{el}}(T, V, \mathcal{E}, N) \equiv -\,\frac{\bar{\varepsilon}_0\,\mathcal{E}^2}{2}\,\chi_e(T, V/N)\,V\,, \tag{21b}$$

that is, the electric pseudo-free energy of the dielectric is the sum of the free energy at null electric field and the electric part $\mathcal{F}^*_{\mathrm{el}}$.

The additivity (factorization) property of the electric pseudo-free energy is transmitted to the non-electric state equations: the entropy, the pressure and the chemical potential (these are sums of the non-electric part, corresponding to vanishing electric field, and the electric part):

$$\mathcal{S}(T, V, \mathcal{E}, N) = -\left(\frac{\partial\,\widetilde{\mathcal{F}}^*}{\partial\,T}\right)_{V,\mathcal{E},N} = \mathcal{S}_0(T, V, N) + \frac{\bar{\varepsilon}_0\,\mathcal{E}^2}{2}\left(\frac{\partial\,\chi_e}{\partial\,T}\right)_{V,N} V\,, \tag{22a}$$

$$\mathfrak{P}(T, V, \mathcal{E}, N) = -\left(\frac{\partial\,\widetilde{\mathcal{F}}^*}{\partial\,V}\right)_{T,\mathcal{E},N} = \mathfrak{P}_0(T, V, N) + \frac{\bar{\varepsilon}_0\,\mathcal{E}^2}{2}\left(\frac{\partial\,(\chi_e V)}{\partial\,V}\right)_{T,N}, \tag{22b}$$

$$\mu(T, V, \mathcal{E}, N) = \left(\frac{\partial\,\widetilde{\mathcal{F}}^*}{\partial N}\right)_{T,V,\mathcal{E}} = \mu_0(T, V, N) - \frac{\bar{\varepsilon}_0\,\mathcal{E}^2}{2}\left(\frac{\partial\,\chi_e}{\partial N}\right)_{T,V} V\,. \tag{22c}$$

---

[8] Because the absence of the correct properties of the convexity, we shall use only the classic definition of the Legendre transformation.

[9] In the strictly sense, $\widetilde{\mathcal{F}}^*$ is a simple Gibbs potential, so that the common terminology is criticizable.

a.2. The electric Gibbs pseudo-potential is defined, analogously to the previous case, as the Legendre transformation on the thermal, volumic and electric degrees of freedom[10]

$$\widetilde{\mathcal{G}}^{*} \equiv \widetilde{\mathcal{U}} - T\,\mathcal{S} + \mathfrak{P}\,V - \mathcal{E}\,\mathcal{P}\,, \tag{23}$$

having the differential form

$$\mathrm{d}\widetilde{\mathcal{G}}^{*} = -\,\mathcal{S}\,\mathrm{d}T + V\,\mathrm{d}\mathfrak{P} - \mathcal{P}\,\mathrm{d}\mathcal{E} + \mu\,\mathrm{d}N\,, \tag{24}$$

so that it allows the deduction of the state equations in the representation $(T, \mathfrak{P}, \mathcal{E}, N)$.
We consider the simplest case (analogously in the former case), when the dielectric behaves as a neutral fluid, having the proper Gibbs potential (the free enthalpy) $\mathcal{G}_0(T, \mathfrak{P}, N)$ in the absence of the electric field and when the electric susceptibility is $\overline{\chi}_e(T, \mathfrak{P})$; then, we can consider as known the electric state equation (the expression of the dipolar electric moment):

$$\left(\frac{\partial\,\widetilde{\mathcal{G}}^{*}}{\partial\,\mathcal{E}}\right)_{T,\mathfrak{P},N} = -\,\mathcal{P}(T, \mathfrak{P}, \mathcal{E}, N) = -\,\bar{\varepsilon}_0\,\overline{\chi}_e(T, \mathfrak{P})\,\mathcal{E}\,N\,.$$

By partial integration, with respect to the electric field, and taking into account that at vanishing electrical field the electric Gibbs pseudo-potential reduces to the proper Gibbs potential:

$$\widetilde{\mathcal{G}}^{*}\Big|_{\mathcal{E}=0} = \mathcal{U}_0 - T\,\mathcal{S}_0 + \mathfrak{P}\,V_0 = \mathcal{G}_0(T, \mathfrak{P}, N)\,,$$

we obtain:

$$\widetilde{\mathcal{G}}^{*}(T, \mathfrak{P}, \mathcal{E}, N) = \mathcal{G}_0(T, \mathfrak{P}, N) + \mathcal{G}^{*}_{\mathrm{el}}(T, \mathfrak{P}, \mathcal{E}, N)\,, \tag{25a}$$

$$\mathcal{G}^{*}_{\mathrm{el}}(T, \mathfrak{P}, \mathcal{E}, N) \equiv -\,\frac{\bar{\varepsilon}_0\,\mathcal{E}^2}{2}\,\overline{\chi}_e(T, \mathfrak{P})\,N\,, \tag{25b}$$

that is, the electric Gibbs pseudo-potential of the dielectric is the sum of the Gibbs potential at null electric field and the electric part $\mathcal{G}^{*}_{\mathrm{el}}$.

The additivity (factorization) property of the electric Gibbs pseudo-potential is transmitted to the non-electric state equations: the entropy, the volume and the chemical potential (these are sums of the non-electric part, corresponding to vanishing electric field, and the electric part):

$$\mathcal{S}(T, \mathfrak{P}, \mathcal{E}, N) = -\left(\frac{\partial\,\widetilde{\mathcal{G}}^{*}}{\partial\,T}\right)_{\mathfrak{P},\mathcal{E},N} = \mathcal{S}_0(T, \mathfrak{P}, N) + \frac{\bar{\varepsilon}_0\,\mathcal{E}^2}{2}\left(\frac{\partial\,\overline{\chi}_e}{\partial\,T}\right)_{\mathfrak{P}} N\,, \tag{26a}$$

$$V(T, \mathfrak{P}, \mathcal{E}, N) = \left(\frac{\partial\,\widetilde{\mathcal{G}}^{*}}{\partial\,\mathfrak{P}}\right)_{T,\mathcal{E},N} = V_0(T, \mathfrak{P}, N) - \frac{\bar{\varepsilon}_0\,\mathcal{E}^2}{2}\left(\frac{\partial\,\overline{\chi}_e}{\partial\,\mathfrak{P}}\right)_{T} N\,, \tag{26b}$$

$$\mu(T, \mathfrak{P}, \mathcal{E}, N) = \left(\frac{\partial\,\widetilde{\mathcal{G}}^{*}}{\partial N}\right)_{T,\mathfrak{P},\mathcal{E}} = \mu_0(T, \mathfrak{P}, N) - \frac{\bar{\varepsilon}_0\,\mathcal{E}^2}{2}\,\overline{\chi}_e(T, \mathfrak{P})\,. \tag{26c}$$

We observe, in addition, that the electric Gibbs pseudo-potential is a maximal Legendre transformation, so that with the Euler relation we obtain

$$\widetilde{\mathcal{G}}^{*}(T, \mathfrak{P}, \mathcal{E}, N) = \mu(T, \mathfrak{P}, \mathcal{E})\,N\,. \tag{27}$$

---

[10] Because the absence of the correct properties of the convexity, we shall use only the classic definition of the Legendre transformation (like in the preceding case).

2.2.2 The method of modified potentials

We use the expression (15) for the electric work, without extracting terms with total exact differential type from the internal energy of the dielectric; then, the differential form (16) can be written in the following explicit manner:

$$\begin{aligned} \mathrm{d}\mathcal{U} &= T\,\mathrm{d}\mathcal{S} - \left(\mathfrak{P} + \frac{\bar{\varepsilon}_0\,\mathcal{E}^2}{2}\right)\mathrm{d}V + \mathcal{E}\,\mathrm{d}(\mathcal{D}\,V) + \mu\,\mathrm{d}N \\ &= T\,\mathrm{d}\mathcal{S} - \pi\,\mathrm{d}V + \mathcal{E}\,\mathrm{d}\mathfrak{D} + \mu\,\mathrm{d}N\,. \end{aligned} \tag{28}$$

We observe that in this case the electric work has contributions on two thermodynamic degrees of freedom, so that we must redefine the electric and volumic state parameters:

$$X_V = V\,, \qquad P'_V = -\,\pi \equiv -\left(\mathfrak{P} + \frac{\bar{\varepsilon}_0\,\mathcal{E}^2}{2}\right), \tag{29a}$$

$$X'_{\mathcal{E}} = \mathfrak{D} \equiv \mathcal{D}\,V\,, \qquad P_{\mathcal{E}} = \mathcal{E}\,. \tag{29b}$$

In this last case it appears some peculiarities of the electric state parameters (both for the extensive and for the intensive), so that there are needed cautions when it is used this method:

- V and $\mathfrak{D} \equiv \mathcal{D}\,V$ must be considered as independent variables,
- the effective pressure has an supplementary electric contribution $\bar{\varepsilon}_0\,\mathcal{E}^2/2$.

Although the modified potential method implies the employment of some unusual state parameters, however it has the major advantage that $\mathcal{U}(\mathcal{S}, V, \mathfrak{D}, N)$ is the true fundamental energetic thermodynamic equation, and it is a *convex* and *homogeneous of degree* 1 function; thus, it is valid the Euler equation:

$$\mathcal{U} = T\,\mathcal{S} - \pi\,V + \mathcal{E}\,\mathfrak{D} + \mu\,N\,, \tag{30}$$

and it is possible to define true thermodynamic potential with Legendre transformations. From the Euler relation (30) and passing to the common variables, it results

$$\mathcal{U} = T\,\mathcal{S} - \mathfrak{P}\,V + \mathcal{E}\,\mathcal{P}\,V + \frac{\bar{\varepsilon}_0\,\mathcal{E}^2}{2}\,V + \mu\,N\,,$$

so that it is ensured that $\widetilde{\mathcal{U}} \equiv \mathcal{U} - W_{\mathcal{E}}$ is a homogeneous function of degree 1 with respect to the variables $(\mathcal{S}, V, \mathcal{P}, N)$.

In order to compare the results of the method of modified potentials with those of the method of the pseudo-potentials we shall present only *the electric free energy* and *the electric Gibbs potential* as energetic thermodynamic potentials.

b.1. The electric free energy is the Legendre transformation on the thermal and electric degrees of freedom

$$\mathcal{F}^*(T, V, \mathcal{E}, N) \equiv \inf_{\mathcal{S},\mathfrak{D}} \left\{ \mathcal{U}(\mathcal{S}, V, \mathfrak{D}, N) - T\,\mathcal{S} - \mathcal{E}\,\mathfrak{D} \right\}, \tag{31}$$

and it has the following differential form[11]:

$$\mathrm{d}\mathcal{F}^* = -\,\mathcal{S}\,\mathrm{d}T - \pi\,\mathrm{d}V - \mathfrak{D}\,\mathrm{d}\mathcal{E} + \mu\,\mathrm{d}N\,. \tag{32}$$

[11] In the strictly sense, $\mathcal{F}^*$ is a simple Gibbs potential, so that the common terminology is criticizable.

We shall emphasize some important properties of the above defined electric free energy $\mathcal{F}^*(T,V,\mathcal{E},N)$.

1. When the electric field vanishes it becomes the proper free energy (the Helmholtz potential)

$$\mathcal{F}^*(T,V,0,N) = \mathcal{U}_0(T,V,N) - T\,\mathcal{S}_0(T,V,N) = \mathcal{F}_0(T,V,N)\,.$$

2. The electric state equation is

$$\left(\frac{\partial \mathcal{F}^*}{\partial \mathcal{E}}\right)_{T,V,N} = -\,\mathfrak{D}(T,V,\mathcal{E},N) = -\,\bar{\varepsilon}_0\left[1+\chi_e(T,V/N)\right]\mathcal{E}\,V\,.$$

3. By partial integration with respect to the electric field and the use of the condition of null field, we obtain the general expression of the electric free energy (for a linear and homogeneous dielectric)

$$\mathcal{F}^*(T,V,\mathcal{E},N) = \mathcal{F}_0(T,V,N) - \frac{\bar{\varepsilon}_0\,\mathcal{E}^2}{2}\left[1+\chi_e(T,V/N)\right]V\,. \tag{33}$$

4. $\mathcal{F}^*(T,V,\mathcal{E},N)$ is a function *concave* in respect to the variables T and $\mathcal{E}$; as a result we get the relation

$$\left(\frac{\partial^2 \mathcal{F}^*}{\partial \mathcal{E}^2}\right)_{T,V,N} = -\,\bar{\varepsilon}_0\left[1+\chi_e\right]V < 0\,,$$

and it follows "the stability condition" $\chi_e > -1$. (We note that actually is realized a more strong condition $\chi_e > 0$, but this has no thermodynamic reasons).

5. The state equations, deduced from Eq. (33) are:

$$\mathcal{S}(T,V,\mathcal{E},N) = -\left(\frac{\partial \mathcal{F}^*}{\partial T}\right)_{V,\mathcal{E},N} = \mathcal{S}_0(T,V,N) + \frac{\bar{\varepsilon}_0\,\mathcal{E}^2}{2}\left(\frac{\partial \chi_e}{\partial T}\right)_{V,N} V\,, \tag{34a}$$

$$\pi(T,V,\mathcal{E},N) = -\left(\frac{\partial \mathcal{F}^*}{\partial V}\right)_{T,\mathcal{E},N} = \mathfrak{P}_0(T,V,N) + \frac{\bar{\varepsilon}_0\,\mathcal{E}^2}{2}\left[1+\left(\frac{\partial(\chi_e V)}{\partial V}\right)_{T,N}\right], \tag{34b}$$

$$\mu(T,V,\mathcal{E},N) = \left(\frac{\partial \mathcal{F}^*}{\partial N}\right)_{T,V,\mathcal{E}} = \mu_0(T,V,N) - \frac{\bar{\varepsilon}_0\,\mathcal{E}^2}{2}\left(\frac{\partial \chi_e}{\partial N}\right)_{T,V} V\,. \tag{34c}$$

Because $\pi = \mathfrak{P} + \bar{\varepsilon}_0\,\mathcal{E}^2/2$, it results that the state equations (34) are identical with Eqs. (22), and this shows that $\widetilde{\mathcal{F}}^*$ (the correspondent pseudo-potential to $\mathcal{F}^*$) gives correct state equations.

From Eq. (33) it result that the free energy (Helmholtz potential) is

$$\mathcal{F} = \mathcal{F}^* + \mathcal{E}\,\mathfrak{D} = \mathcal{F}_0 + \frac{\mathcal{E}\,\mathcal{D}}{2}\,V\,,$$

so that the electric part of the volumic density of free energy is:

$$\mathfrak{f}_{\mathrm{el}} \equiv \frac{\mathcal{F}-\mathcal{F}_0}{V} = \frac{\mathcal{E}\,\mathcal{D}}{2}\,.$$

We emphasize that in many books the previous expression for the electric part of the free energy density is erroneously considered as electric part of the internal energy density.

Correctly, the internal energy has the expression

$$\mathcal{U} = \mathcal{F}^* + T\,\mathcal{S} + \mathcal{E}\,\mathfrak{D} = (\mathcal{F}_0 + T\,\mathcal{S}_0) + \left(\frac{\mathcal{E}\,\mathcal{D}}{2} + \frac{\bar{\varepsilon}_0\,\mathcal{E}^2}{2}\,T\,\frac{\partial \chi_e}{\partial T}\right)V\,,$$

so that the electric part of the volumic density of internal energy is

$$\mathfrak{u}_{\mathrm{el}} = \frac{\mathcal{U} - \mathcal{U}_0}{V} = \frac{\mathcal{E}\,\mathcal{D}}{2} + \frac{\bar{\varepsilon}_0\,\mathcal{E}^2}{2}\,T\,\frac{\partial \chi_e}{\partial T} \neq \frac{\mathcal{E}\,\mathcal{D}}{2}\,.$$

b.2. The electric Gibbs potential is defined analogously, as the Legendre transformation on the thermal, volumic and electric degrees of freedom

$$\mathcal{G}^*(T,\pi,\mathcal{E},N) \equiv \inf_{\mathcal{S},V,\mathfrak{D}} \left\{\mathcal{U}(\mathcal{S},V,\mathfrak{D},N) - T\,\mathcal{S} + \pi\,V - \mathcal{E}\,\mathfrak{D}\right\}, \tag{35}$$

and it has the differential form

$$\mathrm{d}\mathcal{G}^* = -\,\mathcal{S}\;\mathrm{d}T + V\;\mathrm{d}\pi - \mathfrak{D}\;\mathrm{d}\mathcal{E} + \mu\;\mathrm{d}N\,. \tag{36}$$

According to the definition, $\mathcal{G}^*$ is a maximal Legendre transform, so that the Euler relation leads to:

$$\mathcal{G}^* = \mu\,N\,. \tag{37}$$

On the other side, by replacing the variables π and $\mathfrak{D}$, accordingly to the definitions (29), we obtain the *the electric Gibbs potential is equal to the electric Gibbs pseudo-potential*[12] (but they have different variables):

$$\mathcal{G}^*(T,\pi,\mathcal{E},N) = \widetilde{\mathcal{G}}^*(T,\mathfrak{P},\mathcal{E},N)$$

$$\pi = \mathfrak{P} + \frac{\bar{\varepsilon}_0\,\mathcal{E}^2}{2}\,.$$

From the preceding properties it follows that the equations deduced from the potential $\mathcal{G}^*$ are identical with Eqs. (22); we observe, however, that it is more convenient to use the pseudo-potential $\widetilde{\mathcal{G}}^*$, because this has more natural variables than the corresponding potential $\mathcal{G}^*$.

2.2.3 Thermodynamic potentials for open systems

Previously we have shown that the electric work implies two methods for treating the dielectrics: either as a closed subsystem of a compound system (this situation was discussed above), or as an open system located in a fixed volume (and the electric field is different from zero only inside the domain with fixed volume).

[12] The equality $\mathcal{G}^* = \widetilde{\mathcal{G}}^*$ (as quantities, but not as functions) can be obtained directly by comparing the consequences of the Euler equation (27) and (37).

If we use the second method, then the electric work has the expression (11) and the dielectric system has only 3 thermodynamic degrees of freedom: thermal, electric and chemical (the volumic degree of freedom is frozen); then, the fundamental differential form is

$$\mathrm{d}\mathcal{U} = T\,\mathrm{d}\mathcal{S} + \mathcal{E}\,\mathrm{d}(V_0\,\mathcal{D}) + \mu\,\mathrm{d}N\,. \tag{38}$$

Among the thermodynamic potentials, that can be obtained by Legendre transformations of the energetic fundamental thermodynamic equation, denoted as $\mathcal{U}(\mathcal{S}, V_0\,\mathcal{D}, N) \equiv \mathcal{U}(\mathcal{S}, \mathcal{D}, N; V_0)$, we shall present only the electric free energy:

$$\mathcal{F}^*(T,\mathcal{E},N;V_0) \equiv \inf_{\mathcal{S},\mathfrak{D}} \left\{ \mathcal{U}(\mathcal{S},\mathfrak{D},N;V_0) - T\,\mathcal{S} - \mathcal{E}\,\mathfrak{D} \right\}, \tag{39}$$

which has the following properties:

1. the differential form:

$$\mathrm{d}\mathcal{F}^* = -\,\mathcal{S}\,\mathrm{d}T - V_0\,\mathcal{D}\,\mathrm{d}\mathcal{E} + \mu\,\mathrm{d}N\,; \tag{40}$$

2. it reduces to the free energy (the Helmholtz potential) at vanishing electric field

$$\mathcal{F}^*(T,0,N;V_0) = \mathcal{U}_0(T,N;V_0) - T\,\mathcal{S}_0(T,N;V_0) = \mathcal{F}_0(T,N;V_0)\,;$$

3. by integrating the electric state equation $V_0\,\mathcal{D}(T,\mathcal{E},N) = V_0\,[1+\chi_e]\,\bar{\varepsilon}_0\,\mathcal{E}$, we obtain

$$\mathcal{F}^*(T,\mathcal{E},N;V_0) = \mathcal{F}_0(T,N;V_0) - V_0\,\frac{\bar{\varepsilon}_0\,\mathcal{E}^2}{2}\,\left[1+\chi_e(T,V/N)\right]. \tag{41}$$

We note that the results are equivalent to those obtained by the previous method, but the situation is simpler because the volumic degree of freedom is frozen.

2.3 Thermodynamic coefficients and processes

2.3.1 Definitions for the principal thermodynamic coefficients

Because the dielectric has 4 thermodynamic degrees of freedom (in the simplest case, when it is fluid), there are a great number of simple thermodynamic coefficients; taking into account this complexity, we shall present only the common coefficients, corresponding to closed dielectric systems (N = constant). In this case it is convenient to use reduced extensive parameters with respect to the particle number; thus, we shall use the specific entropy $s = \dfrac{\mathcal{S}}{N}$ and the specific volume $v = \dfrac{V}{N}$.

a.1. The sensible specific heats are defined for non-isothermal processes "φ":

$$c_\varphi \equiv \frac{C_\varphi}{N} = \frac{1}{N}\,T\left(\frac{\partial\,\mathcal{S}}{\partial\,T}\right)_\varphi = T\left(\frac{\partial\,s}{\partial\,T}\right)_\varphi\,. \tag{42}$$

In the case, when the process φ is simple, we obtain the following specific isobaric/isochoric and iso-polarization/iso-field heats: $c_{V,\mathfrak{p}}$, $c_{V,\mathcal{E}}$, $c_{P,\mathfrak{p}}$, $c_{P,\mathcal{E}}$.

a.2. The latent specific heats are defined for isothermal processes "ψ":

$$\lambda_\psi^{(a)} = \frac{1}{N}\,T\left(\frac{\partial\,\mathcal{S}}{\partial\,a}\right)_\psi = T\left(\frac{\partial\,s}{\partial\,a}\right)_\psi\,. \tag{43}$$

The most important cases (for "ψ" and a) are the isothermal-isobaric process with $a = \mathcal{E}$ when we have *the isobaric electro-caloric coefficient* λ and the conjugated isothermal-isofield process with $a = \mathfrak{P}$, when we have *the iso-field piezo-caloric coefficient* $\lambda_{\mathcal{E}}^{(P)}$:

$$\lambda = T\left(\frac{\partial s}{\partial \mathcal{E}}\right)_{T,\mathfrak{P}}, \qquad \lambda_{\mathcal{E}}^{(P)} = T\left(\frac{\partial s}{\partial \mathfrak{P}}\right)_{T,\mathcal{E}}. \tag{44}$$

a.3. The thermodynamic susceptibilities are of two types: for the volumic degree of freedom (in this case they are called *compressibility coefficients*) and for the electric degree of freedom (these are called *electric susceptibilities*):

$$\varkappa_{\varphi} = \frac{-1}{V}\left(\frac{\partial V}{\partial \mathfrak{P}}\right)_{\varphi} = \frac{-1}{v}\left(\frac{\partial v}{\partial \mathfrak{P}}\right)_{\varphi}, \tag{45}$$

$$\chi_{\psi}^{(\mathrm{el})} = \frac{1}{V}\left(\frac{\partial \mathcal{P}}{\partial \mathcal{E}}\right)_{\psi} = \left(\frac{\partial P}{\partial \mathcal{E}}\right)_{\psi}. \tag{46}$$

In the simple cases "φ" is an isothermal/adiabatic and iso-polarization/iso-field processes; it results the following simple compressibility coefficients: $\varkappa_{T,\mathcal{P}}$, $\varkappa_{T,\mathcal{E}}$, $\varkappa_{s,\mathcal{P}}$ and $\varkappa_{s,\mathcal{E}}$. Analogously "$\psi$" as simple process is isothermal/adiabatic and isobaric/isochoric, resulting the following simple electric susceptibilities: $\chi_{T,v}^{(\mathrm{el})}$, $\chi_{T,\mathfrak{P}}^{(\mathrm{el})}$, $\chi_{s,v}^{(\mathrm{el})}$ and $\chi_{s,\mathfrak{P}}^{(\mathrm{el})}$.

From Eq. (6b) we observe that the isothermal electric susceptibility is proportional to the susceptibility used in the electrodynamics:

$$\chi_{T,v}^{(\mathrm{el})} = \left(\frac{\partial P}{\partial \mathcal{E}}\right)_{T,v} = \bar{\varepsilon}_0\, \chi_e(T,v)\,.$$

a.4. The thermal coefficients are of two types, corresponding to the two non-thermal and non-chemical degrees of freedom (the volumic and the electric ones). If we consider only thermal coefficients for extensive parameters, then we can define the following types of simple coefficients:

- the *isobaric thermal expansion coefficients* (also iso-polarization/iso-field)

$$\alpha_y = \frac{1}{V}\left(\frac{\partial V}{\partial T}\right)_{\mathfrak{P},y,N} = \frac{1}{v}\left(\frac{\partial v}{\partial T}\right)_{\mathfrak{P},y}, \tag{47}$$

  where the index y is $\mathcal{P}$ or $\mathcal{E}$;

- the *pyro-electric coefficients* (also isochoric/isobaric)

$$\pi_a = \frac{1}{N}\left(\frac{\partial \mathcal{P}}{\partial T}\right)_{\mathcal{E},a,N} = \left(\frac{\partial \mathfrak{p}}{\partial T}\right)_{\mathcal{E},a}, \tag{48}$$

  where the index a is V (for the volume V) or P (for the pressure $\mathfrak{P}$).

a.5. The mixed coefficients express correlations between the volumic and the electric degrees of freedom; we mention the following simple coefficients:

- the *electro-strictive coefficient*

$$\gamma = \frac{1}{V}\left(\frac{\partial V}{\partial \mathcal{E}}\right)_{T,\mathfrak{P},N} = \frac{1}{v}\left(\frac{\partial v}{\partial \mathcal{E}}\right)_{T,\mathfrak{P}} ; \tag{49}$$

- the *piezo-electric coefficient*

$$\varrho = \frac{1}{N}\left(\frac{\partial \mathcal{P}}{\partial \mathfrak{P}}\right)_{T,\mathcal{E},N} = \left(\frac{\partial \mathfrak{p}}{\partial \mathfrak{P}}\right)_{T,\mathcal{E}} . \tag{50}$$

2.3.2 Relations between simple coefficients

Because the great number of thermodynamic coefficients, corresponding to the dielectrics, we must select among all possible relations between the simple coefficients; therefore, we shall present only the most important relations: the symmetry relations (consequences of some Maxwell relations) and special relations (of the type Reech or Mayer).

In order to emphasize symmetry relations expressed by the temperature, the pressure and the electric field $(T, \mathfrak{P}, \mathcal{E})$, as variables, for an closed system (N = constant), we use the differential form of the reduced Gibbs pseudo-potential $\widetilde{g}^*(T, \mathfrak{P}, \mathcal{E}) \equiv \mathcal{G}^*/N$, which is obtained with the general reduction formulae from Eqs. (23) – (24):

$$\mathrm{d}\widetilde{g}^* = -\, s\, \mathrm{d}T + v\, \mathrm{d}\mathfrak{P} - \mathfrak{p}\, \mathrm{d}\mathcal{E} . \tag{51}$$

From the above differential form it results 3 Maxwell relations, which can be expressed by simple coefficients, resulting symmetry relations between these coefficients:

$$\left(\frac{\partial s}{\partial \mathfrak{P}}\right)_{T,\mathcal{E}} = -\left(\frac{\partial v}{\partial T}\right)_{\mathfrak{P},\mathcal{E}} \quad \Longrightarrow \quad \lambda_{\mathcal{E}}^{(P)} = -\, T\, v\, \alpha_{\mathcal{E}} , \tag{52a}$$

(the relation between the iso-field piezo-caloric coefficient and the isobaric-isofield thermal expansion coefficient);

$$\left(\frac{\partial s}{\partial \mathcal{E}}\right)_{T,\mathfrak{P}} = \left(\frac{\partial \mathfrak{p}}{\partial T}\right)_{\mathfrak{P},\mathcal{E}} \quad \Longrightarrow \quad \lambda = T\, \pi_P , \tag{52b}$$

(the relation between the isobaric electro-caloric coefficient and the isobaric pyro-electric coefficient);

$$\left(\frac{\partial v}{\partial \mathcal{E}}\right)_{T,\mathfrak{P}} = -\left(\frac{\partial \mathfrak{p}}{\partial \mathfrak{P}}\right)_{T,\mathcal{E}} \quad \Longrightarrow \quad v\gamma = -\, \varrho , \tag{52c}$$

(the relation between the electro-strictive coefficient and the piezo-electric coefficient).

Relations of Reech type can be obtained from the general relation (see [16], Eq. (3.25)), resulting the equality between the ratios of the isobaric specific heats, of the compressibility coefficients, and of the electric susceptibilities (isothermal, and respectively adiabatic):

$$\frac{c_{P,\mathfrak{p}}}{c_{P,\mathcal{E}}} = \frac{\varkappa_{s,\mathcal{E}}}{\varkappa_{T,\mathcal{E}}} = \frac{\chi_{s,\mathfrak{P}}^{(\mathrm{el})}}{\chi_{T,\mathfrak{P}}^{(\mathrm{el})}} . \tag{53}$$

Analogously, from the general Mayer relation for the specific heats (see [16] Eq. (3.28)) we obtain in this case

$$c_{P,\mathcal{E}} - c_{P,\mathfrak{p}} = T\, v\, \frac{\alpha_{\mathcal{E}}^2}{\chi_{T,\mathfrak{P}}^{(\mathrm{el})}} \,. \tag{54}$$

Similar relations with Eqs. (53) – (54) can be obtained for the coefficients associated to another sets of simple processes (*e.g.* isochoric, iso-polarization processes).

2.3.3 The factorization of some simple coefficients

An important characteristics of some thermodynamic coefficients is *the factorization property*: the expression of the considered coefficient is the sum of the part corresponding to the absence of the electric field (similarly as for the neutral fluid) and the "electric part", and this result comes from the factorization of the state equations.

We shall present the factorization of some coefficients using the variables of the electric Gibbs representation[13] $(T, \mathfrak{P}, \mathcal{E}, N)$; in this case the state equations are Eqs. (26).

The entropy is given by Eq. (26a), and here we write it without the variables (for simplicity), and for using later the convenient variables

$$\mathcal{S} = \mathcal{S}_0 + \mathcal{S}_{\mathrm{el}} \,, \tag{55}$$

where $\mathcal{S}_0$ is the entropy of the dielectric as a neutral fluid, in the absence of the electric field, and $\mathcal{S}_{\mathrm{el}}$ is the electric part of the entropy:

$$\mathcal{S}_{\mathrm{el}} = \frac{\bar{\varepsilon}_0\, \mathcal{E}^2}{2} \left(\frac{\partial \overline{\chi}_e}{\partial T}\right)_{\mathfrak{P}} N = \frac{\bar{\varepsilon}_0\, \mathcal{E}^2}{2} \left(\frac{\partial \chi_e}{\partial T}\right)_{v} V \,. \tag{56}$$

Accordingly to the general definition (42), we obtain a factorization of the specific heats:

$$c_\varphi = c_\varphi^{(0)} + c_\varphi^{(\mathrm{el})} \,, \tag{57}$$

where $c_\varphi^{(0)} = T(\partial s^{(0)}/\partial T)_\varphi$ is the specific heat of the dielectric in the absence of the electric field, and $c_\varphi^{(\mathrm{el})} = T(\partial s^{(\mathrm{el})}/\partial T)_\varphi$ is the electric part of the specific heat.

For the isobaric processes there are the specific heat at constant electric field or at constant polarization. From Eq. (56) it results

$$c_{P,\mathcal{E}}^{(\mathrm{el})} = T \left(\frac{\partial \overline{\chi}_e}{\partial T}\right)_{\mathfrak{P}} \frac{\bar{\varepsilon}_0\, \mathcal{E}^2}{2} \,. \tag{58}$$

To obtain $c_{P,\mathfrak{p}}^{(\mathrm{el})}$ we express the electric entropy $\mathcal{S}_{\mathrm{el}}$ in terms of the dipolar electric moment (instead the electric field), using Eq. (6b):

$$\mathcal{S}_{\mathrm{el}} = \frac{\mathfrak{p}^2}{2\,\bar{\varepsilon}_0\, \overline{\chi}_e^2} \left(\frac{\partial \overline{\chi}_e}{\partial T}\right)_{\mathfrak{P}} N = \frac{-\mathfrak{p}^2}{2\,\bar{\varepsilon}_0} \left(\frac{\partial\, (\overline{\chi}_e)^{-1}}{\partial T}\right)_{\mathfrak{P}} N \,;$$

[13] We remark that some coefficients need the use of other thermodynamic representations.

then it results for the electric part of the isobaric-isopolarization specific heat the expression:

$$c_{P,\mathfrak{p}}^{(\mathrm{el})} = \frac{-\mathfrak{p}^2}{2\,\bar{\varepsilon}_0}\, T \left(\frac{\partial^2 (\overline{\chi}_e)^{-1}}{\partial T^2} \right)_{\mathfrak{P}} . \tag{59}$$

We observe that for an ideal dielectric $(\overline{\chi}_e)^{-1} \sim T$, so that we obtain $c_{P,\mathfrak{p}}^{(\mathrm{el})} = 0$, that is $c_{P,\mathfrak{p}} = c_P^{(0)}$ (the iso-polarization specific heat is independent of the electric field)[14].
It is interesting to emphasize that for the ideal dielectrics the internal energy has also particular properties. The electric part of the volumic density of internal energy for an arbitrary dielectric has the expression

$$\mathfrak{u}^{(\mathrm{el})} = \frac{\bar{\varepsilon}_0\, \mathcal{E}^2}{2} \left(1 + \chi_e + T\, \frac{\partial \chi_e}{\partial T} \right) .$$

For an ideal dielectric we obtain that this energy density is equal to the energy density of the electric field $\mathfrak{u}^{(\mathrm{el})} = \bar{\varepsilon}_0\, \mathcal{E}^2/2 = w_{\mathrm{el}}$, that is the whole electric energy is given only by the electric field, without any contribution from the processes of the electric polarization. The behavior of the iso-polarization specific heats and of the internal energy are similar to the neutral fluids which satisfy the Clapeyron - Mendeleev equations, so that it is justified the terminology "ideal" for the dielectrics which have Curie susceptibility.

In contrast with the specific heats, the isobaric electro-caloric coefficient has contribution only from the electric part of the entropy:

$$\lambda = \bar{\varepsilon}_0\, \mathcal{E}\, T \left(\frac{\partial \overline{\chi}_e}{\partial T} \right)_{\mathfrak{P}} .$$

The volume is given by Eq. (26b), that is it can be expressed in the form:

$$V = V_0 + V_{\mathrm{el}} ,$$

where V is the volume of the dielectric as neutral fluid, in the absence of the electrical field, and V_{el} is the electric part of the volume:

$$V_{\mathrm{el}} = -\frac{\bar{\varepsilon}_0\, \mathcal{E}^2}{2} \left(\frac{\partial \overline{\chi}_e}{\partial \mathfrak{P}} \right)_T N . \tag{60}$$

Accordingly to the general definitions (45) and respectively (47), the isothermal compressibility coefficients $\varkappa_{T,\mathcal{E}}$ and the isobaric thermal expansion coefficient $\alpha_{\mathcal{E}}$ (both of them at constant field) factorize in non-electric part (corresponding to null electric field, when the dielectric behaves as a neutral fluid) and electric part:

$$\varkappa_{T,\mathcal{E}} = \varkappa_T^{(0)} + \varkappa_{T,\mathcal{E}}^{(\mathrm{el})} , \tag{61a}$$

$$\alpha_{\mathcal{E}} = \alpha^{(0)} + \alpha_{\mathcal{E}}^{(\mathrm{el})} , \tag{61b}$$

[14] For the corresponding isochoric specific heat we obtain the same result $c_{V,\mathfrak{P}}^{(\mathrm{el})} = c_V$.

where[15]

$$\varkappa^{(\mathrm{el})}_{T,\mathcal{E}} = \frac{\bar{\varepsilon}_0\,\mathcal{E}^2}{2\,v_0}\left(\frac{\partial^2\overline{\chi}_e}{\partial\mathfrak{P}^2}\right)_T , \tag{62a}$$

$$\alpha^{(\mathrm{el})}_{\mathcal{E}} = -\frac{\bar{\varepsilon}_0\,\mathcal{E}^2}{2\,v_0}\,\frac{\partial^2\overline{\chi}_e}{\partial T\,\partial\mathfrak{P}} . \tag{62b}$$

In contrast with the previous coefficients, the electro-strictive coefficient is obtained only from the electric part of the volumic state equation[16]

$$\gamma \approx -\frac{\bar{\varepsilon}_0\,\mathcal{E}}{v_0}\left(\frac{\partial\overline{\chi}_e}{\partial\mathfrak{P}}\right)_T . \tag{63}$$

2.3.4 Thermodynamic processes

Using the previous results we shall present the most significant thermodynamic processes for the dielectrics as closed system (we shall choose the variable set $T, \mathfrak{P}, \mathcal{E}, N = \text{constant}$).

d.1. The isothermal electrization: we consider that initially the dielectric is in null electric field $(T, \mathfrak{P}, \mathcal{E}_i = 0, N)$ and we apply the electric field with an isothermal-isobaric process, so that the final state has the parameters: $(T, \mathfrak{P}, \mathcal{E}_f = \mathcal{E}, N)$.

Using Eq. (26a) for the entropy, the heat transfered in this process is

$$\begin{aligned} \mathcal{Q}_{if} &= T\,\Delta\,\mathcal{S}_{if} = T\left\{\,\mathcal{S}(T,\mathfrak{P},\mathcal{E},N) - \mathcal{S}(T,\mathfrak{P},0,N)\,\right\} \\ &= \frac{\bar{\varepsilon}_0\,\mathcal{E}^2}{2}\,T\left(\frac{\partial\overline{\chi}_e}{\partial T}\right)_{\mathfrak{P}} N . \end{aligned} \tag{64}$$

Since $\overline{\chi}_e(T,\mathfrak{P})$ is in general a decreasing function with respect to the temperature, it results that in the electrization process the dielectric yields heat: $\mathcal{Q}_{if} < 0$.

d.2. The adiabatic-isobaric depolarization: we consider that initially the dielectric is in the presence of the electric field $\mathcal{E}$ and it has the temperature T_i; then, by a quasi-static adiabatic-isobaric process the electric field is decreasing to vanishing value.

Since the equation of this process is $\mathcal{S}(T,\mathfrak{P},\mathcal{E},N) = \text{constant}$, with the supplementary conditions $\mathfrak{P} = \text{constant}$ and $N = \text{constant}$, then by using Eq. (26a), we obtain the equation of the temperature:

$$\mathcal{S}(T_i,\mathfrak{P},\mathcal{E},N) + \mathcal{S}(T_f,\mathfrak{P},0,N) ;$$

that is, after simple algebraical operations, we get:

$$s^0(T_i,\mathfrak{P}) + \frac{\bar{\varepsilon}_0\,\mathcal{E}^2}{2}\left(\frac{\partial\overline{\chi}_e}{\partial T}\right)_{\mathfrak{P}} = s^0(T_f,\mathfrak{P}) . \tag{65}$$

Because the electric susceptibility is in general an decreasing function in respect to the temperature $(\partial\overline{\chi}_e/\partial T)_{\mathfrak{P}} < 0$ and the entropy $s^0(T,\mathfrak{P})$ is an increasing function of

[15] In fact, the factorization is obtained only if we consider small electric effects, so that we could approximate $v \approx v_0$ at the denominators.

[16] We consider that the electro-strictive effects are small, so that we can use the approximation $v \approx v_0$, at the denominator.

temperature, it results $s_i < s_f$, that is the dielectric gets cool during the adiabatic depolarization: $T_f < T_i$.

d.3. The electro-strictive and piezo-electric effects

The *electro-strictive effect* means the variation of the volume (of the dielectric) due to the variation of the electric field, in conditions isothermal-isobaric (also the dielectric is a closed system)[17]; and the electro-strictive coefficient is defined by Eq. (49).

The *piezo-electric effect* means the variation of the dipolar electric moment (of the dielectric), due to the variation of the pressure, in conditions isothermal and at constant electric field (also the dielectric is a closed system)[18] and the piezo-electric effect coefficient is defined by Eq. (50).

Between coefficients of the two effects it is the symmetry relation (53), and the corresponding expressions can be put in explicit forms using the electric susceptibility:

$$\gamma = -\frac{1}{v}\,\varrho = \frac{-\bar{\varepsilon}_0\,\mathcal{E}}{v}\left(\frac{\partial\,\overline{\chi}_e}{\partial\mathfrak{P}}\right)_T . \tag{66}$$

We observe that the necessary condition to have an electro-strictive effect and an piezo-electric effect is that $\overline{\chi}_e$ depends on the pressure; accordingly to Eqs. (9), it results that only the non-ideal dielectrics can have these effects.

Using the volumic equation of state (26b), we can evaluate the global electro-strictive effect, that is the variation of the volume (of the dielectric) at the isothermal-isobaric electrization:

$$\Delta\,V_{if}(\mathcal{E}) = V(T,\mathfrak{P},\mathcal{E},N) - V(T,\mathfrak{P},0,N) = -\,\frac{\bar{\varepsilon}_0\,\mathcal{E}^2}{2}\left(\frac{\partial\,\overline{\chi}_e}{\partial\mathfrak{P}}\right)_T N\,. \tag{67}$$

From the previous expression it results that when the electric susceptibility is a decreasing function of the pressure $(\partial\,\overline{\chi}_e/\partial\mathfrak{P})_T < 0$, then it follows a contraction (a reduction of the volume) at the electrization of the dielectric.

3. Magnetic systems

The thermodynamics of magnetic systems has many formal similitude with the thermodynamics of electric systems; in fact, we shall show that it is possible to obtain the most of the results for magnetic systems by simple replacements from the corresponding relations for dielectric systems.

Because this similitude is only formal, and there are physical differences, and on the other side, in order to have an autonomy with respect to the previous section, we shall present briefly the thermodynamics of the magnetic systems independently of the results obtained for the dielectrics. However, for emphasizing the formal similitude between the electric and the magnetic systems, we shall do this presentation analogously to the previous one (which corresponds to dielectrics).

---

[17] In other words, the electro-strictive effect can be considered as the volumic response of the dielectric to an electric perturbation.

[18] We observe that the piezo-electric effect can be considered as the electric response of the dielectric to a volumic perturbation, being conjugated to the electro-strictive effect.

3.1 General electrodynamic results

Accordingly to the electrodynamics, the magneto-static field created by a distribution of stationary electric currents, in an magnetic medium is characterized by the vectorial fields *the intensity of magnetic induction* $\mathcal{B}(\mathbf{r})$ and *the intensity of the magnetic field* $\mathcal{H}(\mathbf{r})$ (called also the magnetic excitation) which satisfy the magneto-static Maxwell equations [17-20]:

$$\text{rot}\ \mathcal{H}(\mathbf{r}) = \jmath(\mathbf{r})\,, \tag{68a}$$

$$\text{div}\ \mathcal{B}(\mathbf{r}) = 0\,, \tag{68b}$$

where $\jmath(\mathbf{r})$ is the volumic density of the conduction currents (there are excluded the magnetization currents).

Under the influence of the magneto-static field the medium becomes magnetized (it appears magnetization currents), and it is characterized by the *magnetic dipolar moment* $\mathcal{M}$, respectively by the *magnetization* (the volumic density of dipolar magnetic moment) $M(\mathbf{r})$:

$$M(\mathbf{r}) \equiv \lim_{\delta V \to 0} \frac{\delta \mathcal{M}(\mathbf{r})}{\delta V(\mathbf{r})} \quad \Longleftrightarrow \quad \mathcal{M} = \int_V \mathrm{d}^3\mathbf{r}\; M(\mathbf{r})\,. \tag{69}$$

Using the magnetization it results the relation between the characteristic vectors of the magneto-static field:

$$\mathcal{H}(\mathbf{r}) = \frac{1}{\overline{\mu}_0}\,\mathcal{B}(\mathbf{r}) - M(\mathbf{r})\,, \tag{70}$$

where $\overline{\mu}_0$ is the magnetic permeability of the vacuum (it is an universal constant depending on the system of units).

The general relation between the intensity of the magnetic field $\mathcal{H}(\mathbf{r})$ and the magnetization $M(\mathbf{r})$ is

$$M(\mathbf{r}) = \hat{\chi}_{\mathrm{m}}(\mathcal{H},\mathbf{r}) : \mathcal{H}(\mathbf{r}) + M_0(\mathbf{r})\,, \tag{71a}$$

where M_0 is the spontaneous magnetization, and $\hat{\chi}_{\mathrm{m}}$ is the magnetic susceptibility tensor (generally it is dependent on the magnetic field).

For simplicity, we shall consider only the particular case when there is no spontaneous magnetization (that is the absence of ferromagnetic phenomena) $M_0 = 0$, and the magnetic medium is linear and isotropic (then $\hat{\chi}_{\mathrm{m}}$ is reducible to a scalar which is independent on the magnetic field); in this last case, Eq. (71a) becomes:

$$M = \chi_{\mathrm{m}}\,\mathcal{H}\,, \tag{71b}$$

and Eq. (70) allows a parallelism and proportionality relation between the field vectors:

$$\mathcal{B} = \overline{\mu}_0(1 + \chi_{\mathrm{m}})\,\mathcal{H}\,. \tag{72}$$

In the common cases the susceptibility (for the specified types of magnetic media) depends on the temperature and of the particle density (or of the pressure) in the form:

$$\chi_{\mathrm{m}} = \mathfrak{n}\,\overline{\chi}_{\mathrm{m}}(T,\mathfrak{P})\,, \tag{73}$$

where $\mathfrak{n} \equiv N/V$ is the particle density, and $\overline{\chi}_{\mathrm{m}}$ is the specific (per particle) susceptibility[19].

[19] We can define the dipolar magnetic moment per particle as $\mathcal{M} = N\,\mathfrak{m}$, resulting the susceptibility per particle with an analogous relation to (71b) $\mathfrak{m} = \overline{\chi}_{\mathrm{m}}\,\mathcal{H}$.

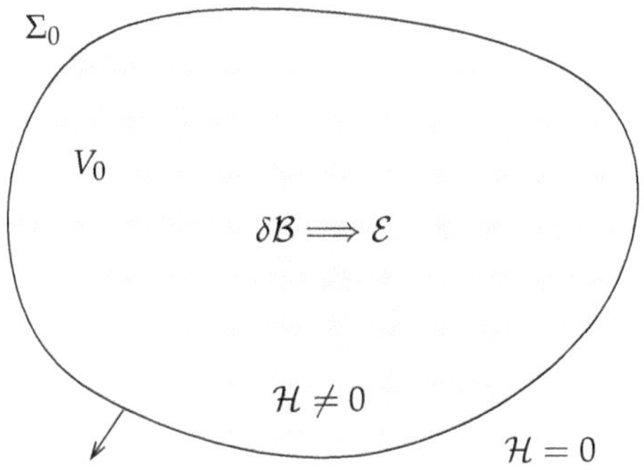

Fig. 4. The system chosen for the evaluation of the magnetic work.

In the strictly sense, the concrete expression of the magnetic susceptibility per particle is an empirical information of the thermal state equation type. On the basis of the type of specific susceptibility, the linear magnetic media are divided in two classes:

a. the *diamagnetic* media, which have negative specific susceptibilities, depending very little of the temperature and the pressure[20]

$$\overline{\chi}_{\mathrm{m}}(T,\mathfrak{P}) \approx \text{constant} \; < \; 0\,; \tag{74a}$$

b. the *paramagnetic* media, which have positive specific susceptibilities, with small values; in addition, there are two types of para-magnets:

– *ideal* para-magnets, having susceptibilities of the Curie type and independent of the pressure

$$\overline{\chi}_{\mathrm{m}}(T) = \frac{K}{T}\,, \tag{74b}$$

– *non-ideal* para-magnets, having susceptibilities of the Curie - Weiss type

$$\overline{\chi}_{\mathrm{m}}(T,\mathfrak{P}) = \frac{K}{T - \Theta(\mathfrak{P})}\,, \tag{74c}$$

where K is a constant depending on the paramagnet, called *Curie constant*), and $\Theta(\mathfrak{P})$ is a function of the pressure, having the dimension of a temperature.

Using the general relations of the magneto-statics, we can deduce the expression of the infinitesimal *magnetic work*, as the energy given to the thermo-isolated magnetic medium when the magnetic field varies:

$$\delta\mathcal{L}_{\mathrm{m}} = \int_{V_0} \mathrm{d}^3\mathbf{r}\, \boldsymbol{\mathcal{H}} \cdot \delta\boldsymbol{\mathcal{B}}\,, \tag{75}$$

under the condition that the system is located in the domain with the volume $V_0 = \text{constant}$, so that outside to this domain the magnetic field *is null*.

[20] In this case the magnetization is in the opposite direction to the magnetic field [see Eq. (71)].

Proof:

The magneto-static field from the magnetic medium is created by conduction electric currents; we consider the situation illustrated in Fig. 4, where inside the domain with the volume V_0 and fixed external surface Σ_0, there are magnetic media and electric conductors[21].

In contrast to the electric case, the magnetic work cannot be evaluated directly, since the magnetic forces are not conservative, and also because the variation of the magnetic field produces an electric field through the electro-magnetic induction, accordingly to the Maxwell – Faraday equation

$$\mathrm{rot}\,\boldsymbol{\mathcal{E}}(\mathbf{r}) = -\frac{\partial\,\boldsymbol{\mathcal{B}}(\mathbf{r},t)}{\partial t}\,. \tag{76}$$

For this reasons, the magnetic work, as the variation of the energy corresponding to the magnetic field inside the considered domain, will be evaluated from the electric work on the currents (as sources of the magnetic field) produced by the electric field which was induced at the variation of the magnetic field.

We consider an infinitesimal variation of the magnetic field $\delta\boldsymbol{\mathcal{B}}(\mathbf{r})$ produced in the infinitesimal time interval δt; the induced electric field performs in the time interval δt, on the currents, the work

$$\delta\mathcal{L}_{\mathrm{el}} = \delta t \int_{V_0} \mathrm{d}^3\mathbf{r}\, \boldsymbol{j}\cdot\boldsymbol{\mathcal{E}}\,. \tag{77}$$

Using Eq. (68a) we transform the integrand in the following form:

$$\boldsymbol{j}\cdot\boldsymbol{\mathcal{E}} = \boldsymbol{\mathcal{E}}\cdot\mathrm{rot}\,\boldsymbol{\mathcal{H}} = \mathrm{div}\,(\boldsymbol{\mathcal{H}}\times\boldsymbol{\mathcal{E}}) + \boldsymbol{\mathcal{H}}\cdot\mathrm{rot}\,\boldsymbol{\mathcal{E}}\,,$$

and it results for the electric work the expression

$$\delta\mathcal{L}_{\mathrm{el}} = \delta t \int_{V_0} \mathrm{d}^3\mathbf{r}\ \mathrm{div}\,(\boldsymbol{\mathcal{H}}\times\boldsymbol{\mathcal{E}}) + \delta t \int_{V_0} \mathrm{d}^3\mathbf{r}\, \boldsymbol{\mathcal{H}}\cdot\mathrm{rot}\,\boldsymbol{\mathcal{E}}\,.$$

We transform the first term in a surface integral, using the Gauss' theorem, and this integral vanishes, because the hypothesis that the magnetic field is null on the frontier of the domain

$$\int_{V_0} \mathrm{d}^3\mathbf{r}\ \mathrm{div}\,(\boldsymbol{\mathcal{H}}\times\boldsymbol{\mathcal{E}}) = \oint_{\Sigma_0} \mathrm{d}\mathcal{A}\, \boldsymbol{n}_0\cdot(\boldsymbol{\mathcal{H}}\times\boldsymbol{\mathcal{E}}) = 0\,;$$

in the second term we use the Maxwell - Faraday equation (76) and we introduce the variation of the magnetic induction with the relation $(\partial\boldsymbol{\mathcal{B}}/\partial t)\,\delta t = \delta\boldsymbol{\mathcal{B}}$, so that it results

$$\delta t \int_{V_0} \mathrm{d}^3\mathbf{r}\, \boldsymbol{\mathcal{H}}\cdot\mathrm{rot}\,\boldsymbol{\mathcal{E}} = -\int_{V_0} \mathrm{d}^3\mathbf{r}\, \boldsymbol{\mathcal{H}}\cdot\frac{\partial\boldsymbol{\mathcal{B}}}{\partial t}\,\delta t = -\int_{V_0} \mathrm{d}^3\mathbf{r}\, \boldsymbol{\mathcal{H}}\cdot\delta\boldsymbol{\mathcal{B}}\,.$$

We observe that the electric work, determined previously, must be compensated by an additional work supplied from outside; therefore the magnetic work given to the system for an infinitesimal variation of the magnetic field is $\delta\mathcal{L}_{\mathrm{m}} = -\,\delta\mathcal{L}_{\mathrm{el}}$, and it results Eq. (75). □

We note that the expression (75) for the magnetic work, implies a domain for integration with a fixed volume (V_0), and in addition the magnetic field must vanish outside this domain.

[21] In order to have the general situation, we do not suppose particular properties for the magnetic medium inside the chosen domain, so that we consider the non-homogeneous case.

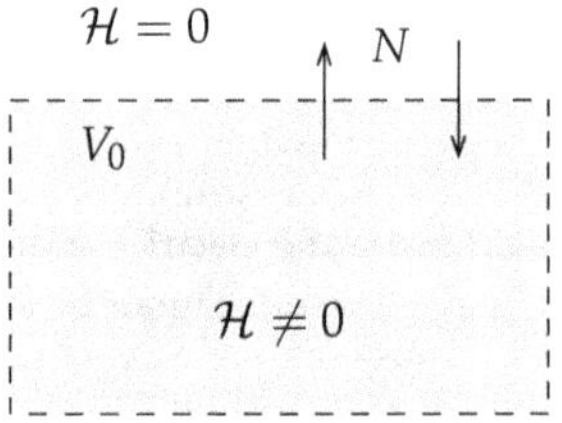

Fig. 5. The open system model.

By comparing the expressions of the magnetic work and electric work (10) we observe the formal similitude between these formula, and this leads to the following substitution rule to obtain the magnetic results from the electric ones:

$$\begin{cases} \mathcal{E} \longrightarrow \mathcal{H} \\ \mathcal{D} \longrightarrow \mathcal{B} \, . \end{cases}$$

However this similitude is only formal, since from the physical point of view, the correspondence between the electrical field vectors and the magnetic ones is $\boldsymbol{\mathcal{E}} \leftrightarrow \boldsymbol{\mathcal{B}}$ and $\boldsymbol{\mathcal{D}} \leftrightarrow \boldsymbol{\mathcal{H}}$.

In the following we shall present the consequences derived from the expression of the magnetic work (75) in an analog manner as the electric case; therefore we shall consider two methods to deal with the thermodynamics of the magnetic media, based on the expression of the magnetic work (and of the necessary conditions for the validity of this expression).

1. The open system method: we consider a fixed domain (having the volume V = constant) which contains an magneto-static field inside, but outside to this domain the magneto-static field vanishes; the interesting system is the magnetic medium located inside the above specified domain, as an open thermodynamic system (the magnetic medium fills completely the domain, but there is a part of this medium, outside the domain at vanishing magneto-static field, because the frontier is totally permeable).

We note the following characteristic features of this situation:

– the thermodynamic system (the portion of the magnetic medium located in magneto-static field) has a fixed volume ($V = V_0 =$ constant), but it is an *open* thermodynamic system ($N \neq$ constant);

– the magneto-striction effect (this is the variation of the volume produced by the variation of the magnetic field) in this case leads to the variation of the particle number, or in another words, by variation of the particle density $\mathfrak{n} \equiv N/V_0 \neq$ constant;

– in the simplest case, when we consider a homogeneous magneto-static field[22], inside the domain with the volume V_0, infinitesimal magnetic work can be written in the form

$$\delta \mathcal{L}_{\mathrm{m}} = \mathcal{H} \, \delta(V_0 \mathcal{B}) \, , \tag{78}$$

[22] This situation is realized by considering a very long cylindrical solenoid, so that it is possible to consider approximately that the magnetic field vanishes outside the solenoid; the space inside and outside the solenoid is filled with a fluid paramagnet (or diamagnet). Accordingly to the previous definitions, the thermodynamic interesting system is only the part of the magnetic medium which is located inside the solenoid, and the frontier of this system is fictitious.

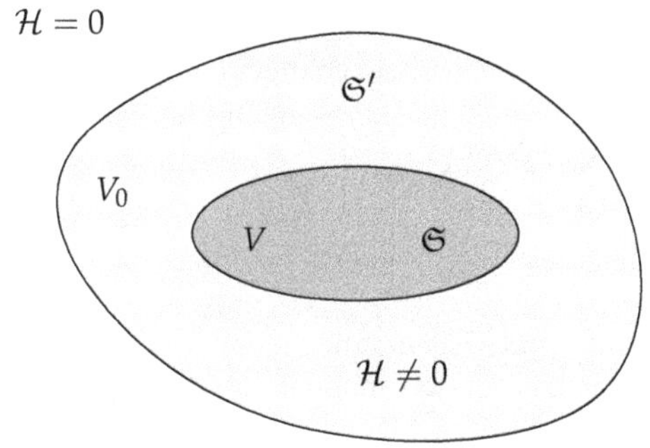

Fig. 6. The model for the closed system.

and this implies the following definition for the magnetic state parameters

$$\begin{cases} X_{\rm m} = V_0 \mathcal{B} \equiv \mathfrak{B} \, , \\ P_{\rm m} = \mathcal{H} \, ; \end{cases} \tag{79}$$

[in this case V = constant, that is the volumic degree of freedom for this system is frozen; but we emphasize that the expression $\delta\mathcal{L}_{\rm m} = \mathcal{H}\ \delta(V\mathcal{B})$ when the volume of the system V can varies is *incorrect*].

2. The closed system method: we consider the magnetic medium surrounded by an non-magnetic medium, so that the magnetic medium does not occupy the whole space where is present the magnetic field.
In this case it is necessary to define the compound system corresponding to the domain with magnetic field

$$\mathfrak{S}^{(\tau)} = \mathfrak{S} \bigcup \mathfrak{S}' \, ,$$

where $\mathfrak{S}$ is the magnetic system with the volume V, and $\mathfrak{S}'$ is an auxiliary non-magnetic system having the volume $V' = V_0 - V$, as it is illustrated in Fig. 6.

We must remark that the auxiliary system (having negligible magnetic properties) is necessary in order to obtain the condition $\mathcal{H} \to 0$ towards the frontier of the domain which have the volume V_0, and also it produces a pressure on the magnetic medium; thus, the volume of the magnetic medium is not fixed and we can distinguish directly magneto-striction effects.

Because the magnetization $\boldsymbol{M}$ is non-vanishing only in the domain V, occupied by the system $\mathfrak{S}$, we transform the expression (75) using Eq. (70), in order to extract the magnetization work on the subsystem $\mathfrak{S}$

$$\begin{aligned} \delta\mathcal{L}_{\rm m}^{(\tau)} = \int_{V_0} {\rm d}^3\mathbf{r}\, \boldsymbol{\mathcal{H}} \cdot \delta\boldsymbol{\mathcal{B}} = \int_{V_0} {\rm d}^3\mathbf{r}\, \boldsymbol{\mathcal{H}} \cdot \overline{\mu}_0\, \delta\boldsymbol{\mathcal{H}} + \int_{V_0} {\rm d}^3\mathbf{r}\, \boldsymbol{\mathcal{H}} \cdot \overline{\mu}_0\, \delta\boldsymbol{M} \\ \equiv \delta W_{\mathcal{H}}^{(\tau)} + \delta\mathcal{L}_{\rm M} \, , \end{aligned}$$

where $\delta W_{\mathcal{H}}^{(\tau)}$ is the variation of the energy for the field inside the total volume V_0, and $\delta\mathcal{L}_{\rm p}$ is the work for magnetize the magnetic medium.

The first term allows the separation of the contributions corresponding the two subsystems when the energy of the magneto-static field changes:

$$\delta W_{\mathcal{H}}^{(\tau)} = \int_{V_0} \mathrm{d}^3\mathbf{r}\,\delta\Big(\frac{\overline{\mu}_0\mathcal{H}^2}{2}\Big) = \delta\Big\{\int_{V} \mathrm{d}^3\mathbf{r}\,\frac{\overline{\mu}_0\mathcal{H}^2}{2}\Big\} + \delta\Big\{\int_{V'} \mathrm{d}^3\mathbf{r}\,\frac{\overline{\mu}_0\mathcal{H}^2}{2}\Big\}$$
$$= \delta W_{\mathcal{H}} + \delta W'_{\mathcal{H}}\,.$$

The second term can be interpreted as magnetization work and it implies only the magnetic medium; in order to include the possible magneto-strictive effects, we shall write this term in the form

$$\delta \mathcal{L}_{\mathrm{M}} = \int_{V_0} \mathrm{d}^3\mathbf{r}\,\overline{\mu}_0\,\boldsymbol{\mathcal{H}}\cdot\delta\boldsymbol{M} = \int_{V_f} \mathrm{d}^3\mathbf{r}\,\overline{\mu}_0\,\boldsymbol{\mathcal{H}}\cdot\boldsymbol{M}_f - \int_{V_i} \mathrm{d}^3\mathbf{r}\,\overline{\mu}_0\,\boldsymbol{\mathcal{H}}\cdot\boldsymbol{M}_i$$
$$= \delta\Big\{\int_{V} \mathrm{d}^3\mathbf{r}\,\overline{\mu}_0\,\boldsymbol{\mathcal{H}}\cdot\boldsymbol{M}\Big\}\Big|_{\mathcal{H}=\mathrm{const.}}$$

that is, the magnetization work implies the variation of the magnetization $\delta\boldsymbol{M}$ and also the variation of the volume δV of the magnetic medium, with the condition of constant magnetic field: $\boldsymbol{\mathcal{H}} = \mathrm{constant}$ (during the process).

On the basis of the previous results we can separate the contribution of the magnetic work on the magnetic medium ($\delta\mathcal{L}_{\mathrm{m}}$) from those on the auxiliary non-magnetic system ($\delta W'_{\mathcal{H}}$):

$$\delta\mathcal{L}_{\mathrm{m}}^{(\tau)} = \delta\mathcal{L}_{\mathrm{m}} + \delta W'_{\mathcal{H}}\,, \tag{80a}$$
$$\delta\mathcal{L}_{\mathrm{m}} = \delta\mathcal{L}_{\mathrm{M}} + \delta W_{\mathcal{H}}\,. \tag{80b}$$

For the magnetic work on the magnetic medium (the subsystem $\mathfrak{S}$) we observe two interpretations in the case when $\mathfrak{S}$ is homogeneous[23]:

1. we take into account only the magnetization work $\delta\mathcal{L}_{\mathrm{M}}$ and we neglect systematically the energy of the magnetic field inside the magnetic medium $\delta W_{\mathcal{H}}$; then the magnetization work in the uniform magnetic field can be expressed with the dipolar magnetic moment

$$\delta\mathcal{L}_{\mathrm{M}} = \overline{\mu}_0\,\boldsymbol{\mathcal{H}}\cdot\delta\int_{V}\mathrm{d}^3\mathbf{r}\,\boldsymbol{M} = \overline{\mu}_0\,\boldsymbol{\mathcal{H}}\cdot\delta\boldsymbol{\mathcal{M}}\,; \tag{81}$$

2. we estimate the contributions of the both terms from Eq. (80b), taking into account the implications due to the homogeneity of the system:

$$đ\mathcal{L}_{\mathrm{M}} = \overline{\mu}_0\,\mathcal{H}\,\mathrm{d}\mathcal{M} = \overline{\mu}_0\,\mathcal{H}\,\mathrm{d}(VM)\,,$$
$$\mathrm{d}W_{\mathcal{H}} = \mathrm{d}\Big(\frac{\overline{\mu}_0\mathcal{H}^2}{2}\,V\Big) = -\,\frac{\overline{\mu}_0\mathcal{H}^2}{2}\,\mathrm{d}V + \mathcal{H}\,\mathrm{d}(V\overline{\mu}_0\mathcal{H})\,.$$

so that we obtain the total magnetic work performed by the magnetic system

$$đ\mathcal{L}_{\mathrm{m}} = đ\mathcal{L}_{\mathrm{M}} + \mathrm{d}W_{\mathcal{H}} = -\,\frac{\overline{\mu}_0\mathcal{H}^2}{2}\,\mathrm{d}V + \mathcal{H}\,\mathrm{d}(V\mathcal{B})\,, \tag{82}$$

[23] The condition of homogeneity implies an uniform magnetic field $\boldsymbol{\mathcal{H}}(\mathbf{r}) = \mathrm{constant}$ in the subsystem $\mathfrak{S}$, and this property is realized only when the magnetic medium is an ellipsoid in an uniform external field.

and the last expression can be interpreted as an work performed on two degrees of freedom (volumic and magnetic).

We observe that for isotropic magnetic media the vectors $\boldsymbol{\mathcal{H}}$, $\boldsymbol{\mathcal{B}}$ and $\boldsymbol{M}$ are colinear; therefore, we shall omit the vector notation, for simplicity.

3.2 Thermodynamic potentials

We shall discuss, for simplicity, only the case when the magnetic system is homogeneous and of fluid type, being surrounded by a non-magnetic environment. Then, the fundamental thermodynamic differential form is:

$$\mathrm{d}\mathcal{U} = đ\,\mathcal{Q} + đ\,\mathcal{L}_V + đ\,\mathcal{L}_\mathrm{m} + đ\,\mathcal{L}_N\,. \tag{83}$$

For the thermodynamic study of the magnetic system there are many methods, depending the choice of the fundamental variables (corresponding to the choice of the concrete expression for the magnetic work $đ\mathcal{L}_\mathrm{m}$).

3.2.1 Pseudo-potentials method

We replace the expression (80b) – (81) for the magnetic work, and also the expression for the other forms of work and for the heat; then, the fundamental thermodynamic differential form has the explicit expression:

$$\mathrm{d}\mathcal{U} = T\,\mathrm{d}\mathcal{S} - \mathfrak{P}\,\mathrm{d}V + \overline{\mu}_0\mathcal{H}\,\mathrm{d}\mathcal{M} + \mathrm{d}W_\mathcal{H} + \mu\,\mathrm{d}N\,. \tag{84}$$

We observe that the preceding differential form contains a term $\mathrm{d}W_\mathcal{H}$ which is a exact total differential (from the mathematical point of view) and it represents the variation of the energy of the magnetic field located in the space occupied by the magnetic medium; we put this quantity in the left side of the above equality we obtain:

$$\mathrm{d}\widetilde{\mathcal{U}} = T\,\mathrm{d}\mathcal{S} - \mathfrak{P}\,\mathrm{d}V + \overline{\mu}_0\mathcal{H}\,\mathrm{d}\mathcal{M} + \mu\,\mathrm{d}N\,, \tag{85}$$

where $\widetilde{\mathcal{U}} \equiv \mathcal{U} - W_\mathcal{H}$ is the *internal pseudo-energy* of the magnetic medium[24].

We present some observations concerning the differential form (85):

• $\widetilde{\mathcal{U}}(\mathcal{S}, V, \mathcal{M}, N)$ is equivalent to the fundamental thermodynamic equation of the system, since it contains the whole thermodynamic information about the system (that is, its derivatives are the state equations); and on the other side, the pseudo-energy has no specified convexity properties (because it was obtained by subtracting a part of the energy from the total internal energy of the system);

• $\widetilde{\mathcal{U}}(\mathcal{S}, V, \mathcal{M}, N)$ is a homogeneous function of degree 1 (because it is obtained as a difference of two homogeneous functions of degree 1);

---

[24] Because we consider in the expression $W_\mathcal{H} = V\overline{\mu}_0\mathcal{H}^2/2$ the intensity of the magnetic field in the presence of the magnetic medium, this energy is due both to the vacuum and to the magnetic medium; thus, $\widetilde{\mathcal{U}}$ *is not the internal energy of the magnetic medium* (without the energy of the magnetic field in vacuum), but it is an artificial quantity.

• by considering the differential form (85) as similar to the fundamental thermodynamic differential form, it follows that the magnetic state parameters are

$$\begin{cases} \widetilde{X}_m = \mathcal{M} = M\,V\,, \\ \widetilde{P}_m = \overline{\mu}_0\,\mathcal{H}\,; \end{cases}$$

• if we perform the Legendre transformations of the function $\widetilde{\mathcal{U}}(\mathcal{S},V,\mathcal{M},N)$, then we obtain objects of the thermodynamic potential types (that is, the derivatives of these quantities give the state equations of the magnetic medium); however, these objects are not true thermodynamic potentials (firstly since they have not the needed properties of convexity-concavity), so that they are usually called *thermodynamic pseudo-potentials*.

In the following we shall present briefly only the most used pseudo-potentials: the *magnetic pseudo-free energy* and the *magnetic Gibbs pseudo-potential*.

a.1. The magnetic pseudo-free energy is the Legendre transformation on the thermal and magnetic degrees of freedom[25]

$$\widetilde{\mathcal{F}}^* \equiv \widetilde{\mathcal{U}} - T\,\mathcal{S} - \overline{\mu}_0\mathcal{H}\,\mathcal{M}\,, \tag{86}$$

having the differential form[26]:

$$\mathrm{d}\widetilde{\mathcal{F}}^* = -\,\mathcal{S}\,\mathrm{d}T - \mathfrak{P}\,\mathrm{d}V - \mathcal{M}\,\mathrm{d}(\overline{\mu}_0\mathcal{H}) + \mu\,\mathrm{d}N\,, \tag{87}$$

so that it allows the deduction of the state equations in the representation $(T,V,\mathcal{H},N)$.
We consider the simplest case, when the magnetic medium behaves as a neutral fluid, having the free energy $\mathcal{F}_0(T,V,N)$, in the absence of the magnetic field and when the magnetic susceptibility is $\chi_m(T,V/N)$; then, we can consider as known the magnetic state equation (the expression of the dipolar magnetic moment):

$$\left(\frac{\partial\,\widetilde{\mathcal{F}}^*}{\partial\,\mathcal{H}}\right)_{T,V,N} = -\,\overline{\mu}_0\,\mathcal{M}(T,V,\mathcal{H},N) = -\,\overline{\mu}_0\,\chi_m(T,V/N)\,\mathcal{H}\,V\,.$$

By partial integration, with respect to the magnetic field, and taking into account that at vanishing magnetic field the magnetic pseudo-free energy reduces to the proper free energy (the Helmholtz potential):

$$\widetilde{\mathcal{F}}^*\Big|_{\mathcal{H}=0} = \mathcal{U}_0 - T\,\mathcal{S}_0 = \mathcal{F}_0(T,V,N)\,,$$

we obtain:

$$\widetilde{\mathcal{F}}^*(T,V,\mathcal{H},N) = \mathcal{F}_0(T,V,N) + \mathcal{F}^*_{\mathrm{mag}}(T,V,\mathcal{H},N)\,, \tag{88a}$$

$$\mathcal{F}^*_{\mathrm{mag}}(T,V,\mathcal{H},N) \equiv -\,\frac{\overline{\mu}_0\mathcal{H}^2}{2}\,\chi_m(T,V/N)\,V\,, \tag{88b}$$

[25] Because the absence of the correct properties of the convexity, we shall use only the classic definition of the Legendre transformation.

[26] In the strictly sense, $\widetilde{\mathcal{F}}^*$ is a simple Gibbs potential, so that the common terminology is criticizable.

that is, the magnetic pseudo-free energy of the magnetic medium is the sum of the free energy at null magnetic field and the magnetic part $\mathcal{F}^*_{\text{mag}}$.

The factorization property of the magnetic pseudo-free energy is transmitted to the non-magnetic state equations: the entropy, the pressure and the chemical potential (these are sums of the non-magnetic part, corresponding to null magnetic field, and the magnetic part):

$$\mathcal{S}(T,V,\mathcal{H},N) = -\left(\frac{\partial \widetilde{\mathcal{F}}^*}{\partial T}\right)_{V,\mathcal{H},N} = \mathcal{S}_0(T,V,N) + \frac{\overline{\mu}_0\mathcal{H}^2}{2}\left(\frac{\partial \chi_m}{\partial T}\right)_{V,N} V\,, \tag{89a}$$

$$\mathfrak{P}(T,V,\mathcal{H},N) = -\left(\frac{\partial \widetilde{\mathcal{F}}^*}{\partial V}\right)_{T,\mathcal{H},N} = \mathfrak{P}_0(T,V,N) + \frac{\overline{\mu}_0\mathcal{H}^2}{2}\left(\frac{\partial (\chi_m V)}{\partial V}\right)_{T,N}\,, \tag{89b}$$

$$\mu(T,V,\mathcal{H},N) = \left(\frac{\partial \widetilde{\mathcal{F}}^*}{\partial N}\right)_{T,V,\mathcal{H}} = \mu_0(T,V,N) - \frac{\overline{\mu}_0\mathcal{H}^2}{2}\left(\frac{\partial \chi_m}{\partial N}\right)_{T,V} V\,. \tag{89c}$$

a.2. The magnetic Gibbs pseudo-potential is defined, analogously to the previous case, as the Legendre transformation on the thermal, volumic and magnetic degrees of freedom[27]

$$\widetilde{\mathcal{G}}^* \equiv \widetilde{\mathcal{U}} - T\,\mathcal{S} + \mathfrak{P}\,V - \overline{\mu}_0\mathcal{H}\,\mathcal{M}\,, \tag{90}$$

having the differential form

$$\mathrm{d}\widetilde{\mathcal{G}}^* = -\,\mathcal{S}\,\mathrm{d}T + V\,\mathrm{d}\mathfrak{P} - \mathcal{M}\,\mathrm{d}(\overline{\mu}_0\mathcal{H}) + \mu\,\mathrm{d}N\,, \tag{91}$$

so that it allows the deduction of the state equations in the representation $(T,\mathfrak{P},\mathcal{H},N)$.
We consider the simplest case (analogously in the former case), when the magnetic medium behaves as a neutral fluid, having the proper Gibbs potential (the free enthalpy) $\mathcal{G}_0(T,\mathfrak{P},N)$ in the absence of the magnetic field and when the magnetic susceptibility is $\overline{\chi}_m(T,\mathfrak{P})$; then, we can consider as known the magnetic state equation (the expression of the dipolar magnetic moment):

$$\left(\frac{\partial \widetilde{\mathcal{G}}^*}{\partial \mathcal{H}}\right)_{T,\mathfrak{P},N} = -\,\overline{\mu}_0\,\mathcal{M}(T,\mathfrak{P},\mathcal{H},N) = -\,\overline{\mu}_0\,\overline{\chi}_m(T,\mathfrak{P})\,\mathcal{H}\,N\,.$$

By partial integration, with respect to the magnetic field, and because at null magnetic field the magnetic Gibbs pseudo-potential reduces to the proper Gibbs potential:

$$\widetilde{\mathcal{G}}^*\Big|_{\mathcal{H}=0} = \mathcal{U}_0 - T\,\mathcal{S}_0 + \mathfrak{P}\,V_0 = \mathcal{G}_0(T,\mathfrak{P},N)\,,$$

we obtain:

$$\widetilde{\mathcal{G}}^*(T,\mathfrak{P},\mathcal{H},N) = \mathcal{G}_0(T,\mathfrak{P},N) + \mathcal{G}^*_{\text{mag}}(T,\mathfrak{P},\mathcal{H},N)\,, \tag{92a}$$

$$\mathcal{G}^*_{\text{mag}}(T,\mathfrak{P},\mathcal{H},N) \equiv -\,\frac{\overline{\mu}_0\mathcal{H}^2}{2}\,\overline{\chi}_m(T,\mathfrak{P})\,N\,, \tag{92b}$$

that is, the magnetic Gibbs pseudo-potential of the magnetic medium is the sum of the Gibbs potential at null magnetic field and the magnetic part $\mathcal{G}^*_{\text{mag}}$.

[27] Because the absence of the correct properties of the convexity, we shall use only the classic definition of the Legendre transformation (like in the preceding case).

The additivity (factorization) property of the magnetic Gibbs pseudo-potential is transmitted to the non-magnetic state equations: the entropy, the volume and the chemical potential (these are sums of the non-magnetic part, corresponding to vanishing magnetic field, and the magnetic part):

$$\mathcal{S}(T,\mathfrak{P},\mathcal{H},N) = -\left(\frac{\partial \widetilde{\mathcal{G}}^*}{\partial T}\right)_{\mathfrak{P},\mathcal{H},N} = \mathcal{S}_0(T,\mathfrak{P},N) + \frac{\overline{\mu}_0\mathcal{H}^2}{2}\left(\frac{\partial \overline{\chi}_m}{\partial T}\right)_{\mathfrak{P}} N\,, \tag{93a}$$

$$V(T,\mathfrak{P},\mathcal{H},N) = \left(\frac{\partial \widetilde{\mathcal{G}}^*}{\partial \mathfrak{P}}\right)_{T,\mathcal{H},N} = V_0(T,\mathfrak{P},N) - \frac{\overline{\mu}_0\mathcal{H}^2}{2}\left(\frac{\partial \overline{\chi}_m}{\partial \mathfrak{P}}\right)_{T} N\,, \tag{93b}$$

$$\mu(T,\mathfrak{P},\mathcal{H},N) = \left(\frac{\partial \widetilde{\mathcal{G}}^*}{\partial N}\right)_{T,\mathfrak{P},\mathcal{H}} = \mu_0(T,\mathfrak{P}) - \frac{\overline{\mu}_0\mathcal{H}^2}{2}\,\overline{\chi}_m(T,\mathfrak{P})\,. \tag{93c}$$

We observe, in addition, that the magnetic Gibbs pseudo-potential is a maximal Legendre transformation, so that with the Euler relation we obtain

$$\widetilde{\mathcal{G}}^*(T,\mathfrak{P},\mathcal{H},N) = \mu(T,\mathfrak{P},\mathcal{H})\,N\,. \tag{94}$$

3.2.2 The method of modified potentials

We use the expression (82) for the magnetic work, without to extract terms of the total exact differential type from the internal energy of the magnetic medium; then, the differential form (83) can be written in the following explicit manner:

$$\begin{aligned} \mathrm{d}\mathcal{U} &= T\,\mathrm{d}\mathcal{S} - \left(\mathfrak{P} + \frac{\overline{\mu}_0\mathcal{H}^2}{2}\right)\mathrm{d}V + \mathcal{H}\,\mathrm{d}(\mathcal{B}\,V) + \mu\,\mathrm{d}N \\ &= T\,\mathrm{d}\mathcal{S} - \pi\,\mathrm{d}V + \mathcal{H}\,\mathrm{d}\mathfrak{B} + \mu\,\mathrm{d}N\,. \end{aligned} \tag{95}$$

We observe that in this case the magnetic work has contributions on two thermodynamic degrees of freedom, so that we must redefine the magnetic and volumic state parameters:

$$X_V = V\,, \qquad P'_V = -\pi \equiv -\left(\mathfrak{P} + \frac{\overline{\mu}_0\mathcal{H}^2}{2}\right), \tag{96a}$$

$$X'_{\mathcal{H}} = \mathfrak{B} \equiv \mathcal{B}\,V\,, \qquad P_{\mathcal{H}} = \mathcal{H}\,. \tag{96b}$$

In this last case it appears the following peculiarities:

– V and $\mathfrak{B} \equiv \mathcal{B}\,V$ must be considered as independent variables,

– the effective pressure has an supplementary magnetic contribution $\overline{\mu}_0\,\mathcal{H}^2/2$.

Although the modified potential method implies the employment of some unusual state parameters, however it has the major advantage that $\mathcal{U}(\mathcal{S},V,\mathfrak{B},N)$ is the true fundamental energetic thermodynamic equation, and it is a *convex* and *homogeneous of degree* 1 function; thus, it is valid the Euler equation:

$$\mathcal{U} = T\,\mathcal{S} - \pi\,V + \mathcal{H}\,\mathfrak{B} + \mu\,N \tag{97}$$

and it is possible to define true thermodynamic potential with Legendre transformations. From the Euler relation (97) and passing to the common variables, it results

$$\mathcal{U} = T\mathcal{S} - \mathfrak{P}V + \overline{\mu}_0\mathcal{H}\,MV + \frac{\overline{\mu}_0\mathcal{H}^2}{2}\,V + \mu N\,,$$

so that it is ensured that $\widetilde{\mathcal{U}} \equiv \mathcal{U} - W_{\mathcal{H}}$ is a homogeneous function of degree 1 with respect to the variables $(\mathcal{S}, V, \mathcal{M}, N)$.

In order to compare the results of the method of modified potentials with those of the method of the pseudo-potentials we shall present only *the magnetic free energy* and *the magnetic Gibbs potential* as energetic thermodynamic potentials.

b.1. The magnetic free energy is the Legendre transformation on the thermal and magnetic degrees of freedom

$$\mathcal{F}^*(T, V, \mathcal{H}, N) \equiv \inf_{\mathcal{S},\mathfrak{B}} \left\{\mathcal{U}(\mathcal{S}, V, \mathfrak{B}, N) - T\mathcal{S} - \mathcal{H}\mathfrak{B}\right\}, \tag{98}$$

and it has the following differential form[28]:

$$\mathrm{d}\mathcal{F}^* = -\mathcal{S}\,\mathrm{d}T - \pi\,\mathrm{d}V - \mathfrak{B}\,\mathrm{d}\mathcal{H} + \mu\,\mathrm{d}N\,. \tag{99}$$

We shall emphasize some important properties of the magnetic free energy $\mathcal{F}^*(T, V, \mathcal{H}, N)$.

1. When the magnetic field vanishes it becomes the proper free energy (Helmholtz potential)

$$\mathcal{F}^*(T, V, 0, N) = \mathcal{U}_0(T, V, N) - T\mathcal{S}_0(T, V, N) = \mathcal{F}_0(T, V, N)\,.$$

2. The magnetic state equation is

$$\left(\frac{\partial \mathcal{F}^*}{\partial \mathcal{H}}\right)_{T,V,N} = -\mathfrak{B}(T, V, \mathcal{H}, N) = -\overline{\mu}_0\left[1 + \chi_m(T, V/N)\right]\mathcal{H}V\,.$$

3. By partial integration with respect to the magnetic field and the use of the condition of null field, we obtain the general expression of the magnetic free energy (for a linear and homogeneous magnetic medium)

$$\mathcal{F}^*(T, V, \mathcal{H}, N) = \mathcal{F}_0(T, V, N) - \frac{\overline{\mu}_0\mathcal{H}^2}{2}\left[1 + \chi_m(T, V/N)\right]V\,. \tag{100}$$

4. $\mathcal{F}^*(T, V, \mathcal{H}, N)$ is a function *concave* in respect to the variables T and $\mathcal{H}$; as a result we get the relation

$$\left(\frac{\partial^2 \mathcal{F}^*}{\partial \mathcal{H}^2}\right)_{T,V,N} = -\overline{\mu}_0\left[1 + \chi_m\right]V < 0\,,$$

and it follows "the stability condition" $\chi_m > -1$. We observe that the thermodynamics allows the existence of negative values of the magnetic susceptibility, that is the diamagnetism; the minimum value $\chi_m = -1$ corresponds to the *perfect diamagnetism*.

[28] In the strictly sense, $\mathcal{F}^*$ is a simple Gibbs potential, so that the common terminology is criticizable.

5. The state equations, deduced from Eq. (100) are:

$$\mathcal{S}(T,V,\mathcal{H},N) = -\left(\frac{\partial \mathcal{F}^*}{\partial T}\right)_{V,\mathcal{H},N} = \mathcal{S}_0(T,V,N) + \frac{\overline{\mu}_0 \mathcal{H}^2}{2}\left(\frac{\partial \chi_m}{\partial T}\right)_{V,N} V\,, \tag{101a}$$

$$\pi(T,V,\mathcal{H},N) = -\left(\frac{\partial \mathcal{F}^*}{\partial V}\right)_{T,\mathcal{H},N} = \mathfrak{P}_0(T,V,N) + \frac{\overline{\mu}_0 \mathcal{H}^2}{2}\left[1+\left(\frac{\partial (\chi_m V)}{\partial V}\right)_{T,N}\right] \tag{101b}$$

$$\mu(T,V,\mathcal{H},N) = \left(\frac{\partial \mathcal{F}^*}{\partial N}\right)_{T,V,\mathcal{H}} = \mu_0(T,V,N) - \frac{\overline{\mu}_0 \mathcal{H}^2}{2}\left(\frac{\partial \chi_m}{\partial N}\right)_{T,V} V\,. \tag{101c}$$

Because $\pi = \mathfrak{P} + \overline{\mu}_0 \mathcal{H}^2/2$, it results that the state equations (101) are identical with Eqs. (89), and this shows that $\widetilde{\mathcal{F}}^*$ (the correspondent pseudo-potential to $\mathcal{F}^*$) gives correct state equations.

From Eq. (100) it result that the free energy (Helmholtz potential) is

$$\mathcal{F} = \mathcal{F}^* + \mathcal{H}\,\mathfrak{B} = \mathcal{F}_0 + \frac{\mathcal{H}\,\mathcal{B}}{2}\,V\,,$$

so that the electric part of the volumic density of free energy is:

$$\mathfrak{f}_{\mathrm{mag}} \equiv \frac{\mathcal{F}-\mathcal{F}_0}{V} = \frac{\mathcal{H}\,\mathcal{B}}{2}\,.$$

We emphasize that in many books the previous expression for the magnetic part of the free energy density is erroneously considered as magnetic part of the internal energy density.

Correctly, the internal energy has the expression

$$\mathcal{U} = \mathcal{F}^* + T\,\mathcal{S} + \mathcal{H}\,\mathfrak{B} = (\mathcal{F}_0 + T\,\mathcal{S}_0) + \left(\frac{\mathcal{H}\,\mathcal{B}}{2} + \frac{\overline{\mu}_0 \mathcal{H}^2}{2}\,T\,\frac{\partial \chi_m}{\partial T}\right) V\,,$$

so that the magnetic part of the volumic density of internal energy is

$$\mathfrak{u}_{\mathrm{mag}} = \frac{\mathcal{U}-\mathcal{U}_0}{V} = \frac{\mathcal{H}\,\mathcal{B}}{2} + \frac{\overline{\mu}_0 \mathcal{H}^2}{2}\,T\,\frac{\partial \chi_m}{\partial T} \neq \frac{\mathcal{H}\,\mathcal{B}}{2}\,.$$

b.2. The magnetic Gibbs potential is defined analogously, as the Legendre transformation on the thermal, volumic and magnetic degrees of freedom

$$\mathcal{G}^*(T,\pi,\mathcal{H},N) \equiv \inf_{\mathcal{S},V,\mathfrak{B}} \left\{\mathcal{U}(\mathcal{S},V,\mathfrak{B},N) - T\,\mathcal{S} + \pi\,V - \mathcal{H}\,\mathfrak{B}\right\}, \tag{102}$$

and it has the differential form

$$\mathrm{d}\mathcal{G}^* = -\,\mathcal{S}\,\mathrm{d}T + V\,\mathrm{d}\pi - \mathfrak{B}\,\mathrm{d}\mathcal{H} + \mu\,\mathrm{d}N\,. \tag{103}$$

According to the definition, $\mathcal{G}^*$ is a maximal Legendre transform, so that the Euler relation leads to:

$$\mathcal{G}^* = \mu\,N\,. \tag{104}$$

On the other side, by replacing the variables π and $\mathfrak{B}$, accordingly to the definitions (96), we obtain the *the magnetic Gibbs potential is equal to the magnetic Gibbs pseudo-potential*[29] (but they have different variables):

$$\mathcal{G}^*(T,\pi,\mathcal{H},N) = \widetilde{\mathcal{G}}^*(T,\mathfrak{P},\mathcal{H},N)$$
$$\pi = \mathfrak{P} + \frac{\overline{\mu}_0\mathcal{H}^2}{2} .$$

From the preceding properties it follows that the equations deduced from the potential $\mathcal{G}^*$ are identical with Eqs. (93); we observe, however, that it is more convenient to use the pseudo-potential $\widetilde{\mathcal{G}}^*$, because this has more natural variables than the corresponding potential $\mathcal{G}^*$.

3.2.3 Thermodynamic potentials for open systems

Previously we have shown that the magnetic work implies two methods for treating the magnetic media: either as a closed subsystem of a compound system (this situation was discussed above), or as an open system located in a fixed volume (and the magnetic field is different from zero only inside the domain with fixed volume).

If we use the second method, then the magnetic work has the expression (78) and the magnetic medium system has only 3 thermodynamic degrees of freedom: thermal, magnetic and chemical (the volumic degree of freedom is frozen); then, the fundamental differential form is

$$\mathrm{d}\mathcal{U} = T\,\mathrm{d}\mathcal{S} + \mathcal{H}\,\mathrm{d}(V_0\,\mathcal{B}) + \mu\,\mathrm{d}N . \tag{105}$$

Among the thermodynamic potentials, obtained by Legendre transformations of the energetic fundamental thermodynamic equation, denoted as $\mathcal{U}(\mathcal{S},V_0\,\mathcal{B},N) \equiv \mathcal{U}(\mathcal{S},\mathfrak{B},N;V_0)$, we shall present only the magnetic free energy:

$$\mathcal{F}^*(T,\mathcal{H},N;V_0) \equiv \inf_{\mathcal{S},\mathfrak{B}} \left\{ \mathcal{U}(\mathcal{S},\mathfrak{B},N;V_0) - T\,\mathcal{S} - \mathcal{H}\,\mathfrak{B} \right\}, \tag{106}$$

which has the following properties:

1. the differential form:

$$\mathrm{d}\mathcal{F}^* = -\,\mathcal{S}\,\mathrm{d}T - V_0\,\mathcal{B}\,\mathrm{d}\mathcal{H} + \mu\,\mathrm{d}N ; \tag{107}$$

2. it reduces to the free energy (the Helmholtz potential) at vanishing magnetic field

$$\mathcal{F}^*(T,0,N;V_0) = \mathcal{U}_0(T,N;V_0) - T\,\mathcal{S}_0(T,N;V_0) = \mathcal{F}_0(T,N;V_0) ;$$

3. by integrating the magnetic state equation, that is written in the form $V_0\,\mathcal{B}(T,\mathcal{H},N) = V_0\,[1+\chi_m]\,\overline{\mu}_0\mathcal{H}$, we obtain

$$\mathcal{F}^*(T,\mathcal{H},N;V_0) = \mathcal{F}_0(T,N;V_0) - V_0\,\frac{\overline{\mu}_0\mathcal{H}^2}{2}\,[1+\chi_m(T,V/N)] . \tag{108}$$

We note that the results are equivalent to those obtained by the previous method, but the situation is simpler because the volumic degree of freedom is frozen.

---

[29] The equality $\mathcal{G}^* = \widetilde{\mathcal{G}}^*$ (as quantities, but not as functions) can be obtained directly by comparing the consequences of the Euler equation (94) and (104).

3.3 Thermodynamic coefficients and processes

3.3.1 Definitions for the principal thermodynamic coefficients

Because the magnetic media has 4 thermodynamic degrees of freedom (in the simplest case, when it is fluid), there are a great number of simple thermodynamic coefficients; taking into account this complexity, we shall present only the common coefficients, corresponding to closed magnetic media systems ($N =$ constant).

a.1. The sensible specific heats are defined for non-isothermal processes "φ":

$$c_{\varphi} = \frac{1}{N} T \left(\frac{\partial \mathcal{S}}{\partial T} \right)_{\varphi} = T \left(\frac{\partial s}{\partial T} \right)_{\varphi} . \tag{109}$$

In the case, when the process φ is simple, we obtain the following specific isobaric/isochoric and iso-magnetization/iso-field heats: $c_{V,\mathrm{m}}, c_{V,\mathcal{H}}, c_{P,\mathrm{m}}, c_{P,\mathcal{H}}$.

a.2. The latent specific heats are defined for isothermal processes "ψ":

$$\lambda_{\psi}^{(a)} = \frac{1}{N} T \left(\frac{\partial \mathcal{S}}{\partial a} \right)_{\psi} = T \left(\frac{\partial s}{\partial a} \right)_{\psi} . \tag{110}$$

The most important cases (for "ψ" and a) are the isothermal-isobaric process with $a = \mathcal{H}$ when we have *the isobaric magnetic-caloric coefficient* λ and the conjugated isothermal-isofield process with $a = \mathfrak{P}$, when we have *the iso-field piezo-caloric coefficient* $\lambda_{\mathcal{H}}^{(P)}$:

$$\lambda = T \left(\frac{\partial s}{\partial \mathcal{H}} \right)_{T,\mathfrak{P}} , \qquad \lambda_{\mathcal{H}}^{(P)} = T \left(\frac{\partial s}{\partial \mathfrak{P}} \right)_{T,\mathcal{H}} . \tag{111}$$

a.3. The thermodynamic susceptibilities are of two types: for the volumic degree of freedom (in this case they are called *compressibility coefficients*) and for the magnetic degree of freedom (these are called *magnetic susceptibilities*):

$$\varkappa_{\varphi} = \frac{-1}{V} \left(\frac{\partial V}{\partial \mathfrak{P}} \right)_{\varphi} = \frac{-1}{v} \left(\frac{\partial v}{\partial \mathfrak{P}} \right)_{\varphi} , \tag{112}$$

$$\chi_{\psi}^{(\mathrm{m})} = \frac{1}{V} \left(\frac{\partial \mathcal{M}}{\partial (\overline{\mu}_0 \mathcal{H})} \right)_{\psi} = \frac{1}{\overline{\mu}_0} \left(\frac{\partial M}{\partial \mathcal{H}} \right)_{\psi} . \tag{113}$$

In the simple cases "φ" is an isothermal/adiabatic and iso-magnetization/iso-field processes; it results the following simple compressibility coefficients: $\varkappa_{T,M}, \varkappa_{T,\mathcal{H}}, \varkappa_{s,M}$ and $\varkappa_{s,\mathcal{H}}$. Analogously "$\psi$" as simple process can be isothermal/adiabatic and isobaric/isochoric, resulting the following simple magnetic susceptibilities: $\chi_{T,v}^{(\mathrm{m})}, \chi_{T,\mathfrak{P}}^{(\mathrm{m})}, \chi_{s,v}^{(\mathrm{m})}$ and $\chi_{s,\mathfrak{P}}^{(\mathrm{m})}$.

From Eq. (71b) we observe that the isothermal magnetic susceptibility is proportional to the susceptibility used in the electrodynamics:

$$\chi_{T,v}^{(\mathrm{m})} = \left(\frac{\partial M}{\partial \mathcal{H}} \right)_{T,v} = \frac{1}{\overline{\mu}_0} \chi_m(T,v) .$$

a.4. The thermal coefficients are of two types, corresponding to the two non-thermal and non-chemical degrees of freedom (the volumic and the magnetic ones). If we consider only thermal coefficients for extensive parameters, then we can define the following types of simple coefficients:

- the *isobaric thermal expansion coefficients* (also iso-magnetization/iso-field)

$$\alpha_y = \frac{1}{V}\left(\frac{\partial V}{\partial T}\right)_{\mathfrak{P},y,N} = \frac{1}{v}\left(\frac{\partial v}{\partial T}\right)_{\mathfrak{P},y}, \tag{114}$$

where the index y is $\mathcal{M}$ or $\mathcal{H}$;

- the *pyro-magnetic coefficients* (also isochoric/isobaric)

$$\pi_a = \frac{1}{N}\left(\frac{\partial \mathcal{M}}{\partial T}\right)_{\mathcal{H},a,N} = \left(\frac{\partial \mathfrak{m}}{\partial T}\right)_{\mathcal{H},a}, \tag{115}$$

where the index a is V (for the volume V) or P (for the pressure $\mathfrak{P}$).

a.5. The mixed coefficients express correlations between the volumic and the magnetic degrees of freedom; we mention the following simple coefficients:

- the *magnetic-strictive coefficient*

$$\gamma = \frac{1}{V}\left(\frac{\partial V}{\partial \mathcal{H}}\right)_{T,\mathfrak{P},N} = \frac{1}{v}\left(\frac{\partial v}{\partial \mathcal{H}}\right)_{T,\mathfrak{P}}; \tag{116}$$

- the *piezo-magnetic coefficient*

$$\varrho = \frac{1}{N}\left(\frac{\partial \mathcal{M}}{\partial \mathfrak{P}}\right)_{T,\mathcal{H},N} = \left(\frac{\partial \mathfrak{m}}{\partial \mathfrak{P}}\right)_{T,\mathcal{H}}. \tag{117}$$

3.3.2 Relations between simple coefficients

Because the great number of thermodynamic coefficients, corresponding to the magnetic media, we must select among all possible relations between the simple coefficients; therefore, we shall present only the most important relations: the symmetry relations (consequences of some Maxwell relations) and special relations (of the type Reech or Mayer).

In order to emphasize symmetry relations expressed by the temperature, the pressure and the magnetic field intensity $(T, \mathfrak{P}, \mathcal{H})$, as variables, for an closed system ($N =$ constant), we use the differential form of the reduced Gibbs pseudo-potential $\widetilde{g}^*(T, \mathfrak{P}, \mathcal{H})$, which is obtained with the general reduction formulae from Eqs. (90) – (91):

$$d\widetilde{g}^* = -\,s\,dT + v\,d\mathfrak{P} - \mathfrak{m}\,d(\overline{\mu}_0 \mathcal{H})\,. \tag{118}$$

From the above differential form it results 3 Maxwell relations, which can be expressed by simple coefficients, resulting symmetry relations between these coefficients:

$$\left(\frac{\partial s}{\partial \mathfrak{P}}\right)_{T,\mathcal{H}} = -\left(\frac{\partial v}{\partial T}\right)_{\mathfrak{P},\mathcal{H}} \quad \Longrightarrow \quad \lambda_{\mathcal{H}}^{(P)} = -\,T\,v\,\alpha_{\mathcal{H}}\,, \tag{119a}$$

(the relation between the iso-field piezo-caloric coefficient and the isobaric-isofield thermal expansion coefficient);

$$\left(\frac{\partial s}{\partial (\overline{\mu}_0 \mathcal{H})}\right)_{T,\mathfrak{P}} = \left(\frac{\partial \mathfrak{m}}{\partial T}\right)_{\mathfrak{P},\mathcal{H}} \quad \Longrightarrow \quad \lambda = \overline{\mu}_0\,T\,\pi_P\,, \tag{119b}$$

(the relation between the isobaric magneto-caloric coefficient and the isobaric pyro-magnetic coefficient);

$$\left(\frac{\partial v}{\partial\left(\overline{\mu}_0 \mathcal{H}\right)}\right)_{T,\mathfrak{P}} = -\left(\frac{\partial \,\mathrm{m}}{\partial \mathfrak{P}}\right)_{T,\mathcal{H}} \qquad \Longrightarrow \qquad \frac{v}{\overline{\mu}_0}\,\gamma = -\varrho\,. \tag{119c}$$

(the relation between the magnetic-strictive coefficient and the piezo-magnetic coefficient).

Relations of Reech type can be obtained from the general relation (see [16] Eq. (3.25)), resulting the equality between the ratios of the isobaric specific heats, of the compressibility coefficients, and of the magnetic susceptibilities (isothermal, and respectively adiabatic):

$$\frac{c_{P,\mathrm{m}}}{c_{P,\mathcal{H}}} = \frac{\varkappa_{S,\mathcal{H}}}{\varkappa_{T,\mathcal{H}}} = \frac{\chi^{(\mathrm{m})}_{S,\mathfrak{P}}}{\chi^{(\mathrm{m})}_{T,\mathfrak{P}}}\,. \tag{120}$$

Analogously, from the general Mayer relation for the specific heats (see [16] Eq. (3.28)) we obtain in this case

$$c_{P,\mathcal{H}} - c_{P,\mathrm{m}} = T\,v\,\frac{\alpha^2_{\mathcal{H}}}{\chi^{(\mathrm{m})}_{T,\mathfrak{P}}}\,. \tag{121}$$

Similar relations with Eqs. (120) – (121) can be obtained for the coefficients associated to another sets of simple processes (*e.g.* isochoric, iso-magnetization processes).

3.3.3 The factorization of some simple coefficients

An important characteristics of some thermodynamic coefficients is *the factorization property*: the expression of the considered coefficient is the sum of the part corresponding to the absence of the magnetic field (like for the neutral fluid) and the "magnetic part", and this result comes from the factorization of the state equations.

We shall present the factorization of some coefficients using the variables of the magnetic Gibbs representation[30] $(T, \mathfrak{P}, \mathcal{H}, N)$; in this case the state equations are Eqs. (93).

The entropy is given by Eq. (93a), and here we write it without the variables, for simplicity and for using later the convenient variables

$$\mathcal{S} = \mathcal{S}_0 + \mathcal{S}_{\mathrm{mag}}\,, \tag{122}$$

where $\mathcal{S}_0$ is the entropy of the magnetic medium as a neutral fluid, in the absence of the magnetic field, and $\mathcal{S}_{\mathrm{mag}}$ is the magnetic part of the entropy:

$$\mathcal{S}_{\mathrm{mag}} = \frac{\overline{\mu}_0 \mathcal{H}^2}{2}\left(\frac{\partial \overline{\chi}_m}{\partial T}\right)_{\mathfrak{P}} N = \frac{\overline{\mu}_0 \mathcal{H}^2}{2}\left(\frac{\partial \chi_m}{\partial T}\right)_{v} V\,. \tag{123}$$

Accordingly to the general definition (109), we obtain a factorization of the specific heats:

$$c_{\varphi} = c^{(0)}_{\varphi} + c^{(\mathrm{mag})}_{\varphi}\,, \tag{124}$$

[30] We remark that some coefficients need the use of other thermodynamic representations.

where $c_{\varphi}^{(0)} = T(\partial s^{(0)}/\partial T)_{\varphi}$ is the specific heat of the magnetic medium in the absence of the magnetic field, and $c_{\varphi}^{(\mathrm{mag})} = T(\partial s^{(\mathrm{mag})}/\partial T)_{\varphi}$ is the magnetic part of the specific heat.

For the isobaric processes there are the specific heat at constant magnetic field or at constant magnetization. From Eq. (123) it results

$$c_{P,\mathcal{H}}^{(\mathrm{mag})} = T\left(\frac{\partial \overline{\chi}_m}{\partial T}\right)_{\mathfrak{P}} \frac{\overline{\mu}_0 \mathcal{H}^2}{2}\,, \tag{125}$$

To obtain $c_{P,\mathfrak{m}}^{(\mathrm{mag})}$ we express the magnetic entropy $\mathcal{S}_{\mathrm{mag}}$ in terms of the dipolar magnetic moment (instead the magnetic field), using Eq. (71b):

$$\mathcal{S}_{\mathrm{mag}} = \frac{\mathfrak{m}^2}{2\,\overline{\mu}_0\,\overline{\chi}_m^2}\left(\frac{\partial \overline{\chi}_m}{\partial T}\right)_{\mathfrak{P}} N = \frac{-\mathfrak{m}^2}{2\,\overline{\mu}_0}\left(\frac{\partial\,(\overline{\chi}_m)^{-1}}{\partial T}\right)_{\mathfrak{P}} N\,;$$

then it results for the magnetic part of the isobaric-isomagnetization specific heat the expression:

$$c_{P,\mathfrak{m}}^{(\mathrm{mag})} = \frac{-\mathfrak{m}^2}{2\,\overline{\mu}_0}\,T\left(\frac{\partial^2\,(\overline{\chi}_m)^{-1}}{\partial T^2}\right)_{\mathfrak{P}}\,; \tag{126}$$

We observe that for an ideal paramagnet $(\overline{\chi}_m)^{-1} \sim T$, so that we obtain $c_{P,\mathfrak{m}}^{(\mathrm{mag})} = 0$, that is $c_{P,\mathfrak{m}} = c_P^{(0)}$ (the iso-magnetization specific heat is independent of the magnetic field)[31].
It is interesting to emphasize that for the ideal para magnets the internal energy has also particular properties. The magnetic part of the volumic density of internal energy for an arbitrary magnetic medium has the expression

$$\mathfrak{u}^{(\mathrm{mag})} = \frac{\overline{\mu}_0 \mathcal{H}^2}{2}\left(1 + \chi_m + T\,\frac{\partial \chi_m}{\partial T}\right)\,.$$

For an ideal paramagnet we obtain that this energy density is equal to the energy density of the magnetic field $\mathfrak{u}^{(\mathrm{mag})} = \overline{\mu}_0 \mathcal{H}^2/2 = w_{\mathrm{mag}}$, that is the whole magnetic energy is given only by the magnetic field, without any contribution from the processes of the magnetization. The behavior of the iso-magnetization specific heats and of the internal energy are similar to the neutral fluids which satisfy the Clapeyron - Mendeleev equations, so that it is justified the terminology "ideal" for the para-magnets which have Curie susceptibility.

We observe, in addition, that the diamagnetic systems, having a constant magnetic susceptibility (approximatively), have null magnetic entropy $\mathcal{S}_{\mathrm{mag}}$, accordingly to Eq. (123); therefore, the diamagnetic systems have caloric properties independent of the magnetic field (or of the magnetization).

In contrast with the specific heats, the isobaric magneto-caloric coefficient has contribution only from the magnetic part of the entropy:

$$\lambda = \overline{\mu}_0\,\mathcal{H}\,T\left(\frac{\partial \overline{\chi}_m}{\partial T}\right)_{\mathfrak{P}}\,.$$

---

[31] For the corresponding isochoric specific heat we obtain the same result $c_{V,\mathfrak{P}}^{(\mathrm{mag})} = c_V^{(0)}$.

The volume is given by Eq. (93b), that is it can be expressed in the form:

$$V = V_0 + V_{\rm mag} \,,$$

where V is the volume of the magnetic medium as neutral fluid, in the absence of the magnetic field, and $V_{\rm mag}$ is the magnetic part of the volume:

$$V_{\rm mag} = -\,\frac{\overline{\mu}_0 \mathcal{H}^2}{2} \left(\frac{\partial \overline{\chi}_m}{\partial \mathfrak{P}}\right)_T N \,. \tag{127}$$

Accordingly to the general definitions (112) and respectively (114), the isothermal compressibility coefficients $\varkappa_{T,\mathcal{H}}$ and the isobaric thermal expansion coefficient $\alpha_{\mathcal{H}}$ (both of them at constant field) factorize in non-magnetic part (corresponding to null magnetic field, when the magnetic medium behaves as a neutral fluid) and magnetic part:

$$\varkappa_{T,\mathcal{H}} = \varkappa_T^{(0)} + \varkappa_{T,\mathcal{H}}^{(\rm mag)} \,, \tag{128a}$$

$$\alpha_{\mathcal{H}} = \alpha^{(0)} + \alpha_{\mathcal{H}}^{(\rm mag)} \,, \tag{128b}$$

where[32]

$$\varkappa_{T,\mathcal{H}}^{(\rm mag)} = \frac{\overline{\mu}_0\, \mathcal{H}^2}{2\, v_0} \left(\frac{\partial^2 \overline{\chi}_m}{\partial \mathfrak{P}^2}\right)_T \,, \tag{129a}$$

$$\alpha_{\mathcal{H}}^{(\rm mag)} = -\,\frac{\overline{\mu}_0\, \mathcal{H}^2}{2\, v_0}\, \frac{\partial^2 \overline{\chi}_m}{\partial T\, \partial \mathfrak{P}} \,. \tag{129b}$$

In contrast with the previous coefficients, the magneto-strictive coefficient is obtained only from the magnetic part of the volumic state equation[33]

$$\gamma \approx -\,\frac{\overline{\mu}_0 \mathcal{H}}{v_0} \left(\frac{\partial \overline{\chi}_m}{\partial \mathfrak{P}}\right)_T \,. \tag{130}$$

3.3.4 Thermodynamic process

Using the previous results we shall present the most significant thermodynamic processes for the magnetic media as closed system (we shall choose the variable set $T, \mathfrak{P}, \mathcal{H}, N = $ constant).

d.1. The isothermal magnetization: we consider that initially the magnetic medium in null magnetic field $(T, \mathfrak{P}, \mathcal{H}_i = 0, N)$ and we apply the magnetic field with an isothermal-isobaric process, so that the final state has the parameters: $(T, \mathfrak{P}, \mathcal{H}_f = \mathcal{H}, N)$.

Using Eq. (93a) for the entropy, the heat transfered in this process is

$$\begin{aligned} \mathcal{Q}_{if} &= T\, \Delta \mathcal{S}_{if} = T\, \{\, \mathcal{S}(T, \mathfrak{P}, \mathcal{H}, N) - \mathcal{S}(T, \mathfrak{P}, 0, N)\, \} \\ &= \frac{\overline{\mu}_0\, \mathcal{H}^2}{2}\, T \left(\frac{\partial \overline{\chi}_m}{\partial T}\right)_{\mathfrak{P}} N \,. \end{aligned} \tag{131}$$

[32] In fact, the factorization is obtained only if we consider small magnetic effects, so that we could approximate $v \approx v_0$ at the denominators.

[33] We consider that the magneto-strictive effects are small, so that we can use the approximation $v \approx v_0$, at the denominator.

Since $\overline{\chi}_m(T,\mathfrak{P})$ is in general a decreasing function with respect to the temperature, it results that in the magnetization process the magnetic medium yields heat: $\mathcal{Q}_{if} < 0$.

d.2. The adiabatic-isobaric demagnetization: we consider that initially the magnetic medium is in the presence of the magnetic field $\mathcal{H}$ and it has the temperature T_i ; then, by a quasi-static adiabatic-isobaric process the magnetic field is decreasing to vanishing value.

Because the equation of this process is $\mathcal{S}(T,\mathfrak{P},\mathcal{H},N) = \text{constant}$, with the supplementary conditions $\mathfrak{P} = \text{constant}$ and $N = \text{constant}$, then by using Eq. (93a), we obtain the equation of the temperature:

$$\mathcal{S}(T_i,\mathfrak{P},\mathcal{H},N) = \mathcal{S}(T_f,\mathfrak{P},0,N) ;$$

that is, after simple algebraical operations, we get:

$$s^0(T_i,\mathfrak{P}) + \frac{\overline{\mu}_0\mathcal{H}^2}{2}\left(\frac{\partial\overline{\chi}_m}{\partial T}\right)_{\mathfrak{P}} = s^0(T_f,\mathfrak{P}) . \tag{132}$$

Because the magnetic susceptibility is in general an decreasing function in respect to the temperature $(\partial\overline{\chi}_m/\partial T)_{\mathfrak{P}} < 0$ and the entropy $s^0(T,\mathfrak{P})$ is an increasing function of temperature, it results $s_i < s_f$, that is the paramagnet gets cool during the adiabatic demagnetization: $T_f < T_i$.

d.3. The magneto-strictive and piezo-magnetic effects

The *magneto-strictive effect* means the variation of the volume (of the magnetic medium) due to the variation of the magnetic field, in conditions isothermal-isobaric (also the magnetic medium is a closed system)[34]; and the magneto-strictive coefficient is defined by Eq. (116).

The *piezo-magnetic effect* means the variation of the dipolar magnetic moment (of the magnetic medium), due to the variation of the pressure, in conditions isothermal and at constant magnetic field (also the magnetic medium is a closed system)[35] and the piezo-magnetic effect coefficient is defined by Eq. (117).

Between the coefficients of the two effects it is the symmetry relation (53), and the corresponding expressions can be put in explicit forms using the magnetic susceptibility:

$$\gamma = -\frac{1}{v}\,\varrho = \frac{-\overline{\mu}_0\mathcal{H}}{v}\left(\frac{\partial\overline{\chi}_m}{\partial\mathfrak{P}}\right)_T . \tag{133}$$

We observe that the necessary condition to have an magneto-strictive effect and an piezo-magnetic effect is that $\overline{\chi}_m$ depends on the pressure; accordingly to Eqs. (76), it results that only the non-ideal para-magnets can have these effects.

Using the volumic equation of state (93b), we can evaluate the global magneto strictive effect, that is the variation of the volume (of the paramagnet) at the isothermal-isobaric magnetization:

$$\Delta V_{if}(\mathcal{H}) = V(T,\mathfrak{P},\mathcal{H},N) - V(T,\mathfrak{P},0,N) = -\frac{\overline{\mu}_0\mathcal{H}^2}{2}\left(\frac{\partial\overline{\chi}_m}{\partial\mathfrak{P}}\right)_T N . \tag{134}$$

From the previous expression it results that when the magnetic susceptibility is a decreasing function of the pressure $(\partial\overline{\chi}_m/\partial\mathfrak{P})_T < 0$, then it follows a contraction (a reduction of the volume) at the magnetization of the diamagnet.

[34] In other words, the magneto-strictive effect can be considered as the volumic response of the magnetic medium to an magnetic perturbation.

[35] We observe that the piezo-magnetic effect can be considered as the magnetic response of the magnetic medium to a volumic perturbation, being conjugated to the magneto-strictive effect.

4. Acknowledgment

The financial support of the JINR Dubna - IFIN-HH Magurele-Bucharest project no. 01-3-1072-2009/2013 is kindly acknowledged.

5. References

A. Sommerfeld, *Thermodynamics and Statistical Mechanics*, (Academic Press, New York 1956).
M. Planck, *Treatise on Thermodynamics*, (Dover Publisher, New York, 1945).
E. A. Guggenheim, *Thermodynamics – An Advanced Treatment for Chemists and Physicists*, (North Holland, Amsterdam, 1967).
A. H. Wilson, *Thermodynamics and Statistical Mechanics*, (Cambridge Univ. Press, 1957).
A. B. Pippard, *Elements of Classical Thermodynamics*, (Cambridge Univ. Press, 1966).
D. ter Haar, H. Wegeland, *Elements of Thermodynamics*, (Addison-Wesley, Reading Mass. 1966).
C. J. Adkins, *Equilibrium Thermodynamics*, (Cambridge Univ. Press, 1983).
G. N. Hatsopoulos, J. H. Keenan, *Principles of General Thermodynamics*, (John Wiley & Sons, New York, 1965).
V. S. Helrich, *Modern Thermodynamics and Statistical Mechanics*, (Springer-Verlag, Berlin, 2009).
F. Schwabl, *Statistical Mechanics*, (Springer-Verlag, Berlin, 2006).
W. Greiner, W. Neise, H. Stöcker, *Thermodynamics and Statistical Mechanics*, (Springer, New York, 2095).
L. D. Landau, E. M. Lifshitz, *Statistical Physics, Part 1* (3rd ed.), (Pergamon Press, Oxford, 1980).
H. B. Callen, *Thermodynamics and an Introduction to Thermostatistics* (2nd ed.), (John Wiley & Sons, New York, 1985).
L. Tisza, *Generalized Thermodynamics*, (The MIT Press, Cambridge Mass. 1966).
E. H. Lieb, J. Yngvason, *The Physics and Mathematics of the Second Law of Thermodynamics*, *Physics Reports* 310, 1-96 (1999).
R. P. Lungu, *Thermodynamics*, `http://www.unescochair-hhf.ro/index.php?page=training.`
J. D. Jackson, *Classical Electrodynamics* (3rd ed.), (John Wiley & Sons, New York, 1999).
W. Greiner, *Classical Electrodynamics*, (Springer-Verlag, Berlin, 1998).
W. H. K. Panofsky, M. Phillips, *Classical Electrodynamics*, (Addison-Wesley, Reading Mass. 1962).
L. D. Landau, E. M. Lifshitz, *Electrodynamics of Continuous Media*, (Pergamon Press, Oxford, 1984).

5

Topological Electromagnetism: Knots and Quantization Rules

Manuel Arrayás[1], José L. Trueba[1] and Antonio F. Rañada[2]
[1]*Universidad Rey Juan Carlos*
[2]*Universidad Complutense de Madrid*
Spain

1. Introduction

In this chapter, we revise the main features of a topological model of electromagnetism, also called the model of electromagnetic knots, that was presented in 1989 (Rañada, 1989) and has been developed in a number of references. Some of them are (Arrayás & Trueba, 2010; 2011; Irvine & Bouwmeester, 2008; Rañada, 1990; 1992; Rañada & Trueba, 1995; 1997; 2001; Rañada, 2003). One of the main characteristics of this model is that it allows to obtain interesting topological quantization rules for the electric charge (Rañada & Trueba, 1998) and the magnetic flux through a superconducting ring (Rañada & Trueba, 2006). We will pay special attention to these features.

An electromagnetic knot is defined as a standard electromagnetic field with the property that any pair of its magnetic lines, or any pair of its electric lines, is a link with linking number ℓ. This number is a measure of how much the force lines curl themselves the ones around the others. These lines coincide with the level curves of a pair of complex scalar fields $\phi(\mathbf{r}, t)$, $\theta(\mathbf{r}, t)$. In the model of electromagnetic knots, the physical space and the complex plane are compactified to S^3 and S^2, so that the scalars can be interpreted as maps $S^3 \mapsto S^2$, which are known to be classified in homotopy classes characterized by the integer value of the Hopf index n, which is related to the linking number ℓ.

The topological model of electromagnetism is locally equivalent to Maxwell's standard theory in the sense that the set of electromagnetic knots coincides locally with the set of the standard radiation fields (radiation fields are electromagnetis fields such that the magnetic field is orthogonal to the electric field at any point and at any instant of time). In other words, standard radiation fields can be understood as patched together electromagnetic knots. This can still be expressed as the statement that, in any bounded domain of space-time, any standard radiation fields can be approximated arbitrarily enough by electromagnetic knots.

It is remarkable that the standard Maxwell's equations are the *exact linearization*, by change of variables *not by truncation*, of a set of nonlinear equations referring to the complex scalar fields $\phi(\mathbf{r}, t)$ and $\theta(\mathbf{r}, t)$. The fact that this change is not completely invertible has the surprising consequence that the linearity of the Maxwell's equations is compatible with the existence of topological constants of the motion which are nonlinear in the magnetic and electric fields. In this chapter we will see how to find some of these topological constants.

2. Electromagnetic knots

As said before, the topological model of electromagnetic knots makes use of two fundamental complex scalar fields $\phi(\mathbf{r},t)$ and $\theta(\mathbf{r},t)$, the level curves of which coincide with the magnetic and electric lines, respectively. This means that each one of these lines are labelled by the constant value of the corresponding scalar. These complex scalar fields are assumed to have only one value at infinity, which is equivalent, from the mathematical point of view, to compactify the three-space to the sphere S^3. Moreover, the complex plane C is also compactified to the sphere S^2. Both compactifications imply that the scalars ϕ and θ can be interpreted (via stereographic projection) as maps $S^3 \rightarrow S^2$, which can be classified in homotopy classes and, as such, be characterized by the value of the Hopf index n. It can be shown that the two scalars have the same Hopf index and that the magnetic (resp. electric) lines are generically linked with the same Gauss linking number ℓ. If μ is the multiplicity of the level curves (i.e. the number of different magnetic (resp. electric) lines that have the same label ϕ (resp. θ)), then $n = \ell\mu^2$; the Hopf index can thus be interpreted as a generalized linking number if we define a line as a level curve with μ disjoint components.

From the dimensionless scalars $\phi(\mathbf{r},t)$ and $\theta(\mathbf{r},t)$, one can construct a magnetic field $\mathbf{B}$ and an electric field $\mathbf{E}$ as

$$\begin{aligned} \mathbf{B}(\mathbf{r},t) &= \frac{\sqrt{a}}{2\pi i} \frac{\nabla\phi \times \nabla\bar{\phi}}{(1+\bar{\phi}\phi)^2}, \\ \mathbf{E}(\mathbf{r},t) &= \frac{\sqrt{a}c}{2\pi i} \frac{\nabla\bar{\theta} \times \nabla\theta}{(1+\bar{\theta}\theta)^2}, \end{aligned} \tag{1}$$

where $\bar{\phi}$ and $\bar{\theta}$ are the complex conjugates of ϕ and θ respectively, i is the imaginary unit, a is a constant introduced so that the magnetic and electric fields have correct dimensions, and c is the speed of light in vacuum. In the SI of units, a can be expressed as a pure number times the Planck constant $\hbar$ times the speed of light c times the vacuum permeability μ_0.

In order to obtain a solution of the Maxwell's equations in vacuum from the fields given by Equations (1), they also have to satisfy

$$\begin{aligned} \mathbf{B}(\mathbf{r},t) &= \frac{\sqrt{a}}{2\pi i c(1+\bar{\theta}\theta)^2} \left(\frac{\partial\bar{\theta}}{\partial t}\nabla\theta - \frac{\partial\theta}{\partial t}\nabla\bar{\theta} \right), \\ \mathbf{E}(\mathbf{r},t) &= \frac{\sqrt{a}}{2\pi i(1+\bar{\phi}\phi)^2} \left(\frac{\partial\phi}{\partial t}\nabla\bar{\phi} - \frac{\partial\bar{\phi}}{\partial t}\nabla\phi \right). \end{aligned} \tag{2}$$

Equations (1) and (2) constitute the definition of an electromagnetic knot, and the magnetic and the electric fields resulting from these equations satisfy exactly Maxwell's equations in vacuum.

It is possible to write Equations (1) and (2) in a more compact way by using the language of differential forms (a nice reference in which Electromagnetism is written in this language is (Hehl & Obukhov, 2003)). If $\mu, \nu = 0,1,2,3$ are space-time indices and $i, j, = 1,2,3$ are purely space indices, $A^\mu = (V/c, \mathbf{A})$ (in which V is the electrostatic potential and $\mathbf{A}$ is the vector potential) is the 4-vector potential of the electromagnetic field, so that the electromagnetic tensor is

$$F_{\mu\nu} = \partial_\mu A_\nu - \partial_\nu A_\mu, \tag{3}$$

in which $x_0 = ct$. From this tensor one finds the components of the electric field as $\mathbf{E}_i = c\,F^{i0}$, and the magnetic field as $\mathbf{B}_i = -\epsilon_{ijk}F^{jk}/2$ as usual. Moreover, the dual to the electromagnetic tensor is defined as

$$G_{\mu\nu} = {}^* F_{\mu\nu} = \frac{1}{2}\,\epsilon_{\mu\nu\alpha\beta}F^{\alpha\beta}, \tag{4}$$

with components $\mathbf{B}_i = G^{0i}$, $\mathbf{E}_i = -c\,\epsilon_{ijk}\,G^{jk}/2$. Now, the Faraday 2-form is defined as

$$\mathcal{F} = \frac{1}{2}\,F_{\mu\nu}dx^\mu \wedge dx^\nu, \tag{5}$$

and its dual 2-form is defined as

$$*\,\mathcal{F} = \frac{1}{2}\,G_{\mu\nu}dx^\mu \wedge dx^\nu. \tag{6}$$

Because of clarity, we will use in this work *natural units*, in which the speed of light c, the Planck constant $\hbar$, the vacuum permittivity ε_0 and the vacuum permeability μ_0 are chosen as $c = \hbar = \varepsilon_0 = \mu_0 = 1$. In this system of units, the constant a in Equations (1) and (2) is a pure number. In the language of differential forms, Equations (1) and (2) simply and remarkably mean that the Faraday form $\mathcal{F}$ and its dual $*\mathcal{F}$ of any electromagnetic knot are the two pull-backs of σ, the area 2-form in S^2, by the maps ϕ and θ from S^3 to S^2, i. e.

$$\begin{aligned}\mathcal{F} &= -\sqrt{a}\,\phi^*\sigma,\\ *\mathcal{F} &= \sqrt{a}\,\theta^*\sigma.\end{aligned} \tag{7}$$

As a consequence the two maps are dual to one another in the sense that

$$*\,(\phi^*\sigma) = -\theta^*\sigma, \tag{8}$$

* being the Hodge or duality operator. The existence of two maps satisfying Equation (8) guarantees that both $\mathcal{F}$ and $*\mathcal{F}$ obey the Maxwell equations in empty space without the need of any other requirement. The electromagnetic fields obtained as in Equations (7) are electromagnetic knots. They are radiation fields , i. e. they verify the condition $\mathbf{E}\cdot\mathbf{B} = 0$. Note that, because of the Darboux theorem, any electromagnetic field in empty space can be expressed locally as the sum of two radiation fields.

As stated before, the model of electromagnetic knots is locally equivalent to Maxwell's standard theory (Rañada & Trueba, 1998; Rañada, 2003). However, its difference from the global point of view has interesting consequences, as are the following topological quantizations:

- The electromagnetic helicity $\mathcal{H}$ is quantized. In natural units,

$$\mathcal{H} = \frac{1}{2}\int_{R^3} (\mathbf{A}\cdot\mathbf{B} + \mathbf{C}\cdot\mathbf{E})\,d^3r = na, \tag{9}$$

where $\mathbf{B} = \nabla \times \mathbf{A}$, $\mathbf{E} = \nabla \times \mathbf{C}$, the integer n being equal to the common value of the Hopf indices of ϕ and θ. Note that $\mathcal{H} = N_R - N_L$, where N_R and N_L are the classical expressions of the number of right- and left-handed photons contained in the field (i.e. $\mathcal{H} = N_R - N_L = \int d^3k(\bar{a}_R a_R - \bar{a}_L a_L)$, $a_R(\mathbf{k}), a_L(\mathbf{k})$ being Fourier transforms of A_μ in the classical theory, but creation and annihilation operator in the quantum version). This implies that, if we take the constant a to be $a = 1$,

$$n = N_R - N_L, \tag{10}$$

which is a curious relation between the Hopf index (i.e. the generalized linking number) of the classical field and the classical limit of the difference $N_R - N_L$. This difference has a clear topological meaning, what is attractive from the intuitive physical point of view.

- The topology of the model of electromagnetic knots implies also the quantization of the electromagnetic energy in a cavity, as studied in Reference (Rañada, 2003). More precisely, the model predicts that its energy $\mathcal{E}$ in a cubic cavity verifies

$$\mathcal{E} = n\omega, \tag{11}$$

with $n = d/4$, d being an integer, equal to the degree of a certain map between two orbifolds, and ω is the angular frequency of the electromagnetic radiation. This rule is different from the Planck-Einstein law but very similar.

- The model of electromagnetic knots explains the discretization of the values of the electric charge and the magnetic flux through a superconducting ring. These properties will be studied in the next sections of this work.

3. The problem of the quantization of the electric charge

It is a experimental fact that electric charge is discrete. The theoretical prediction of this fact has been linked to the existence of magnetic monopoles. So far there is not any evidence of the existence of monopoles, although some modern unified theories of cosmology and fundamental interactions imply the existence of magnetic monopoles.

In the next section we will present a theoretical argument for the quantization of the electric charge where there is not need for the existence of a magnetic charge or quantum mechanics. However, in this section we also present the standard arguments of the electric charge quantization. We advice to consult the bibliography, specially (Jackson, 1998) and (Schwinger et al., 1998) for more details.

3.1 Thomson's calculation of the angular momentum

J. J Thomson considered in (Thomson, 1904) the electromagnetic field of a system consisting in a magnetic pole and an electric charge. He calculated the momentum and the angular momentum of the electromagnetic field. Then, from its conservation he deduced the magnetic part of the Lorentz force.

Let us assume that we have a magnetic pole g at the point A and a electric charge e at the point B both at rest. We have then that at an arbitrary point P of the space the electric field and magnetic fields are given, in natural units, by

$$\mathbf{E} = \frac{e}{4\pi}\frac{\mathbf{r}_1}{r_1^3},$$

$$\mathbf{B} = \frac{g}{4\pi}\frac{\mathbf{r}_2}{r_2^3} \tag{12}$$

where $\mathbf{r}_1 = \mathbf{r}_P - \mathbf{r}_A$ and $\mathbf{r}_2 = \mathbf{r}_P - \mathbf{r}_B$. It is very interesting to see how Thomson assumed in this work that the magnetic field produced by a magnetic pole was of the Coulomb type. In (Thomson, 1904), the author cites Coulomb and Gauss to provide the experimental proof. The fact that he assumed that, in a magnet, the total magnetic charge has to be zero led him to get the right answers.

The linear momentum of the field is

$$\mathbf{P}_f = \int \mathbf{E} \times \mathbf{B}\, d^3r = 0, \tag{13}$$

as the linear momentum field lines are circles with their centres along the line AB and their planes at right angles to it. The angular momentum of the electromagnetic field is defined as

$$\mathbf{L}_f = \int \mathbf{r} \times (\mathbf{E} \times \mathbf{B})\, d^3r. \tag{14}$$

Since the total linear momentum is null, the total angular momentum will be independent of the point chosen to calculate it, according to Classical Mechanics. It will point in the direction of the line AB. To evaluate it, we take origin at the position of the magnetic pole A, the axis z as the line AB, and we have $\mathbf{L}_f = L_z\,\hat{\mathbf{z}}$ with

$$L_z = \frac{eg}{(4\pi)^2}\int \frac{\sin\theta}{r|\mathbf{r} - \mathbf{R}|^2}\,\sin\alpha\, d^3r, \tag{15}$$

where $\mathbf{R} = R\,\hat{\mathbf{z}}$ is the position of the electric charge at B, $\theta = \angle PAB$ and $\alpha = \angle APB$. Using spherical coordinates and the law of sines, it turns out that

$$L_z = \frac{egR}{8\pi}\int_0^\infty \int_0^\pi \frac{r\sin^3\theta}{(r^2 + R^2 - 2Rr\cos\theta)^{3/2}}\, dr\, d\theta. \tag{16}$$

The integral can be calculated by different methods as can be seen in (Adawi, 1976). A change of variables $r = R(\cos\theta + \sin\theta\tan\gamma)$, $\gamma \in [\theta - \pi/2, \pi/2]$ solves the integral and yields $1/2$, so that

$$\mathbf{L}_f = \frac{eg}{4\pi}\frac{\mathbf{R}}{R}. \tag{17}$$

From the conservation of the total linear and angular momenta of the field plus the system of the pole and the charge, Thomson deduces then the magnetic part $e(\mathbf{v} \times \mathbf{B})$ of the Lorentz force over the charge. Note that Jackson follows the converse argument, starting from the Lorentz force between a monopole and a charge, to get the same result.

3.2 The semiclassical quatization rules by Saha and Wilson

Thomson result (17) was used by Saha (Saha, 1949) and independently by Wilson (Wilson, 1949) to get the same quantization condition that Dirac had obtained earlier (we will revise Dirac's argument below). The idea is that, from quantum mechanics, the angular momentum is quantized. Using Saha words, if we apply the *quantum logic*, identifying the angular momentum of the field created by a charge and a monopole with the quantum number for the angular momentum, we get the Dirac result in natural units,

$$eg = 2\pi n, \tag{18}$$

so the existence of a monopole implies the quantization of the charge. For further considerations of the role of the angular momentum and its conservation in the monopole problem, we will refer to the work (Goldhaber, 1965).

3.3 Dirac's argument

Now the turn for the source: Dirac's consideration about the wave function of a particle (Dirac, 1931; 1948). A particle in quantum mechanics is represented by a wave function

$$\psi = Ae^{i\gamma} \tag{19}$$

where A and γ are real functions of $\mathbf{r}$ and t, denoting the amplitude and the phase respectively. The physical meaning of the wave function, according to the quantum postulates, allows for an arbitrary numerical constant coefficient that we can choose to be of modulus unity. So we can add to the phase γ an arbitrary function β. This arbitrary function β does not have to be a unique value in each point $(\mathbf{r}, t)$, as if we go around a closed curve could change, but this change has to be the same for all the wave functions or vary for different wave functions in multiples of 2π, otherwise will have physical consequences such as interference between states. But it has to have definite derivatives as it has to be a solution of a quantum wave equation.

Following Dirac, we will introduce the four vector κ^μ as

$$\kappa_x = \frac{\partial\beta}{\partial x}, \; \kappa_y = \frac{\partial\beta}{\partial y}, \; \kappa_z = \frac{\partial\beta}{\partial z}, \; \kappa_t = \frac{\partial\beta}{\partial t}, \tag{20}$$

and they have to be well defined as stated above. Thus the change in phase round a close curve in the 4-D space, where the vector κ^μ is defined, can be calculated as

$$\oint \kappa_\mu \, ds^\mu = 2\pi n. \tag{21}$$

If we take the close curve very small, the continuity of the wave function imposes the value $n = 0$ for a simple connected domain, as the integration domain reduces to a point. However, if there are points where the wave function vanishes, then the phase would have not meaning. Since the wave function is a complex number, we need two conditions for its vanishing, so we well have in general a nodal line. But now, if we take the closed curve for the integration in (21) around such a line, the continuity considerations are not longer able to tell us that the phase change must be zero. All we can say is that the change will be $2\pi n$ being n an integer, positive or negative depending on the defined orientation.

On the other hand, we can apply Stoke's theorem to the circulation in Equation (21) to write it as

$$\oint \kappa_\mu \, ds^\mu = \int_S (\mathrm{curl}\,\kappa)_j \, dS^j, \tag{22}$$

where the domain is any hypersurface bounded by the closed curve, and the $(\mathrm{curl}\,\kappa)_j$ is a 6-D vector that we can write in three dimensional vector notation as

$$\begin{aligned} \nabla \times \mathbf{k} &= e\,\mathbf{B} \\ \nabla \kappa_0 - \frac{\partial \mathbf{k}}{\partial t} &= e\,\mathbf{E}, \end{aligned} \tag{23}$$

where $\mathbf{k} = (\kappa_x, \kappa_y, \kappa_z)$. We can identify, as the notation in (23) suggests, this curl with an electromagnetic field given by the electromagnetic potentials $(V, \mathbf{A}) = (-\kappa_0, \mathbf{k})/e$. This can be seen clearer calculating the momentum using Equation (19) with the arbitrary phase β,

$$\mathbf{P} = -i\nabla\left(\psi e^{i\beta}\right) = e^{i\beta}(-i\,\nabla\psi + \mathbf{k}) = \mathbf{p} + e\mathbf{A}. \tag{24}$$

The interpretation of the phase curl as an electromagnetic field as far reaching consequences, as Dirac noted. If the close curve is taken in three-dimensional space, only the magnetic flux will come to play so, from Equations (21) and (22), one obtains

$$e\int_S \mathbf{B}\cdot d\mathbf{S} = 2\pi n. \tag{25}$$

So the magnetic flux through any surface bounded by the curve will be equal to the phase shift difference of the wave equation. We have seen that if there is not any nodal line across the surface defined by the curve, the phase difference is equal zero. If we take a closed surface around a nodal line, in the case that the nodal line comes in and out, and that difference should be again zero. But if the nodal line had an end, and we take the close surface around that end, then the phase shift will be nonzero. But that would mean that the there is a net magnetic flux crossing a closed surface, so there is a magnetic charge or monopole inside the surface. The magnetic flux can be written as g, being g the strength of the magnetic pole. Then we get, from Equation (25),

$$eg = 2\pi n, \tag{26}$$

which is the same condition as in Equation (18) that we got with the semiclassical rule. There is a nice account of bibliography related to the monopole problem in Reference (Goldhaber & Trower, 1990).

4. Quantization of the electric charge in the model of electromagnetic knots

A topological mechanism for the quantization of the charge in the model of electromagnetic knots can be seen in Reference (Rañada & Trueba, 1998). Quantization of charge is usually stated by saying that the electric charge of any particle is an integer multiple of a fundamental value e, the electron charge, whose value in the International System of Units is $e = 1.6 \times 10^{-19}$ C. The Gauss theorem allows a different, although fully equivalent, statement of this property:

the electric flux across any closed surface Σ which does not intersect any charge is always an integer multiple of e. This can be written as

$$\int_{\Sigma} \mathbf{E} \cdot \mathbf{n}\, dS = ne, \tag{27}$$

where $\mathbf{n}$ is a unit vector orthogonal to the surface, $\mathbf{E}$ is the electric field and dS the surface element. We could as well write Equation (27) as

$$\int_{\Sigma} *\mathcal{F} = ne, \tag{28}$$

$*\mathcal{F}$ being the dual to the Faraday 2-form $\mathcal{F} = 1/2\, F_{\mu\nu} dx^{\mu} \wedge dx^{\nu}$. Stating in this way the discretization of the charge is interesting because it shows a close similarity with the expression of the topological degree of a map. Assume that we have a regular map θ of Σ on a 2-sphere S^2 and let σ be the normalized area 2-form in S^2. It then happens that

$$\int_{\Sigma} \theta^{*}\sigma = n, \tag{29}$$

$\theta^{*}\sigma$ being the pull-back of σ and n an integer called the degree of the map, which gives the number of times that S^2 is covered when one runs once through Σ (equal to the number of points in Σ in which θ takes any prescribed value). The comparison of Equations (28) and (29) shows that there is a close formal similarity between the dual to the Faraday 2-form and the pull-back of the area 2-form of a sphere S^2.

Suppose that an electromagnetic field is given, such that its form $*\mathcal{F}$ is regular except at the positions of some point charges. Suppose also that we have a map $\theta : R^3 \mapsto S^2$ which is regular except at some point singularities where its level curves converge or diverge. It happens then that Equations (28) and (29) are simultaneously satisfied for all the closed surfaces Σ which do not intersect any charge or singularity. This means that the electric charge will be automatically and topologically quantized in a model in which these two forms $*\mathcal{F}$ and $\theta^{*}\sigma$ are proportional, the fundamental charge being equal to the proportionality coefficient and the number of fundamental charges in a volume having then the meaning of a topological index.

This is exactly what happens in the topological model of electromagnetic knots. In it, the dual to the Faraday 2-form is expressed as

$$*\mathcal{F} = \sqrt{a}\, \theta^{*}\sigma, \tag{30}$$

where a is a normalizing constant, that is a pure number in natural units, or proportional to the product $\hbar c \mu_0$ in the International System of Units. The electric field is then $\mathbf{E} = \sqrt{a}\, c\, (2\pi i)^{-1} (1 + \bar{\theta}\theta)^{-2} \nabla\bar{\theta} \times \nabla\theta$, the electric lines being therefore the level curves of θ. The degree of the map $\Sigma \mapsto S^2$ induced by θ is given by Equation (29). Therefore,

$$\int_{\Sigma} *\mathcal{F} = n\sqrt{a}. \tag{31}$$

As this is equal to the charge Q inside Σ, it does happen that $Q = n\sqrt{a}$, what implies that there is then a fundamental charge $q_0 = \sqrt{a}$, the degree n being the number of fundamental charges inside Σ. This gives a topological interpretation of n, the number of fundamental charges inside any volume.

It is easy to understand that $n = 0$ if θ is regular in the interior of Σ. This is because each level curve of θ, i. e. each electric line, is labeled by its value along it (a complex number) and, in the regular case, any one of these lines enters into this interior as many times as it goes out of it. But assume that θ has a singularity at point P, from which the electric lines diverge or to which they converge. If Σ is a sphere around P, we can identify R^3 except P with $\Sigma \times R$, so that the induced map $\theta : \Sigma \mapsto S^2$ is regular. In this case, n need not vanish and is equal to the number of times that θ takes any prescribed complex value in Σ, with due account to the orientation. Otherwise stated, among the electric lines diverging from or converging to P, there are $|n|$ whose label is equal to any prescribed complex number.

To understand better this mechanism of discretization, let us take the case of a Coulomb potential as in Reference (Rañada & Trueba, 1997): $\mathbf{E} = Q\mathbf{r}/(4\pi r^3)$, $\mathbf{B} = 0$. The corresponding scalar is

$$\theta = \tan\left(\frac{\vartheta}{2}\right) \exp\left(i\frac{Q}{\sqrt{a}}\varphi\right), \tag{32}$$

where φ and ϑ are the azimuth and the polar angle. The scalar (32) is well defined only if $Q = n\sqrt{a}$, n being an integer. The lines diverging from the charge are labeled by the corresponding value of θ, so that there are $|n|$ lines going in or out of the singularity and having any prescribed complex number as their label. If $n = 1$, it turns out that $\theta = (x + iy)/(z + r)$.

This mechanism has a very curious aspect: it does not apply to the source but to the electromagnetic field itself. This is surprising since one would expect that the topology should operate restricting the fields of the charged particles. However, in this model, it is the field who mediates the force the one which is submitted to a topological condition. It must be emphasized furthermore that the maps $S^3 \mapsto S^2$, given by the two scalars ϕ, θ are regular except for singularities at the position of point charges, either electrical or magnetic (if the latter do exist). At these points, the level curves (the electric lines) converge or diverge.

In the case that the value of a in natural units is $a = 1$ (in order to obtain the right quantization of the electromagnetic helicity), the topological model of electromagnetic knots predicts that the fundamental charge has the value

$$q_0 = 1, \tag{33}$$

which is about 3.3 times the electron charge. Note that this applies both to the electron charge and to the hypothetical monopole charge. This property can be stated saying that, in the topological model, the electromagnetic fields can only be coupled to point charges which are integer multiple of the fundamental charge $q_0 = 1$. Note that the same discretization mechanism would apply to the hypothetical magnetic charges (located at singularities of ϕ), their fundamental value being also $q_0 = 1$.

5. Quantization of the magnetic flux in the model of electromagnetic knots

Electromagnetic knots are compatible with the quantization of the magnetic flux of a superconducting ring, which in standard theory is always an integer multiple of $g/2$, where $g = \sqrt{a}$ (or $g = 1$ in natural units) is the value of the magnetic monopole in the topological model of electromagnetic knots. The mechanism of quantization was stablished in Reference (Rañada & Trueba, 2006). To understand how this mechanism of quantization works, let us begin with the case of an infinite solenoid.

5.1 Flux quantization in an infinite solenoid

Consider again the equations for any electromagnetic knot. The Faraday 2-form and its dual generated by the pair of complex scalar fields ϕ and θ can be written, with the constant a fixed to $a = 1$ in natural units, as

$$\begin{aligned} \mathcal{F} &= ds \wedge dp, \quad \text{with } p = 1/(1+|\phi|^2),\ s = \arg(\phi)/2\pi \\ {}^*\mathcal{F} &= dv \wedge du, \quad \text{with } v = 1/(1+|\theta|^2),\ u = \arg(\theta)/2\pi, \end{aligned} \tag{34}$$

so that $\phi = \sqrt{(1-p)/p}\, e^{i2\pi s}$ and $\theta = \sqrt{(1-v)/v}\, e^{i2\pi u}$. This implies that the magnetic and electric fields have the form

$$\begin{aligned} \mathbf{B} &= \nabla p \times \nabla s = (\partial_0 u \nabla v - \partial_0 v \nabla u)\,, \\ \mathbf{E} &= \nabla u \times \nabla v = (\partial_0 s \nabla p - \partial_0 p \nabla s)\,. \end{aligned} \tag{35}$$

The quantities (p, s) and (v, u) are called *Clebsch variables* of the fields **B** and **E**, respectively (or of the scalars ϕ and θ). Note that ϕ and θ are not uniquely determined by the magnetic and electric fields. Indeed, a different pair defines the same fields **E**, **B** if the corresponding Clebsch variables (P, S), (V, U) can be obtained through a canonical transformation $(p, s) \to (P, S)$ or $(v, u) \to (V, U)$. However, the canonical transformation must satisfy two conditions: (i) $0 \le P, V \le 1$, and (ii) S, U must be arguments of complex numbers in units of 2π, i. e. they can be multivalued but their change along a closed curve must be an integer.

Let us turn to our physical problem. Consider an infinite perfect solenoid around the z-axis with N turns per unit length and intensity I (perfect means that no flux escapes through the coils). This can happen exactly only in a superconducting ring. Indeed, from the purposes of the present study, perfect solenoids and superconducting rings can be considered synonymous. The magnetic field vanishes outside and is equal to $B = \mu_0 N I$ inside. Now let us ask what can be the scalar ϕ (which gives a map $S^3 \mapsto S^2$) that corresponds to that magnetic field if we restrict ourselves to the model of electromagnetic knots. With the configuration of the magnetic lines of that solenoid, it is impossible that ϕ be regular in all the sphere S^3. However, we may consider the 3-space as $S^2 \times R$ and require that ϕ be regular in the induced map $S^2 \mapsto S^2$, the first S^2 being the plane (x, y), the second the complex plane, both completed with the point at infinity. If $\phi = |\phi| \exp(2\pi i s)$ and $p = 1/(1+|\phi|^2)$, then

$$\mathbf{B} = \nabla p \times \nabla s. \tag{36}$$

As $B = 0$ outside the solenoid, p and s can not be independent functions there. This may happen in three ways:

1. The first possibility is $s = f(p)$, f being a nontrivial function. We can change s to $s - f(p)$. This is a canonical transformation of the variables (s, p) which does not affect the value of B in view of Equation (36). The new expression of ϕ is real outside the solenoid, but not inside in general. Consequently, the magnetic flux across a section of the solenoid is

topologically quantized, being equal to the area of the set $\phi(S)$ in the sphere S^2, where S is any surface that cuts the solenoid and is bordered by a circuit outside it. Indeed its value is necessarily Flux $= n/2$, because any curve contained in a great circle of a sphere encircles a integer multiple of semispheres.

2. The second possibility is $s = s_0 =$ constant. The situation is similar to and gives the same flux quantization as in the previous case (outside the solenoid, ϕ takes values also in a great circle of S^2). That is Flux $= n/2$.

3. The last possibility is $p = p_0 =$ constant. Let $p = p_0$ outside and s variable. Then the scalar would be

$$\phi = \sqrt{\frac{1-p_0}{p_0}}\, e^{i2\pi s(r,\varphi)}, \tag{37}$$

where $r = (x^2 + y^2)^{1/2}$ and φ is the azimuth. Moreover,

$$\int_0^{2\pi} \frac{\partial s}{\partial \varphi}\, d\varphi = m, \tag{38}$$

where m is an integer number. In order for ϕ to be a regular map, there are two possibilities: $s = s_0 =$ constant, and $s =$ function of φ but with either p_0 (and $\phi = \infty$) or $p_0 = 1$ (and $\phi = 0$). In both cases, it turns out that Flux $= n/2$.

So, in conclusion, in the topological model of electromagnetism based on electromagnetic knots, the magnetic flux in an infinite perfect solenoid is always an semi-integer multiple of the fundamental magnetic charge q_0 (with $q_0 = 1$ in natural units),

$$\text{Flux} = \frac{n}{2}\, q_0. \tag{39}$$

This is interesting, because it says that the flux in the solenoid is necessarily quantized, the fundamental fluxoid being half the fundamental magnetic charge q_0 (as the real fluxoid is half the Dirac monopole). This quantity, however, is $q_0/2 = 1/2$ in natural units, as compared with $g/2 = 10.37$ for the Dirac monopole.

5.2 Flux quantization in a finite solenoid

Let us consider now the case of a superconducting ring, i. e. of a perfect but finite solenoid. We can imagine it as a cylinder around de z-axis between $z = -L/2$ and $z = L/2$ and the radii r_0 and $r_0 + h$, although these magnitudes are quite irrelevant in this case. Since the magnetic field does not enter inside the superconductor, $B = 0$ inside it. If the superconductor is infinitely think (i.e. $h = \infty$), the topology of the problem is the same as in the previous case of infinite solenoid, and all the results are also the same. In the realistic case in which h is finite, there are also three cases. In the two first cases, the result would be the same. However, it is not clear that the same could be said of the third possibility. We have to take a different way for the third possibility.

Consider the quantization of the magnetic flux across a superconducting ring in the standard theory (Feynman et al., 1965). In this case the wave function can be treated as a classical macroscopic field $\psi = \sqrt{\rho}\, e^{i\vartheta}$, the following equation being satisfied

$$\hbar \nabla \vartheta = Q\mathbf{A}, \tag{40}$$

where Q is the charge of a Cooper pair of electrons, equal to $2e$. The flux is thus

$$\text{Flux} = \oint \mathbf{A} \cdot d\mathbf{s} = \frac{2\pi n'}{Q}, \tag{41}$$

where n' is an integer. We see that the fundamental unit of flux is then $2\pi/Q$.

Let us take a finite superconducting ring of cylindrical shape, with axis along the z-axis, between the planes $z = \pm L/2$ and radii r_0 and $r_0 + h$. The interior magnetic field created by the superconducting ring at the central plane $z = 0$ can be written as

$$\mathbf{B} = B(r)\,\hat{\mathbf{z}}, \tag{42}$$

r being the radial coordinate. The magnetic flux across the ring is

$$\text{Flux} = \int_{C_0} B(r)\, r dr d\varphi, \tag{43}$$

where C_0 is the circle of radius r_0. Because of the symmetry of the problem, we can take a scalar $\phi(r, \varphi)$, with $p = 1/(1 + |\Phi|^2)$ and $s = \arg(\Phi)/2\pi$, such that

$$\mathbf{B} = \frac{1}{r}\left(\frac{\partial p}{\partial r}\frac{\partial s}{\partial \varphi} - \frac{\partial p}{\partial \varphi}\frac{\partial s}{\partial r}\right)\hat{\mathbf{z}}. \tag{44}$$

It is convenient to define the dimensionless radial coordinate

$$R = \frac{r}{r_0}, \tag{45}$$

so that, in each plane (R, φ), ϕ can be taken as a map $\phi : C_1 \to S^2$, where C_1 is the circle of unit radius, and

$$\mathbf{B} = \frac{1}{r_0^2}\frac{1}{R}\left(\frac{\partial p}{\partial R}\frac{\partial s}{\partial \varphi} - \frac{\partial s}{\partial \varphi}\frac{\partial s}{\partial R}\right)\hat{\mathbf{z}}. \tag{46}$$

The magnetic flux across the superconductor results

$$\text{Flux} = \int_{C_1}\left(\frac{\partial p}{\partial R}\frac{\partial s}{\partial \varphi} - \frac{\partial p}{\partial \varphi}\frac{\partial s}{\partial R}\right) dR d\varphi. \tag{47}$$

The quantity between brackets in (47) is the Jacobian of the change of variables $(p, s) \to (R, \varphi)$, so that

$$\text{Flux} = \int_{\phi(C_1)} dp ds, \tag{48}$$

where $\phi(C_1)$ is the image in S^2 of the unit circle C_1.

In the framework of London's theory of type II superconductors (the case in which the magnetic flux is quantized), the magnetic field in the superconductor satisfies a phenomenological equation in the transition layer in which the magnetic field goes to zero.

This is the second London equation,

$$\mathbf{A} = -\lambda^2 \nabla \times \mathbf{B}, \tag{49}$$

where $\mathbf{A}(\mathbf{r})$ obeys Coulomb gauge and λ is the penetration length of the magnetic field inside the superconductor material (in practice, λ is about ten Angstroms, much shorter than the inner radius of the superconductor ring r_0).

From Equation (44), the vector potential $\mathbf{A}(\mathbf{r})$ for the magnetic field $\mathbf{B}(\mathbf{r})$ in the Coulomb gauge ($\nabla \cdot \mathbf{A} = 0$) can be written as

$$\mathbf{A} = \frac{p}{r} \frac{ds}{d\varphi} \mathbf{u}_{\varphi}. \tag{50}$$

It follows that $s = s(\varphi)$ and $p = p(r)$. Furthermore, the quantity $\int_0^{2\pi} A_{\varphi} r \, d\varphi$ has to be independent of r inside the superconductor. From these considerations one obtains

$$p = p_0, \; s = n \frac{\varphi}{2\pi}. \tag{51}$$

Inserting Equation (50) into London Equation (49), we obtain the following ordinary differential equation for $p(r)$,

$$\lambda^2 \left(\frac{d^2 p}{dr^2} - \frac{1}{r} \frac{dp}{dr} \right) - p = 0. \tag{52}$$

Up to first order in λ / r_0, we can neglect the first term in (52) to obtain

$$p(r) = 0, \; r \geq r_0, \tag{53}$$

characterizing p inside the superconductor. As the Clebsch variable p has to be continuous and constant inside the superconductor, with a value $p = p_0$, we obtain $p_0 = 0$, i.e. $\phi = \infty$. In the model of electromagnetic knots, if an electromanetic field is generated by the scalar field ϕ and the Clebsch variables (p, s), it is also generated by the scalar $1/\bar{\phi}$ and the Clebsch variables $(1 - p, -s)$. In the latter case, Equation (53) would be $1 - p(r) = 0$, $r \geq r_0$, so that $p_0 = 1$ and $\phi = 0$ inside the superconductor.

Consequently, the value of the scalar field ϕ inside the superconductor is $\phi = \infty$ or $\phi = 0$. In both cases, the magnetic flux is

$$\text{Flux} = \int_{C_1} A(R) \, r_0 R \, d\varphi = \int_0^{2\pi} \frac{n}{2\pi} d\varphi = n. \tag{54}$$

If we consider the solutions given by the families 1 and 2 at the begining of this subsection, it results that the magnetic flux is quantized, being always an integer multiple of $1/2$.

The previous argument relies on London's equation. However, the same conclusion can be reached considering the following. The radial derivative of p is in general discontinuous at r_0. However, this irregularity in the map ϕ is vanished if either $\phi = 0$ or $\phi = \infty$ inside the superconductor. Therefore, the requirement that the map is regular leads to the topological quantization of the flux, without taking into account the London equation.

5.3 The fine structure constant at infinite energy equal to $1/4\pi$?

Because the topological model here presented is classical, the electric charge $q_0 = \sqrt{\hbar c \epsilon_0}$ must be interpreted as the fundamental bare charge, both electric and magnetic (remember that we are using natural units). The corresponding fine-structure constant $\alpha = q_0^2/4\pi\hbar c\epsilon_0$ is clearly equal to $\alpha_0 = 1/4\pi$, which is certainly a nice and simple number. We will argue now that $1/4\pi$ is an appealing and interesting value for the non-renormalized fine-structure constant (i. e. neglecting the effect of the quantum vacuum). As we show now, the topological model seems to describe the electromagnetic field at infinite energy.

The argument goes as follows. Let us combine this topological quantization of the charge with the appealing and plausible idea that, in the limit of very high energies, the interactions of charged particles could be determined by their bare charges, i. e. the values of that their charges would have if they were not renormalized by the quantum vacuum; see e. g. Section 11.8 of Reference (Milonni, 1994). A warning is necessary, however. As the concept of bare charge is not simple, it is convenient to speak instead of charge at a certain scale. To avoid confusion and be precise, the expression "bare charge" will be used here as synonymous or equivalent of "infinite energy limit of the charge" or, more correctly, "charge at infinite momentum transfer Q", defined as $e_\infty = \sqrt{4\pi\hbar c\epsilon_0\alpha_\infty}$, where $\alpha_\infty = \lim \alpha(Q^2)$ when $Q^2 \to \infty$.

The possibility of a finite value for α_∞ is an interesting idea worth of consideration. In fact, it was discusseed by Gell-Mann and Low in their classical and seminal paper "QED at small distances", Reference (Gell-Mann & Low, 1954), in which they showed that it is something to be seriously studied. However, they could not decide from their analysis whether e_∞ is finite or infinite. The current wisdom idea that it is infinite was established later on the basis of perturbative calculations, but the alternative posed by Gell-Mann and Low has not been really settled. It is still open.

The infinite energy charge e_∞ of an electron is partially screened by the sea of virtual pairs that are continuously being created and destroyed in empty space. It is hence said that it is renormalized. Because the pairs are polarized, as are the molecules in a dielectric, a polarization cloud is formed around any charged particle, with the result that the observed value of the electron charge is smaller than e_∞. Moreover, the apparent electron charge increases as any probe goes deeper into the polarization cloud and is therefore less screened. This effect is difficult to measure, since it can be appreciated only at extremely short distances. However it has been observed in experiments of electron-positron scattering at high energies Reference (Levine *et al.*, 1997). This means that the vacuum is dielectric. On the other hand, it is paramagnetic because the effect of the magnetic field is due to the spin of the pairs. The consequence is that the hypothetical magnetic charge would be observed with a greater value at low energy than at very high energy, contrary to the electron charge.

It is easy to understand the reason for the expression "bare charge" to denote e_∞. When two electrons interact with very high momentum transfer, each one is located so deeply inside the polarization cloud around the other that very little space is left between them to screen the charges, so that the bare charges, namely e_∞, interact directly. As unification is is assumed to happen at very high energy, it is an appealing idea that $\alpha_\infty = \alpha_{GUT}$ (GUT stands for "Grand Unified Theories", that include weak and gravitation ones. This suggests that a unified theory could be a theory of bare particles (i. e. in the sense that it neglects the effect of the vacuum.)

If this were the case, nature would have provided us with a natural cut-off, $\alpha_{\text{GUT}} = \alpha_\infty$. As a consequence, it can be argued that the topological model implies that $\alpha_{\text{GUT}} = \alpha_\infty = 1/4\pi$. The argument goes as follows.

1. The value of the fundamental charge predicted by this topological quantization, $e_0 = \sqrt{\hbar c \epsilon_0} = 5.28 \times 10^{-19}$ C is in the right interval to verify $e_0 = e_\infty = g_\infty$, in other words to be equal to the common value of both the fundamental electric and magnetic bare charges. This is so because, as the quantum vacuum is dielectric but paramagnetic, the following inequalities must be satisfied: $e < e_0 < g$, as it happens since $e = 0.3028$, $e_0 = 1$, $g = e/2\alpha = 20.75$, in natural units. Note that it is impossible to have a completely symmetry between electricity and magnetism simultaneously at low and high energies. The lack of symmetry between the charges of the electron and the Dirac monopole would be due to the vacuum polarization: according to the topological model, the electric and magnetic infinite energy charges are equal and verify $e_\infty g_\infty = e_0^2 = 1$, but they would be decreased and increased, respectively, by the sea of virtual pairs, until their current values that verify $eg = 2\pi$. This qualitative picture seems nice and appealing.

2. Let us admit as a working hypothesis that two charged particles interact with their bare charges at high energies. There could be then a conflict between (a) a unified theory of electroweak and strong forces in which $\alpha = \alpha_s$ and (b) an infinite value of α_∞. The reason is that unification implies that the curves of the running constants $\alpha(Q^2)$ and $\alpha_s(Q^2$ must converge asymptotically to the same value α_{GUT}. It could be argued that, to have unification at a certain scale, it would suffice that these two curves be close in an energy interval, even if they cross and separate afterwards. However in that case the unified theory would be just an approximate accident. On the other hand, the assumption that both running constants go asymptotically to the same finite value gives a much deeper meaning to the idea of unified theory. In that case, e_∞ must expected to be finite, and the equality $\alpha_{\text{GUT}} = \alpha_\infty$ must be satisfied.

3. The value $\alpha_0 = e_0^2/4\pi\hbar c\epsilon_0 = 1/4\pi = 0.0796$ for the infinite energy fine-structure constant is thought-provoking and fitting, since α_{GUT} is believed to be in the interval $(0.05, 0.1)$. This reaffirms the assertion that the fundamental value of the charge given by this topological mechanism e_0 could be equal to e_∞, the infinite energy electron charge (and the infinite energy monopole charge, as well). It also supports the statement that $\alpha_{\text{GUT}} = \alpha_0 = 1/4\pi$. All this is certainly curious and intriguing: indeed, the topological mechanism for the quantization of the charge here described is obtained simply by putting some topology in elementary classical low energy electrodynamics.

We believe, therefore, that the following three ideas must be studies carefully: (1) the complete symmetry between electricity and magnetism at the level of the infinite energy charges, where both are equal to $\sqrt{\hbar c \epsilon_0}$, this symmetry being broken by the dielectric and paramagnetic quantum vacuum; (2) That the topological model on which the topological mechanism of quantization is based could give a theory of high- energy electromagnetism at the unification scale and (3) that the value that this model predicts for the fine-structure constant $\alpha_0 = 1/4\pi$ could be equal to the infinite energy limit α_∞ and also to α_{GUT}, the constant of the unified theory of strong and electroweak interactions.

6. Conclusions

The topological model of electromagnetism constructed with electromagnetic knots is based on the existence of a topological structure which underlies the Maxwell's standard theory, in such a way that the Maxwell's equations in empty space are the exact linearization of some nonlinear equations with topological properties and constants of the motion. Although the model is classical, it embodies the topological quantizations of the helicity and the energy inside a cavity, which suggest that it offers a way to understand better the relation between the classical and quantum aspects of the electromagnetic theory. The model is locally equivalent to Maxwell's standard theory in empty space (but globally non-equivalent). This means that it can not enter in conflict with Maxwell's theory in experiments of local nature.

In the model of electromagnetic knots, the electric charge which is topologically quantized, its fundamental value being $q_0 = 1$ in natural units (or $q_0 = \sqrt{\hbar c \epsilon_0} = 5.28 \times 10^{-19}$ C in the International System of Units). Furthermore, the number of fundamental charges inside a volume is equal to the degree of a map between two spheres. It turns out that there are exactly $|m|$ electric lines going out or coming into a point charge $q = mq_0$, for which a complex scalar field is equal to any prescribed complex number (taking into account the orientation of the map).

The topological model is completely symmetric between electricity and magnetism, in the sense that it predicts that the fundamental hypothetical magnetic charge would be also q_0. Note that $q_0 = 3.3\,e$, where e is the electron charge, and that the corresponding fine structure constant is $\alpha_0 = 1/4\pi$. Hence, q_0 could be interpreted as the bare electron and monopole charge. As the quantum vacuum is dielectric but paramagnetic, the observed electric charge must be smaller than q_0, but the Dirac magnetic monopole must be greater (it is equal to $20.75\,q_0$). This suggests that α_0 could be the fine structure constant at infinite energy and, consequently, that the coupling constant of the Grand Unified Theory could be also $\alpha_s = \alpha_0 = 1/4\pi$.

The model of electromagnetic knots also predicts that the magnetic flux is quantized, the fundamental flux unit being $1/2$ in natural units. Consequently, the relation between the fundamental magnetic flux and electric charge in this model is the same as that between the Dirac monopole and the electron charge in standard theory. The quantum vacuum increases the value of the magnetic fields by a factor $2\pi/e$, according to the above mentioned interpretation. This can be represented by a relative permeability $\mu_r = 2\pi/e = 20.75$ (with respect to the state in which there is neither matter nor radiation and the effect of the zero point radiation has been discounted). The renormalized magnetic flux must be therefore equal to $\mu_r \times \text{Flux}$, where $\text{Flux} = 1/2$ is the bare value. This implies that the flux is a multiple integer of π/e, either in standard theory or in the topological theory after multiplying by the permeability μ_r to take care of the effect of the quantum vacuum. Hence, the topological quantization of the magnetic flux coincides with the standard one after introducing a relative permeability to account for the effect of the quantum vacuum. This is fully coherent with the interpretation given in the previous paragraph that the topological model of electromagnetic knots gives a theory of high energy electromagnetism at the unification scale or a theory of bare electromagnetism.

7. References

Adawi, I. (1976). Thomson's monopoles. *American Journal of Physics*, Vol. 44, (762-765).

Arrayás, M. & Trueba, J. L. (2010). Motion of charged particles in a knotted electromagnetic field. *Journal of Physics A: Mathematical and Theoretical*, Vol. 43, (235-401).

Arrayás, M. & Trueba, J. L. (2011). Exchange of helicity in a knotted electromagnetic field. *Annalen der Physik*, doi: 10.1002/andp.201100119.

Dirac, P. A. M. (1931). Quantised singularities in the electromagnetic field. *Proceedings of the Royal Society of London A*, Vol. 133, (60-72).

Dirac, P. A. M. (1948). The theory of magnetic poles. *Physical Review*, Vol. 74, (817-830).

Feynman, R. P.; Leighton, R. B. & Sands, M. (1965). *The Feynman lectures on Physics Vol. 2*, Addison-Wesley, Reading.

Gell-Mann, M. & Low, F. (1954). QED at small distances. *Physical Review*, Vol. 95, (1300-1312).

Goldhaber, A. S. (1965). Role of spin in the monopole problem. *Physical Review B*, Vol. 140, (1407-1414).

Goldhaber, A. S. & Trower, W. P. (1990). Magnetic monopoles. *American Journal of Physics*, Vol. 58, (429-439).

Hehl, F. W. & Obukhov, Y. N. (2003). *Foundations of Classical Electrodynamics: Charge, Flux and Metric*, Birkhauser, Boston.

Irvine, W. T. & Bouwmeester, D. (2008). Linked and knotted beams of light. *Nature Physics*, Vol. 4, (716-720).

Jackson, J. D. (1998). *Classical Electrodynamics*, John Wiley and Sons, 3rd Edition, New York.

Levine, I. *et al.* (TOPAZ Collaboration) *Physical Review Letters*, Vol 78, (424-427).

Milonni, P. W. (1994), *The Quantum Vacuum. An Introduction to Quantum Electrodynamics*, Academic, Boston.

Rañada, A. F. (1989). A topological theory of the electromagnetic field. *Letters in Mathematical Physics*, Vol. 18, (97-106).

Rañada, A. F. (1990). Knotted solutions of Maxwell equations in vacuum. *Journal of Physics A: Mathematical and General*, Vol. 23, (L815-820).

Rañada, A. F. (1992). Topological electromagnetism. *Journal of Physics A: Mathematical and General*, Vol. 25, (1621-1641).

Rañada, A. F. & Trueba, J. L. (1995). Electromagnetic knots. *Physics Letters A*, Vol. 202, (337-342).

Rañada, A. F. & Trueba, J. L. (1997). Two properties of electromagnetic knots. *Physics Letters A*, Vol. 235, (25-33).

Rañada A. F. & Trueba, J. L. (1998). A topological mechanism of discretization for the electric charge. *Physics Letters B*, Vol. 422, (196-200).

Rañada, A. F. & Trueba, J. L. (2001). Topological electromagnetism with hidden nonlinearity. In: *Modern Nonlinear Optics, Part 3*, Edited by M. W. Evans, John Wiley & Sons, New York. (197-253).

Rañada, A. F. (2002). Is the fine structure constant at the unification scales equal to $1/4\pi$? *Annales de la Fondation Louis de Broglie* (Paris), Vol. 27, (505-510).

Rañada, A. F. (2003). Interplay of topology and quantization: topological energy quantization in a cavity. *Physics Letters A*, Vol. 310, (434-444).

Rañada A. F. & Trueba, J. L. (2006). Topological quantization of the magnetic flux. *Foundations of Physics*, Vol. 36, (427-436).

Saha, M. N. (1949). Note on Dirac's theory of magnetic monopoles. *Physical Review*, Vol. 75, (1968-1968).

Schwinger, J.; Deraad Jr., L. L.; Milton, K. A.; Tsai, W. Y. & Norton, J. (1998). *Classical Electrodynamics*, Westview Press, Cambridge.

Thomson, J. J. (1904). *Elements of the Mathematical Theory of Electricity and Magnetism*, Cambridge University Press, Cambridge.

Wilson, H. A. (1949). Note on Dirac's theory of magnetic poles. *Physical Review*, Vol. 75, (309-309).

6

Waveguides, Resonant Cavities, Optical Fibers and Their Quantum Counterparts

Victor Barsan
Department of Theoretical Physics, National Institute of Physics and Nuclear Engineering, Bucharest, Magurele
Romania

1. Introduction

In this chapter, we shall expose several analogies between oscillatory phenomena in mecahanics and optics. The main subject will be the analogy between propagation of electromagnetic waves in dielectrics and of electrons in various time-independent potentials. The basis of this analogy is the fact that both wave equations for electromagnetic monoenergetic waves (i.e. with well-defined frequency), obtained directly from the Maxwell equations, and the time-independent Schrodinger equation are Helmholtz equations; when specific restrictions - like behaviour at infinity and boundary conditions - are imposed, they generate similar eigenvalues problems, with similar solutions. The benefit of such analogies is twofold. First, it could help a researcher, specialized in a specific field, to better understand a new one. For instance, they might efficiently explain the fiber-optics properties to people already familiar with quantum mechanics. Also, even if such researchers work frequently with the quantum mechanical wave function, the electromagnetic modal field may provide an interesting vizualization of quantum probability density field [1]. Second, it provides the opportunity of cross-fertilization between (for instance) electromagnetism and optoelectronics, through the development of ballistic electron optics in two dimensional (2D) electron systems (2DESs); transferring concepts, models of devices and experiments from one field to another stimulate the progress in both domains.

Even if in modern times the analogies are not credited as most creative approaches in physics, in the early days of developement of science the perception was quite different. "Men's labour ... should be turned to the investigation and observation of the resemblances and analogies of things... for these it is which detect the unity of nature, and lay the fundation for the constitution of the sciences.", considers Francis Bacon, quoted by [1]. Some two centuries later, Goethe was looking for the "ultimate fact" - the Urphänomenon - specific to every scientific discipline, from botanics to optics [2], and, in this investigation, attributed to analogies a central role. However, if analogies cannot be considered anymore as central for the scientific investigation, thay could still be pedagogically useful and also inspiring for active scientific research.

Let us describe now shortly the structure of this chapter. The next two sections are devoted to a general description of the two fields to be hereafter investigated: metallic and dielectric waveguides (Section 2) and 2DESs (Section 3). The importance of these topics for the development of optical fibers, integrated optics, optoelectronics, transport phenomena in mesoscopic and nanoscopic systems, is explained. In Section 4, some very general considerations about the physical basis of analogies between mechanical (classical or quantum) and electromagnetic phenomena, are outlined. Starting from the main experimental laws of electromagnetism, the Maxwell's equations are introduced in Section 5. In the next one, the propagation of electromagnetic waves in metalic and dielectric structures is studied, and the transverse solutions for the electric and magnetic field are obtained. These results are applied to metalic waveguides and cavities in Section 7. The optical fibers are described in Section 8, and the behaviour of fields, including the modes in circular fibers, are presented. Although the analogy between wave guide- and quantum mechanical- problems is treated in a huge number of references, the subject is rarely discussed in full detail. This is why, in Section 9, the analogy between the three-layer slab optical waveguide and the quantum rectangular well is mirrored and analyzed with utmost attention. The last part of the chapter is devoted to transport phenomena in 2DESs and their electromagnetic counterpart. In Section 10, the theoretical description of ballistic electrons is sketched, and, in Section 11, the transverse modes in electronic waveguides are desctibed. A rigorous form of the effective mass approach for electrons in semiconductors is presented in Section 12, and a quantitative analogy between the electronic wave function and the electric or magnetic field is established. Section 13 is devoted to optics experiments made with ballistic electrons. Final coments and conclusions are exposed in Section 14.

2. Metallic and dielectric waveguides; optical fibers

Propagation of electromagnetic waves through metallic or dielectric structures, having dimensions of the order of their wavelength, is a subject of great interest for applied physics. The only practical way of generating and transmitting radio waves on a well-defined trajectory involves such metallic structures [3]. For much shorter wavelengths, i.e. for infrared radiation and light, the propagation through dielectric waveguides has produced, with the creation of optical fibers, a huge revolution in telecommunications. The main inventor of the optical fiber, C. Kao, received the 2009 Nobel Prize in Physics (together with W. S. Boyle and G. E. Smith). As one of the laurees remarks, "it is not often that the awards is given for work in applied science". [4]

The creation of optical fibers has its origins in the efforts of improving the capabilities of the existing (at the level of early '60s) communication infrastructure, with a focus on the use of microwave transmission systems. The development of lasers (the first laser was produced in May 16, 1960, by Theodore Maiman) made clear that the coherent light can be an information carrier with 5-6 orders of magnitude more performant than the microwaves, as one can easily see just comparing the frequencies of the two radiations. In a seminal paper, Kao and Hockham [5] recognized that the key issue in producing "a successful fiber waveguide depends... on the availability of suitable low-loss dielectric material", in fact - of a glass with very small $\left(< 10^{-6}\right)$ concentration of impurities, particularly of transition elements (Fe, Cu, Mn). Besides telecommunication applications, an appropriate bundle of optical fibers can transfer an image - as scientists sudying the insect eye realized, also in the early '60s. [6]

Another domain of great interest which came to being with the development of dielectric waveguides and with the progress of thin-film technology is the integrated optics. In the early '70s, thin films dielectric waveguides have been used as the basic element of all the components of an optical circuit, including lasers, modulators, detectors, prisms, lenses, polarizers and couplers [7]. The transmission of light between two optical components became a problem of interconnecting of two waveguides. So, the traditional optical circuit, composed of separate devices, carefully arranged on a rigid support, and protected against mechanical, thermal or atmospheric perturbations, has been replaced with a common substrate where all the thin-film optical components are deposited [7].

3. 2DESs and ballistic electrons

Electronic transport in conducting solids is generally diffusive. Its flow follows the gradient in the electrochemical potential, constricted by the physical or electrostatic edges of the specimen or device. So, the mean free path of electrons is very short compared to the dimension of the specimen. [8] One of the macroscopic consequences of this behaviour is the fact that the conductance of a rectangular 2D conductor is directly proportional to its width (W) and inversely proportional to its length (L). Does this ohmic behaviour remain correct for arbitrary small dimensions of the conductor? It is quite natural to expect that, if the mean free path of electrons is comparable to W or L - conditions which define the ballistic regime of electrons - the situation should change. Although the first experiments with ballistic electrons in metals have been done by Sharvin and co-workers in the mid '60s [9] and Tsoi and co-workers in the mid '70s [10], the most suitable system for the study of ballistic electrons is the two-dimensional electron system (2DES) obtained in semiconductors, mainly in the $GaAs - Al_xGa_{1-x}As$ heterostructures, in early '80s. In such 2DESs, the mobility of electrons are very high, and the ballistic regime can be easily obtained. The discovery of quantum conductance is only one achievement of this domain of mesoscopic physics, which shows how deep is the non-ohmic behaviour of electrical conduction in mesoscopic systems. In the ballistic regime, the electrons can be described by a quite simple Schrodinger equation, and electron beams can be controlled via electric or magnetic fields. A new field of research, the classical ballistic electron optics in 2DESs, has emerged in this way. At low temperatures and low bias, the current is carried only by electrons at the Fermi level, so manipulating with such electrons is similar to doing optical experiments with a monochromatic source [11].

The propagation of ballistic electrons in mesoscopic conductors has many similarities with electromagnetic wave propagation in waveguides, and the ballistic electron optics opened a new domain of micro- or nano-electronics. The revealing of analogies between ballistic electrons and guided electromagnetic waves, or between optics and electric field manipulation of electron beams, are not only useful theoretical exercises, but also have a creative potential, stimulating the transfer of knowledge and of experimental techniques from one domain to another.

4. Mechanical and electrical oscillations

It is useful to begin the discussion of the analogies presented in this chapter with some very general considerations [12]. The most natural starting point is probably the comparison between the mechanical equation of motion of a mechanical oscillator having the mass m and the stiffness k:

$$m\frac{d^2x}{dt^2} + kx^2 = 0 \tag{1}$$

and the electrectromagnetical equation of motion of a LC circuit [12]:

$$L\frac{d^2q}{dt^2} + \frac{q}{C} = 0 \tag{2}$$

which provides immediately an analogy between the mechanical energy:

$$\frac{1}{2}m\left(\frac{dx}{dt}\right)^2 + \frac{1}{2}kx^2 = \mathcal{E} \tag{3}$$

and the electromagnetic one:

$$\frac{1}{2}L\left(\frac{dq}{dt}\right)^2 + \frac{1}{2}\frac{q^2}{C} = \mathcal{E} \tag{4}$$

The analogy between these equations reveals a much deeper fact than a simple terminological dictionary of mechanical and electromagnetic terms: it shows the inertial properties of the magnetic field, fully expressed by Lenz's law. Actually, magnetic field inertia (defined by the inductance L) controls the rate of change of current for a given voltage in a circit, in exactly the same way as the inertial mass controls the change of velocity for a given force. Magnetic inertial or inductive behaviour arises from the tendency of the magnetic flux threading a circuit to remain constant, and reaction to any change in its value generates a voltage and hence a current which flows to oppose the change of the flux. ([5], p.12) Even if, in the previous equations, the mechanical oscillator is a classical one, its deep connections with its quantum counterpart are wellknown ([13], vol.1, Ch. 12). Also, understanding of classical waves propagation was decissive for the formulation of quantum-wave theory [13], so the classical form of (1) and (3) is not an obstacle in the development of our arguments.

These basic remarks explain the similarities between the propagation of elastic and mechanical waves. The velocity of waves through a medium is determined by the inertial and elastical properties of the medium. They allow the storing of wave energy in the medium, and in the absence of energy dissipation, they also determine the impedance presented by the medium to the waves. In addition, when there is no loss mechanism, a plane wave solution will be obtained, but any resistive or loss term, will produce a decay with time or distance of the oscillatory solution.

Referring now to the electromagnetic waves, the magnetic inertia of the medium is provided by the inductive property of that medum, i.e. permeability μ, allowing storage of magnetic energy, and the elasticity or capacitive property - by the permittivity ϵ, allowing storage of the potential or electric field energy. ([12], p.199)

5. Maxwell's equations

The theory of electromagnetic phenomena can be described by four equations, two of them independent of time, and two - time-varying. The time-independent ones express the fact that the electric charge is the source of the electric field, but a "magnetic charge" does not exist:

$$\nabla \cdot (\epsilon \mathbf{E}) = \rho \tag{5}$$

$$\nabla \cdot (\mu \mathbf{H}) = 0 \tag{6}$$

One time-varying equation expresses Faraday's (or Lenz's) law [12], relating the time variation of the magnetic induction, $\mu\mathbf{H} = \mathbf{B}$, with the space variation of $\mathbf{E}$:

$$\frac{\partial}{\partial t}(\mu\mathbf{H}) \text{ is connected with } \frac{\partial \mathbf{E}}{\partial z} \text{ (say)}$$

More exactly,

$$\nabla \times \mathbf{E} = -\mu \frac{\partial \mathbf{H}}{\partial t} \tag{7}$$

The other one expresses Ampere's law [12], relating that the time variation of $\epsilon \mathbf{E}$ defines the space variation of $\mathbf{H}$:

$$\frac{\partial}{\partial t}(\epsilon \mathbf{E}) \text{ is connected with } \frac{\partial \mathbf{H}}{\partial z} \text{ (say)}$$

More exactly,

$$\nabla \times \mathbf{H} = \epsilon \frac{\partial \mathbf{E}}{\partial t} \tag{8}$$

assuming that no free chages or electric current are present - a natural assumption for our approach, as we shall use Maxwell's equations only for studying the wave propagation. In this context, the only role played by (5) and (6) will be to demonstrate the transverse character of the vectors $\mathbf{E}, \mathbf{H}$.

6. Propagation of electromagnetic waves in waveguides and cavities

The propagation of electromagnetic waves in hollow metalic cylinders is an interesting subject, both for theoretical and practical reasons - e.g., for its applications in telecommunications. We shall consider that the metal is a perfect conductor; if the cylinder is infinite, we shall call this metallic structure *waveguide*; if it has end faces, we shall call it *cavity*. The transversal section of the cylinder is the same, along the cylinder axis. With a time dependence $\exp(-i\omega t)$, the Maxwell equations (5)-(8) for the fields inside the cylinder take the form [3]:

$$\nabla \times \mathbf{E} = i\omega \mathbf{B}, \quad \nabla \cdot \mathbf{B} = 0, \quad \nabla \times \mathbf{B} = -i\mu\epsilon\omega \mathbf{E}, \quad \nabla \cdot \mathbf{E} = 0 \tag{9}$$

For a cylinder filled with a uniform non-dissipative medium having permittivity ϵ and permeability μ,

$$\left(\nabla^2 + \mu\epsilon\omega^2\right) \begin{Bmatrix} \mathbf{E} \\ \mathbf{B} \end{Bmatrix} = 0 \tag{10}$$

The specific geometry suggests us to single out the spatial variation of the fields in the z direction and to assume

$$\left.\begin{matrix} \mathbf{E}(x,y,z,t) \\ \mathbf{B}(x,y,z,t) \end{matrix}\right\} = \left\{\begin{matrix} \mathbf{E}(x,y)\exp(\pm ikz - i\omega t) \\ \mathbf{B}(x,y)\exp(\pm ikz - i\omega t) \end{matrix}\right. \tag{11}$$

The wave equation is reduced to two variables:

$$\left[\nabla_t^2 + \left(\mu\epsilon\omega^2 - k^2\right)\right] \begin{Bmatrix} \mathbf{E} \\ \mathbf{B} \end{Bmatrix} = 0 \tag{12}$$

where ∇_t^2 is the transverse part of the Laplacian operator:

$$\nabla_t^2 = \nabla - \frac{\partial^2}{\partial z^2} \tag{13}$$

It is convenient to separate the fields into components parallel to and transverse the oz axis:

$$\mathbf{E} = \mathbf{E}_z + \mathbf{E}_t, \text{ with } \mathbf{E}_z = \hat{\mathbf{z}} E_z, \quad \mathbf{E}_t = (\hat{\mathbf{z}} \times \mathbf{E}) \times \hat{\mathbf{z}} \tag{14}$$

$\widehat{z}$ is as usual, a unit vector in the $z-$direction. Similar definitions hold for the magnetic field $\mathbf{B}$. The Maxwell equations can be expressed in terms of transverse and parallel fields as [3]:

$$\frac{\partial \mathbf{E}_t}{\partial z} + i\omega\widehat{\mathbf{z}} \times \mathbf{B}_t = \boldsymbol{\nabla}_t E_z, \ \ \widehat{\mathbf{z}} \cdot (\boldsymbol{\nabla}_t \times \mathbf{E}_t) = i\omega B_z \tag{15}$$

$$\frac{\partial \mathbf{B}_t}{\partial z} - i\mu\epsilon\omega\widehat{\mathbf{z}} \times \mathbf{E}_t = \boldsymbol{\nabla}_t B_z, \ \ \widehat{\mathbf{z}} \cdot (\boldsymbol{\nabla}_t \times \mathbf{B}_t) = -i\mu\epsilon\omega E_z \tag{16}$$

$$\boldsymbol{\nabla}_t \cdot \mathbf{E}_t = -\frac{\partial E_z}{\partial z}, \ \ \boldsymbol{\nabla}_t \cdot \mathbf{B}_t = -\frac{\partial B_z}{\partial z} \tag{17}$$

According to the first equations in (15) and (16), if E_z and B_z are known, the transverse components of $\mathbf{E}$ and $\mathbf{B}$ are determined, assuming the z dependence is given by (11). Considering that the propagation in the positive z direction (for the opposite one, k changes it sign) and that at least one E_z and B_z have non-zero values, the transverse fields are

$$\mathbf{E}_t = \frac{i}{\mu\epsilon\omega^2 - k^2}\left[k\boldsymbol{\nabla}_t E_z - \omega\widehat{\mathbf{z}} \times \boldsymbol{\nabla}_t B_z\right] \tag{18}$$

$$\mathbf{B}_t = \frac{i}{\mu\epsilon\omega^2 - k^2}\left[k\boldsymbol{\nabla}_t B_z + \omega\epsilon\omega\widehat{\mathbf{z}} \times \boldsymbol{\nabla}_t E_z\right] \tag{19}$$

Let us notice the existence of a special type of solution, called the *transverse electromagnetic* (TEM) wave, having only field components transverse to the direction of propagation [6]. From the second equation in (15) and the first in (16), results that $E_z = 0$ and $B_z = 0$ implies that $\mathbf{E}_t = \mathbf{E}_{ETM}$ satisfies

$$\boldsymbol{\nabla}_t \times \mathbf{E}_{ETM} = 0, \ \ \boldsymbol{\nabla}_t \cdot \mathbf{E}_{ETM} = 0 \tag{20}$$

So, $\mathbf{E}_{ETM}$ is a solution of an *electrostatic* problem in 2D. There are 4 consequences:

1. the axial wave number is given by the infinite-medium value,

$$k = k_0 = \omega\sqrt{\mu\epsilon} \tag{21}$$

as can be seen from (12).

2. the magnetic field, deduced from the first eq. in (16), is

$$\mathbf{B}_{ETM} = \pm\sqrt{\mu\epsilon}\widehat{\mathbf{z}} \times \mathbf{E}_{ETM} \tag{22}$$

for waves propagating as $\exp(\pm ikz)$. The connection between $\mathbf{B}_{ETM}$ and $\mathbf{E}_{ETM}$ is just the same as for plane waves in an infinite medium.

3. the TEM mode cannot exist inside a single, hollow, cylindrical conductor of infinite conductivity. The surface is an equipotential; the electric field therefore vanishes inside. It is necessary to have two or more cylindrical surfaces to support the TEM mode. The familiar coaxial cable and the parallel-wire transmission line are structures for which this is the dominant mode.

4. the absence of a cutoff frequency (see below): the wave number (21) is real for all ω.

In fact, two types of field configuration occur in hollow cylinders. They are solutions of the eigenvalue problems given by the wave equation (12), solved with the following boundary conditions, to be fulfilled on the cylinder surface:

$$\mathbf{n} \times \mathbf{E} = 0, \ \ \mathbf{n} \cdot \mathbf{B} = 0 \tag{23}$$

where $\mathbf{n}$ is a normal unit at the surface S. From the first equation of (23):

$$\mathbf{n} \times \mathbf{E} = \mathbf{n} \times (-\mathbf{n}E_t + \widehat{\mathbf{z}}E_z) = \mathbf{n} \times \widehat{\mathbf{z}}E_z = 0$$

so:

$$E_z\,|_S = 0 \tag{24}$$

Also, from the second one:

$$\mathbf{n} \cdot \mathbf{B} = \mathbf{n} \cdot (-\mathbf{n}B_t + \widehat{\mathbf{z}}B_z) = -B_t = 0$$

With this value for B_t in the component of the first equation (16) parallel to $\mathbf{n}$, we get:

$$\frac{\partial B_z}{\partial n}\,|_S = 0 \tag{25}$$

where $\partial/\partial n$ is the normal derivative at a point on the surface. Even if the wave equation for E_z and B_z is the same ((eq. (12)), the boundary conditions on E_z and B_z are different, so the eigenvalues for E_z and B_z will in general be different. The fields thus naturally divide themselves into two distinct categories:

Transverse magnetic (TM) waves:

$$B_z = 0 \text{ everywhere; boundary condition, } E_z\,|_S = 0 \tag{26}$$

Transverse electric (TE) waves:

$$E_z = 0 \text{ everywhere; } \quad \text{boundary condition, } \frac{\partial B_z}{\partial n}\,|_S = 0 \tag{27}$$

For a given frequency ω, only certain values of wave number k can occur (typical waveguide situation), or, for a given k, only certain ω values are allowed (typical resonant cavity situation).

The variuos TM and TE waves, plus the TEM waves if it can exist, constitute a complete set of fields to describe an arbitrary electromagnetic disturbance in a waveguide or cavity [3].

7. Waveguides

For the propagation of waves inside a hollow waveguide of uniform cross section, it is found from (18) and (19) that the transverse magnetic fields for both TM and TE waves are related by:

$$H_t = \pm\frac{1}{Z}\widehat{\mathbf{z}} \times \mathbf{E}_t \tag{28}$$

where Z is called the *wave impedance* and is given by

$$Z = \begin{cases} \frac{k}{\epsilon\omega} = \frac{k}{k_0}\sqrt{\frac{\mu}{\epsilon}} & (TM) \\ \frac{\mu\omega}{k} = \frac{k_0}{k}\sqrt{\frac{\mu}{\epsilon}} & (TE) \end{cases} \tag{29}$$

and k_0 is given by (21). The $\pm$ sign in (28) goes with z dependence, $\exp(\pm ikz)$ [3]. The transverse fields are determined by the longitudinal fields, according to (18) and (19):

TM waves:

$$E_t = \pm \frac{ik}{\gamma^2} \nabla_t \psi \tag{30}$$

TE waves:

$$H_t = \pm \frac{ik}{\gamma^2} \nabla_t \psi \tag{31}$$

where $\psi \exp(\pm ikz)$ is E_z (H_z) for TM (TE) waves, and γ^2 is defined below. The scalar function ψ satisfies the 2D wave eq (12):

$$\left(\nabla_t + \gamma^2 \right) \psi = 0 \tag{32}$$

where

$$\gamma^2 = \mu\epsilon\omega^2 - k^2 \tag{33}$$

subject to the boundary condition,

$$\psi |_S = 0 \text{ or } \frac{\partial \psi}{\partial n} |_S = 0 \tag{34}$$

for TM (TE) waves.

Equation (32) for ψ, together with boundary condition (34), specifies an eigenvalues problem. The similarity with non-relativistic quantum mechanics is evident.

7.1 Modes in a rectangular waveguide

Let us illustrate the previous general theory by considering the propagation of TE waves in a rectangular waveguide (the corners of the rectangle are situated in $(0,0), (a,0), (a,b), (0,b)$). In this case, is easy to obtain explicit solutions for the fields [3]. The wave equation for $\psi = H_z$ is

$$\left(\frac{\partial^2}{\partial x^2} + \frac{\partial^2}{\partial y^2} + \gamma^2 \right) \psi = 0 \tag{35}$$

with boundary conditions $\partial\psi/\partial n = 0$ at $x = 0,\ a$ and $y = 0,\ b$. The solution for ψ is easily find to be:

$$\psi_{mn}(x,y) = H_0 \cos\left(\frac{m\pi x}{a} \right) \cos\left(\frac{n\pi y}{b} \right) \tag{36}$$

with γ givem by:

$$\gamma_{mn}^2 = \pi^2 \left(\frac{m^2}{a^2} + \frac{n^2}{b^2} \right) \tag{37}$$

with m, n - integers. Consequently, from (33),

$$k_{mn}^2 = \mu\epsilon\omega^2 - \gamma_{mn}^2 = \mu\epsilon \left(\omega^2 - \omega_{mn}^2 \right),\ \omega_{mn}^2 = \frac{\gamma_{mn}^2}{\mu\epsilon} \tag{38}$$

As only for $\omega > \omega_{mn}$, k_{mn} is real, so the waves propagate without attenuation; ω_{mn} is called cutoff frequency. For a given ω, only certain values of k, namely k_{mn}, are allowed.

For TM waves, the equation for the field $\psi = E_z$ will be also (39), but the boundary condition will be different: $\psi = 0$ at $x = 0,\ a$ and $y = 0,\ b$. The solution will be:

$$\psi_{mn}(x,y) = E_0 \sin\left(\frac{m\pi x}{a} \right) \sin\left(\frac{n\pi y}{b} \right) \tag{39}$$

with the same result for k_{mn}.

In a more general geometry, there will be a spectrum of eigenvalues γ_λ^2 and corresponding solutions ψ_λ, with λ taking discrete values (which can be integers or sets of integers, see for instance (37)). These different solutions are called the modes of the guide. For a given frequency ω, the wave number k is determined for each value of λ :

$$k_\lambda^2 = \mu\epsilon\omega^2 - \gamma_\lambda^2 \tag{40}$$

Defining a cutoff frequency ω_λ,

$$\omega_\lambda = \frac{\gamma_\lambda}{\sqrt{\mu\epsilon}}\sqrt{\omega^2 - \omega_\lambda^2} \tag{41}$$

then the wave number can be written:

$$k_\lambda = \sqrt{\mu\epsilon}\sqrt{\omega^2 - \omega_\lambda^2} \tag{42}$$

7.2 Modes in a resonant cavity

In a resonant cavity - i.e., a cylinder with metallic, perfect conductive ends perpendicular to the oz axis - the wave equation is identical, but the eigenvalue problem is somewhat different, due to the restrictions on k. Indeed, the formation of standing waves requires a $z-$dependence of the fields having the form

$$A\sin kz + B\cos kz \tag{43}$$

So, the wavenumber k is restricted to:

$$k = p\frac{\pi}{d}, \quad p = 0, 1, ... \tag{44}$$

and the condition a(35) impose a quantization of ω :

$$\mu\epsilon\omega_{p\lambda}^2 = \left(p\frac{\pi}{d}\right)^2 + \gamma_\lambda^2 \tag{45}$$

So, the existence of quantized values of k implies the quantization of ω.

8. Electromagnetic wave propagation in optical fibers

Optical fibers belong to a subset (the most commercially significant one) of dielectric optical waveguides [6]. Although the first study in this subject was published in 1910 [14], the explosive increase of interest for optical fibers coincides with the technical production of low loss dielectrics, some six decenies later. In practice, they are highly clindrical flexible fibers made of nearly transparent dielectric material. These fibers - with a diameter comparable to a human hair - are composed of a central region, the *core* of radius a and reffractive index n_{co}, surrounded by the *cladding*, of refractive index $n_{cl} < n_{co}$, covered with a protective jacket [15]. In the core, n_{co} may be constant - in this case, one says that the refractive-index profile is a step profile (as also $n_{cl} = const.$), or may be graded, for instance:

$$n_{co}(r) = n_{co}(0)\left[1 - \Delta\left(\frac{r}{a}\right)^\alpha\right], \quad r < a \tag{46}$$

For $\alpha = 2$, the profile is called parabolic. One of the main parameters characterizing an optical fiber is the profile hight parameter Δ,

$$\Delta = \frac{1}{2}\left(1 - \frac{n_{cl}^2}{n_{co}^2}\right) \simeq 1 - \frac{n_{cl}}{n_{co}}, \quad n_{co} = \max n_{co}(r)\,|_{r \leq a} \tag{47}$$

Besides Δ, one usually also defines the fiber parameter V :

$$V = ka\sqrt{2\Delta} \tag{48}$$

Assimilating the propagating light with a geometric ray, it must be incident on the core-cladding interface at an angle smaller than the critical angle θ_c :

$$\theta_c = \arcsin \frac{n_{cl}}{n_{co}} \tag{49}$$

in order to be totally reflected at this interface, and therefore to remain inside the core. However, due to the wave character of light, it must satisfy a self-interference condition, in order to be trapped in the waveguide [6]. There are only a finite number of paths which satisfy this condition, and therefore a finite number of modes which propagate through the fiber. The fiber is multimode if $12.5\mu m < r < 100\mu m$ and $0.01 < \Delta < 0.03$, and single-mode if $2\mu m < r < 5\mu m$ and $0.003 < \Delta < 0.01$ [15]. By far the most popular fibers for long distance telecommunications applications allow only a single mode of each polarization to propagate [6].

8.1 Modes in circular fibers

We consider a fiber of uniform cross section with relative magnetic permeability = 1 and n varying only on transverse directions [3]. Assuming a $z-$ and $t-$ dependence $\exp(ik_z z - i\omega t)$, the Maxwell equations can be combined, to yield the Helmholtz wave equations for $\mathbf{H}$ and $\mathbf{E}$:

$$\nabla^2 \mathbf{H} + \frac{n^2\omega^2}{c^2}\mathbf{H} = i\omega\epsilon_0\left(\nabla n^2\right) \times \mathbf{E} \tag{50}$$

$$\nabla^2 \mathbf{E} + \frac{n^2\omega^2}{c^2}\mathbf{E} = -\nabla\left[\frac{1}{n^2}\left(\nabla n^2\right) \cdot \mathbf{E}\right] \tag{51}$$

where we have written $\epsilon = n^2\epsilon_0$. Just as in Sect. 6, the transverse components of E and H can be expressed in terms of the longitudinal fields E_z, H_z, *i.e.*

$$E_t = \frac{i}{\gamma^2}\left[k_z \nabla_t E_z - \omega\mu_0 \widehat{\mathbf{z}} \times \boldsymbol{\nabla}_t H_z\right] \tag{52}$$

and

$$H_t = \frac{i}{\gamma^2}\left[k_z \nabla_t H_z + \omega\epsilon_0 n^2 \widehat{\mathbf{z}} \times \boldsymbol{\nabla}_t E_z\right] \tag{53}$$

where $\gamma^2 = n^2\omega^2/c^2 - k_z^2$ is the radial propagation constant, as for metallic waveguides. If we take the z component of the eqs (54), (55) and use (52) to eliminate the transverse field components, assuming that $\partial n^2/\partial z = 0$, we find generalizations of the 2D scalar wave equation (32):

$$\nabla_t^2 H_z + \gamma^2 H_z - \left(\frac{\omega}{\gamma c}\right)^2 \left(\nabla_t n^2\right) \cdot \nabla_t H_z = -\frac{\omega k_z \epsilon_0}{\gamma^2} \hat{\mathbf{z}} \cdot \left[\nabla_t n^2 \times \nabla_t E_z\right] \quad (54)$$

and

$$\nabla_t^2 E_z + \gamma^2 E_z - \left(\frac{k_z^2}{\gamma n}\right)^2 \left(\nabla_t n^2\right) \cdot \nabla_t E_z = \frac{\omega k_z \mu_0}{\gamma^2 n^2} \hat{\mathbf{z}} \cdot \left[\nabla_t n^2 \times \nabla_t H_z\right] \quad (55)$$

In contrast to (32) for ideal metallic guides, the equations for E_z, H_z are coupled. In general, thee is no separation into purely TE and TM modes. The only simplification occurs in the case of a step-profile refractive index, where we can solve the equation (54) or (55) in each domain of constant refractive index, and match the two solutions, using appropriate boundary conditions. In this case, the radial part of the electric field (for the first mode) in the core is [6]:

$$R(r) = \frac{J_0\left(\sqrt{n_{co}^2 k_0^2 - \beta^2}\,(r/a)\right)}{J_0\left(\sqrt{n_{co}^2 k_0^2 - \beta^2}\right)}, \quad r < a \quad (56)$$

and in the cladding:

$$R(r) = \frac{K_0\left(\sqrt{n_{cl}^2 k_0^2 - \beta^2}\,(r/a)\right)}{K_0\left(\sqrt{n_{cl}^2 k_0^2 - \beta^2}\right)}, \quad r > a \quad (57)$$

These solutions are identical (using an appropriate "dictionary") with the solution of the Schrodinger equation for a particle moving in a potential with cylindrical symmetry, the radial part of the potential being a rectangular well of finite depth. However, this kind of analogies can be more easely developed for planar dielectric waveguides, namely for "step-index" dielectrics, consisting of a central slab of finite thickness and of higher refractive index (core), and two lateral, half-space medium of lower refractive index (cladding). Indeed, in such a situation, the quantum counterpart of the dielectric guide is much more extensively studied, in almost any textbook of quantum mechanics.

9. An optical-quantum analogy: the three-layer slab optical waveguide and the quantum rectangular well

We shall calculate in detail the TE modes of a three-layer slab optical waveguide, with a 1D structure, and the bound states of a particle in a rectangular well, and we shall find that these problems have identical solutions. Of course, the physical meaning of the parameters entering in each solution are different, but the mathematical structure of the solutions is identical.

9.1 The optical problem

We consider a three-layer slab optical waveguide, with a 1D structure [16]. The electromagnetic wave propagates along the x axis, and the slabs are: a semi-infinite medium of refractive index n_1, having as right border the yz plane; a slab of refractive index n_2, having as left border the plane yz and as left border a plane paralel to it, cutting the ox axis at $x_0 = W$;

and a semi-infinite medium of refractive index n_3, for the remaining space. The inner slab corresponds to the core, and the outer ones - to the cladding.

It is instructive to obtain the wave equation for the electric and magnetic field in this simple geometry, starting directly from the Maxwell equations (7), (8). Assuming that $\mu = \mu_0$ throughout the entire system and that the $t-$dependence is:

$$\mathbf{E} = \mathbf{E}\left(t=0\right)\exp\left(-i\omega t\right), \quad \mathbf{H} = \mathbf{H}\left(t=0\right)\exp\left(-i\omega t\right)$$

the equations for the field components are:

$$\frac{\partial E_z}{\partial y} - \frac{\partial E_y}{\partial z} = i\mu_0\omega H_x \tag{58}$$

$$\frac{\partial E_x}{\partial z} - \frac{\partial E_z}{\partial x} = i\mu_0\omega H_y \tag{59}$$

$$\frac{\partial E_y}{\partial x} - \frac{\partial E_x}{\partial y} = i\mu_0\omega H_z \tag{60}$$

Also,

$$\frac{\partial H_z}{\partial y} - \frac{\partial H_y}{\partial z} = -i\epsilon\omega E_x \tag{61}$$

$$\frac{\partial H_x}{\partial z} - \frac{\partial H_z}{\partial x} = -i\epsilon\omega E_y \tag{62}$$

$$\frac{\partial H_y}{\partial x} - \frac{\partial H_x}{\partial y} = -i\epsilon\omega E_z \tag{63}$$

TE mode

We shall look for the TE mode. By definition, in this mode there is no electric field in longitudinal direction, $E_z = 0$, there is no space variation in the y direction, so $\partial/\partial y \to 0$, and the z-dependence is $\exp\left(-i\beta z\right)$, so $\partial/\partial z \to -i\beta$. The Maxwell equations (58)-(63) become:

$$H_x = \frac{\beta}{\mu_0\omega}E_y \tag{64}$$

$$-\beta E_x = \mu_0\omega H_y \tag{65}$$

$$H_z = -\frac{i}{\mu_0\omega}\frac{\partial E_y}{\partial x} \tag{66}$$

$$\beta H_y = -\epsilon\omega E_x \tag{67}$$

$$-i\beta H_x - \frac{\partial H_z}{\partial x} = -i\epsilon\omega E_y \tag{68}$$

$$\frac{\partial H_y}{\partial x} = 0 \tag{69}$$

From (69), $H_y = const$ and we can put $H_y = 0$, so from (65), (67), $E_x = 0$. So,

$$E_x = E_z = H_y = 0 \tag{70}$$

With (64), (66) in (68):

$$\frac{d^2E_y}{dx^2} + \left(\epsilon\mu_0\omega^2 - \beta^2\right) E_y = 0 \tag{71}$$

Defining:

$$n_{eff} = \frac{\beta}{k_0} \to \beta = n_{eff}k_0, \ k_0 = \omega\sqrt{\epsilon_0\mu_0} \tag{72}$$

we have:

$$\frac{d^2E_y}{dx^2} + k_0^2\left(\epsilon_r - n_{eff}^2\right) E_y = 0 \tag{73}$$

With $k_0 = 2\pi/\lambda$, (73) becomes:

$$\frac{d^2E_y}{dx^2} + \frac{4\pi^2}{\lambda^2}\left(\epsilon_r - n_{eff}^2\right) E_y = 0 \tag{74}$$

It is interesting to compare (74) with the Schrodinger equation for a particle of mass m moving in a potential V :

$$\frac{d^2\psi}{dx^2} + \frac{4\pi^2}{h^2}\left(-2mV + 2m\mathcal{E}\right)\psi = 0 \tag{75}$$

For bound states, $\mathcal{E} = -|\mathcal{E}| < 0$. In (75), the energy is subject of quantization, similar to n_{eff}^2 in (73) - with appropriate boundary conditions, see below. So, the quantum-mechanical energy is proportional to ϵ_r, confirming the analogy stated in Sect.4. The opposite of the potential is proportional to the square of the refractive index - the so-called "upside-down correspondence" [1] between optical and mechanical propagation: a light wave tends to concentrate in the area with maximum refractive index, while a particle tends to propagate on the bottom of the potential. Also, the wavelength λ corresponds to the Planck constant h : when $\lambda \to 0$, the wave optics is replaced by geometrical optics, similarly with transition ftrom quantum to classical mechanics.

Let us discuss now the boundary conditions. In the absence of charges and current flow on surfaces, the boundary conditions for the electromagnetic fields are:

1. the tangential components of the electric field are continuous while crossing the border
2. the tangential components of the magnetic field are continuous
3. the normal components of the electric flux density $\mathbf{D} = \varepsilon\mathbf{E}$ are continuous
4. the normal components of the magnetic flux density $\mathbf{B} = \mu\mathbf{H}$ are continuous

The tangential electric field at the boundary is $E_y\widehat{\mathbf{y}} + E_z\widehat{\mathbf{z}} = E_y\widehat{\mathbf{y}}$ and the tangential magnetic field is $H_y\widehat{\mathbf{y}} + H_z\widehat{\mathbf{z}} = H_z\widehat{\mathbf{z}}$. But $H_z = -\frac{i}{\mu_0\omega}\frac{dE_y}{dx}$, so the continuity of H_z is equivalent to the continuity of $\frac{dE_y}{dx}$. So, the conditions (1) and (2) impose the continuity of E_y and of its derivative, $\frac{dE_y}{dx}$. The normal component of the electric field E_x is identically zero, according to (70), so the condition (3) is automatically fulfilled. As $\mu = \mu_0$, condition (4) claims the continuity of the normal components of the magnetic flux, $H_x = \frac{\beta}{\mu_0\omega}E_y$, so this condition coincides with (1).

Consequently, in our case, the boundary conditions request the continuity of E_y and of its derivative dE_y/dx , at the slab boundaries. The equation (73), together with these boundary conditions, define a Sturm-Liouville problem, which determines the eigenvalues of n_{eff} or, equivalently, of β.

For the physics of optical fibers, the most interesting situation is that corresponding to an oscillatory solution inside the core and exponentially small ones outside the core (in the cladding):

$$E_y(x) = C_1 \exp(\gamma_1 x), \quad \gamma_1 = k_0\sqrt{n_{eff}^2 - n_1^2} \tag{76}$$

$$= C_2 \sin(\gamma_2 x + \alpha), \qquad \gamma_2 = k_0\sqrt{n_2^2 - n_{eff}^2} \tag{77}$$

$$= C_3 \exp(-\gamma_3 (x - W)), \quad \gamma_3 = k_0\sqrt{n_{eff}^2 - n_3^2} \tag{78}$$

As we just have seen, the boundary conditions are equivalent to the continuity of $E_y(x)$ and of its derivative, $dE_y(x)/dx$:

$$\frac{dE_y(x)}{dx} = \gamma_1 C_1 \exp(\gamma_1 x), \quad \gamma_1 = k_0\sqrt{n_{eff}^2 - n_1^2} \tag{79}$$

$$= \gamma_2 C_2 \cos(\gamma_2 x + \alpha) \tag{80}$$

$$= -\gamma_3 C_3 \exp(-\gamma_3 (x - W)) \tag{81}$$

So, the continuity at $x = 0$ means:

$$C_1 = C_2 \sin\alpha \tag{82}$$

$$\gamma_1 C_1 = \gamma_2 C_2 \cos\alpha \tag{83}$$

Similarily, at $x = W$:

$$C_2 \sin(\gamma_2 W + \alpha) = C_3 \tag{84}$$

$$\gamma_2 C_2 \cos(\gamma_2 W + \alpha) = -\gamma_3 C_3 \tag{85}$$

Dividing (83) by (82), we get:

$$\frac{\gamma_1}{\gamma_2} = \cot\alpha, \tag{86}$$

$$\alpha = \operatorname{arccot}\frac{\gamma_1}{\gamma_2} + q_1\pi, \quad q_1 = 0, 1, 2, ... \tag{87}$$

Dividing (85) by (81), we get:

$$\frac{\gamma_3}{\gamma_2} = -\cot(\gamma_2 W + \alpha),$$

$$-\operatorname{arccot}\frac{\gamma_3}{\gamma_2} - \alpha + q_2\pi = \gamma_2 W \tag{88}$$

Substitution of α from (87) into (88) gives:

$$-\operatorname{arccot}\frac{\gamma_3}{\gamma_2} - \operatorname{arccot}\frac{\gamma_1}{\gamma_2} + q\pi = \gamma_2 W \tag{89}$$

Written in the form:

$$-\arctan\frac{\gamma_2}{\gamma_3} - \arctan\frac{\gamma_2}{\gamma_1} + q\pi = \gamma_2 W \tag{90}$$

it coincides with eq. (20) Ch.III, vol.1, [13]. It is the energy eigenvalue equation for the Schrodinger equation of a particle of mass m, moving in the potential $V(x)$:

$$\left[\frac{d^2}{dx^2}+\frac{2m}{\hbar^2}\left(-V(x)+\mathcal{E}\right)\right]\psi(x)=0 \tag{91}$$

where $V(x)$ is a piecewise-defined function ([13], III.1.6):

$$V(x)=\begin{cases} V_1, & x>a \\ V_2, & a>x>b \\ V_3, & b>x \end{cases} \tag{92}$$

$$V_2<V_1<V_3$$

It is useful to consider a particular situation, when $n_1 = n_3$ in the optical waveguide, respectively when $V_1 = V_3 = 0$, $V_2 < 0$, in the quantum mechanical problem. In (91), $-V(x) \geqslant 0$ is given, and we have to find the eigenvalues of the energy $\mathcal{E} < 0$. For the optical waveguide (71), $\epsilon\mu_0\omega^2$ is given, and the eigenvalues of the quantity $-\beta^2$ (essentially, the propagation constant β) must be obtained. Let us note once again that the refractive index in the optical waveguide corresponds to the opposite of the potential, in the quantum mechanical problem.

Let us investigate in greater detail the consequences of the particular situation just mentioned, $n_1 = n_3$. With $q \to -q$, eq. (90) becomes:

$$\arctan\frac{\gamma_2}{\gamma_1}+\frac{q\pi}{2}=-\frac{\gamma_2 W}{2} \tag{93}$$

It gives, for q odd:

$$\frac{\gamma_1}{\gamma_2}=\tan\frac{\gamma_2 W}{2} \tag{94}$$

and for q even:

$$\frac{\gamma_2}{\gamma_1}=-\tan\frac{\gamma_2 W}{2} \tag{95}$$

Putting:

$$\frac{W}{2}=a,\ \gamma_1 a=ak_0\sqrt{n_{eff}^2-n_1^2}=\Gamma_1,\ \gamma_2 a=ak_0\sqrt{n_2^2-n_{eff}^2}=\Gamma_2 \tag{96}$$

we get, instead of (94), (95):

$$\frac{\Gamma_1}{\Gamma_2}=\tan\Gamma_2 \quad (q\ odd) \tag{97}$$

$$\frac{\Gamma_2}{\Gamma_1}=-\tan\Gamma_2 \quad (q\ even) \tag{98}$$

Defining K through the equation:

$$\Gamma_1^2=K^2-\Gamma_2^2 \tag{99}$$

the eigenvalue conditions (97), (98) take the form:

$$\frac{\sqrt{K^2-\Gamma_2^2}}{\Gamma_2}=\tan\Gamma_2 \tag{100}$$

$$\frac{\Gamma_2}{\sqrt{K^2 - \Gamma_2^2}} = -\tan\Gamma_2 \tag{101}$$

So, the eigenvalue equation (90) splits into two simpler conditions (100), (101), carracterizing states with well defined parity, as we shall see further on (of course, the parity of q, mentioned just after (93), has nothing to do with the parity of states).

We shall analyze now the same problem, starting from the quantum mechanical side.

9.2 The quantum mechanical problem: the particle in a rectangular potential well

We discuss now the Schrodinger equation for a particle in a rectangular potential well ([17], v.1, pr.25), one of the simplest problems of quantum mechanics:

$$\left[-\frac{\hbar^2}{2m}\frac{d^2}{dx^2} + V(x)\right]\psi(x) = \mathcal{E}\psi(x) \tag{102}$$

$$V(x) = \begin{cases} -U, & 0 < x < a \\ 0, & elsewere \end{cases} \tag{103}$$

Let be:

$$\mathcal{E} = -\frac{\hbar^2\varkappa^2}{2m}; \; U = \frac{\hbar^2 k_0^2}{2m}; \; k^2 = k_0^2 - \varkappa^2 \tag{104}$$

We are looking for bound states inside the well:

$$u_1(x) = A\exp(\varkappa x), \quad x < 0 \tag{105}$$

$$u_2(x) = B\sin(kx + \alpha), \quad 0 < x < a \tag{106}$$

$$u_3(x) = D\exp(\varkappa(a - x)), \quad a < x \tag{107}$$

The wavefunction and its derivatives:

$$u_1'(x) = \varkappa A\exp(\varkappa x), \quad x < 0 \tag{108}$$

$$u_2'(x) = kB\cos(kx + \alpha), \quad 0 < x < a \tag{109}$$

$$u_3'(x) = -\varkappa D\exp(\varkappa(a - x)), \quad a < x \tag{110}$$

must be continuous in $x = 0$:

$$A = B\sin\alpha \tag{111}$$

$$\varkappa A = kB\cos\alpha \tag{112}$$

and in $x = a$:

$$B\sin(ka + \alpha) = C \tag{113}$$

$$kB\cos(ka + \alpha) = -\varkappa C \tag{114}$$

Dividing (111), (112):

$$\frac{1}{\varkappa} = \frac{1}{k}\tan\alpha, \qquad \alpha = \arctan\frac{k}{\varkappa} + n\pi \tag{115}$$

and (113), (114):

$$\frac{1}{k}\tan(ka + \alpha) = -\frac{1}{\varkappa}, \qquad ka + \alpha = -\arctan\frac{k}{\varkappa} + n_1\pi \tag{116}$$

and substituting (115) in (116), we get:

$$\frac{k}{\varkappa} = \tan\left(-\frac{ka}{2} + \frac{n_2}{2}\pi\right) \tag{117}$$

For n_2 even:

$$\frac{k}{\varkappa} = -\tan\frac{ka}{2}, \qquad \arctan\frac{k}{\varkappa} = -\frac{ka}{2} \tag{118}$$

and for n_2 odd:

$$\frac{k}{\varkappa} = \tan\left(-\frac{ka}{2} + \frac{\pi}{2}\right), \qquad \arctan\frac{k}{\varkappa} = -\frac{ka}{2} + \frac{\pi}{2} \tag{119}$$

Putting

$$\xi = \frac{ka}{2}, \ C = \frac{k_0 a}{2} \tag{120}$$

the conditions (118), (119) become respectively:

$$\tan\xi = -\frac{\xi}{\sqrt{C^2 - \xi^2}} \tag{121}$$

$$\cot\xi = \frac{\xi}{\sqrt{C^2 - \xi^2}} \tag{122}$$

Let us write now the wavefunction (106) using the expression (115) for α:

$$u_2(x) = B\sin\left(kx + \arctan\frac{k}{\varkappa} + n\pi\right), \qquad 0 < x < a \tag{123}$$

With (118) and (119), the equation (123) splits in two equations:

$$u_2^*(x) = B\sin\left(kx - \frac{ka}{2} + n\pi\right) \tag{124}$$

$$u_2^{**}(x) = B\sin\left(kx - \frac{ka}{2} + \frac{\pi}{2} + n\pi\right) \tag{125}$$

We translate now the coordinate x, so that the origin of the new axis is placed in the center of the well:

$$x = y + \frac{a}{2} \tag{126}$$

and we get:

$$U_2^*(y) = u_2^*\left(y + \frac{a}{2}\right) = B\sin(ky + n\pi) = (-1)^n B\sin ky \tag{127}$$

$$U_2^{**}(y) = u_2^{**}\left(y + \frac{a}{2}\right) = B\sin\left(ky + \frac{\pi}{2} + n\pi\right) = (-1)^n B\cos ky \tag{128}$$

So, the wavefunctions corresponding to the eigenvalues obtained from (121), (122) have well defined parity.

Let us stress once more that the core of the optical-quantum analogy consists in eqs. (74), (75), which can be formulated as follows: the refractive index for the propagation of light plays a similar role to the potential, for the propagation of a quantum non-relativistic particle, and both the electric (or magnetic) field and the wave function are the solution of essentially the same (Helmholtz) equation. So, if the dynamics of a particle, given by the Schrodinger equation, can be considered as the central aspect of quantum mechanics, the scattering of light by a medium with refractive index $n\left(\vec{r}\right)$ can be considered as the central aspect of optics, at least when the Maxwell equations can be reduced to a Helmholtz equation. Remembering Goethe's opinion, that the "Urphänomenon" of light science is the scattering of light on a "turbid" medium, one could remark that his theory of colours is not always as unrealistic as it was generally considered. [18]

10. Ballistic electrons in 2DESs

As already mentioned, the 2DES, formed at the interface of two semiconductors might play a central role in mesoscopic physics. The thin 2D conduction layer formed in the GaAs/AlGaAs heterojunction may reach a carrier concentration of $2 \times 10^{12}\ cm^{-2}$ and can be depleted by applying a negative voltage to a metalic gate deposited on the surface [Datta]. The mobility can be as high as $10^6 cm^2/Vs$, two order of magnitude higher than in bulk semiconductors. The Fermi wavelength λ_M is about 35 nm, and the electron mean free path may be as long as $\lambda_m = 30\mu$m– the same order of magnitude as the liniar dimension of the sample; the ballistic regime of electrons is therefore easily reached.

At low temperature, the conduction in mesoscopic semiconductor is mainly due to electrons in the conduction band. Their dynamics can be described by an equation of the form:

$$\left[\mathcal{E}_c + \frac{(i\hbar\nabla + e\mathbf{A})^2}{2m} + U(\mathbf{r})\right]\Psi(\mathbf{r}) = \mathcal{E}\Psi(\mathbf{r}) \tag{129}$$

where $\mathcal{E}_c$ referrs to the conduction band energy, $U(\mathbf{r})$ is the potential energy due to space-charge etc., $\mathbf{A}$ is the vector potential and m is the effective mass. Any band discontinuity $\Delta\mathcal{E}_c$ at heterojunctions is incorporated by letting $\mathcal{E}_c$ be position-dependent [11].

In the case of a homogenous semiconductor, $U(\mathbf{r}) = 0$, assuming $\mathbf{A} = 0$ and $\mathcal{E}_c = const.$, the solution of (129) is given by plane waves, $\exp(i\mathbf{k}\cdot\mathbf{r})$, and not by Bloch functions, $u_k(\mathbf{r})\cdot\exp(i\mathbf{k}\cdot\mathbf{r})$. So, the solutions of (129) are not true wavefunctions, but wavefunctions smoothed out over a mesoscopic distance, so any rapid variation at atomic scale is suppressed; eq. (129) is called single-band effective mass equation.

Let us consider a 2DES contained mainly in the xy plane. This means that, in the absence of any external potential, the electrons can move freely in the xy plane, but they are confined in the z-direction by some potential $V(z)$, so their wavefunctions will have the form:

$$\Psi(\mathbf{r}) = \phi_n(z)\exp(ik_x x)\exp(ik_y y) \tag{130}$$

The quantization on the $z-$direction, expressed by the functions $\phi_n(z)$, generate several subbands, with cut-off energy ε_n. At low temperature, only the first subband, corresponding to $n = 1$, is occupied, so, instead of (129), the electrons of the 2DES are described by:

$$\left[\mathcal{E}_s + \frac{(i\hbar\nabla + e\mathbf{A})^2}{2m} + U(\mathbf{r})\right]\Psi(\mathbf{r}) = \mathcal{E}\Psi(\mathbf{r}) \quad (102) \tag{131}$$

where the subband energy is $\mathcal{E}_s = \mathcal{E}_c + \varepsilon_1$; so, the ‡−dimension enters in this equation only through ε_1, which depends on the confining potential $V(z)$. The eq. (131) correctly describes the 2DESs formed in semiconductor heterostructures, but is inappropriate for metallic thin films, where the electron density is much higher, and even at nanoscopic scale, there are tens of occupied bands; so, the system is merely 3D. Consequently, the dimensionality of a system depends not only on its geometry, but also on its electron concentration. Let us remind that the conductive / dielectric properties of a sample depends on frequency of electromagnetic waves: so even basic classification of materials is not necessarely intrinsec, but it might depend of the value of some parameters.

11. Transverse modes (or magneto-electric subbands)

We shall discuss now the concept of transverse modes or subbands, which are analogous to the transverse modes of electromagnetic waveguides [11]. In narrow conductors, the different transverse modes are well separated in energy, and such conductors are often called electron waveguides.

We consider a rectangular conductor that is uniform in the $x-$direction and has some transverse confining potential $U(y)$. The motion of electrons in such a conductor is described by the effective mass eq (131):

$$\left[\mathcal{E}_s + \frac{(i\hbar\nabla + e\mathbf{A})^2}{2m} + U(y)\right]\Psi(x,y) = \mathcal{E}\Psi(x,y) \tag{132}$$

We assume a constant magnetic field B in the $z-$direction, perpendicular to the plane of the conductor, which can be represented by a vector potential defined by:

$$A_x = -By, \ \ A_y = 0 \tag{133}$$

so that the effective-mass equationcan be rewritten as:

$$\left[\mathcal{E}_s + \frac{(p_x + eBy)^2}{2m} + \frac{p_y^2}{2m} + U(y)\right]\Psi(x,y) = \mathcal{E}\Psi(x,y) \tag{134}$$

Writing

$$\Psi(x,y) = \frac{1}{\sqrt{L}}\exp(ikx)\,\chi(y) \tag{135}$$

we get for the transverse function the equation:

$$\left[\mathcal{E}_s + \frac{(\hbar k + eBy)^2}{2m} + \frac{p_y^2}{2m} + U(y)\right]\chi(y) = \mathcal{E}\chi(y) \tag{136}$$

We are interested in the nature of the transverse eigenfunctions and eigenenergies for different combinations of the confining potential $U$and the magnetic field B. A parabolic potential

$$U(y) = \frac{1}{2}m\omega_0^2 y^2 \tag{137}$$

is often a good description of the actual potential in many electron wave guides.

Let us consider the case of confined electrons ($U \neq 0$) in zero magnetic field ($B = 0$). Eq. (134) becomes:

$$\left[\mathcal{E}_s + \frac{\hbar^2 k^2}{2m} + \frac{p_y^2}{2m} + \frac{1}{2} m\omega_0^2 y^2\right] \chi(y) = \mathcal{E}\chi(y) \tag{138}$$

with solutions:

$$\chi_{n,k}(y) = u_n(q), \quad q = \sqrt{\frac{m\omega_0}{\hbar}} y \tag{139}$$

$$E(n,k) = E_s + \frac{(\hbar k)^2}{2m} + \left(n + \frac{1}{2}\right) \hbar\omega_0 \tag{140}$$

where:

$$u_n(q) = \exp\left(-\frac{1}{2}q^2\right) H_n(q) \tag{141}$$

with H_n - Hermite polynomials. The velocity is obtained from the slope of the dispersion curve:

$$v(n,k) = \frac{1}{\hbar}\frac{\partial E(n,k)}{\partial k} = \frac{\hbar k}{m} \tag{142}$$

States with different index n are said to belong to different subbands; the situation is similar to that described in Sect. 8, where we have discussed the confinement due to the potential $V(z)$. However, the confinement in the $y-$direction is somewhat weaker, and several subbands are now normally occupied. The subbands are often referred to as *transverse modes*, in analogy with the modes of an electromagnetic waveguide [11].

12. Effective mass approximation revisited

A more attentive investigation of the effective-mass approximation for electrons in a semiconductor, introduced in a simplified form in Sect.10, will allow quantitative analogies between propagation ballistic electrons and guided electromagnetic waves past abrupt interfaces [19].

According to Morrow and Brownstein [20], out of the general class of Hamiltonians suggested by von Ross [21], only those of the form:

$$H\psi = -\frac{\hbar^2}{2}\left\{m(\vec{r})^\alpha \nabla \cdot \left[m(\vec{r})^\beta \nabla \left(m(\vec{r})^\alpha \psi\right)\right]\right\} + V(\vec{r})\psi = \mathcal{E}\psi \tag{143}$$

with the constraint

$$2\alpha + \beta = -1 \tag{144}$$

(where α and β have specific values for specific substances) can be used in the study of refraction of ballistic electrons at the interface between dissimilar semiconductors.

For the Hamiltonian (143), the boundary conditions for an electron wave at an interface are:

$$m^\alpha \psi = continous \tag{145}$$

and

$$m^{\alpha+\beta} \nabla\psi \cdot \hat{n} = continuous \tag{146}$$

with $\widehat{n}$ – the unit vector normal to the interface. Analogously, the boundary conditions for an electromagnetic waves at an interface between two dielectrics require the continuity of tangential components E_t, H_t across the interface. Based on these considerations, it is reasonable to look for analogies between

$$\Phi = m^{\alpha}\psi \tag{146}$$

and either E or H.

For bulk propagation in a homogenous medium, an exact analogy can be drawn between Φ and both E or H. In this case, eq. (143) reduces to a Helmholtz equation of the form:

$$\nabla^2\Phi = -k^2\Phi,\ k^2 = \frac{2m\,(E-V)}{\hbar^2} \tag{147}$$

The wave equation (148) is exactly analogous to the Helmholtz equation for an electromagnetic wave propagating in a homogenous dielectric of permittivity ϵ and permeability μ, with Φ replaced by E or H, and

$$k^2 = \omega^2\mu\epsilon \tag{148}$$

So, an exact analogy can be drawn between Φ and both E or H. With these analogies, one can define a phase-refractive index for electron waves as:

$$n_{ph}^{EW} = m_r^{1/2}\,(E-V)_r^{1/2} \tag{149}$$

where $m_r = m/m_{ref}$ is the relative effective mass and

$$(\mathcal{E}-V)_r = \frac{\mathcal{E}-V}{\mathcal{E}-V_{ref}} \tag{150}$$

is the relative kinetic energy, where m_{ref}, V_{ref} are the effective mass and the potential energy in a reference region. This electron wave phase-refractive index is analogous to the phase-refracting index for electromagnetic waves,

$$n_{ph}^{EM} = \sqrt{\mu_r\epsilon_r} \tag{151}$$

With these results, phase-propagation effects, such as interference, can be analyzed using standard em results, where E (or H), n_{ph}^{EM} is replaced by Φ, n_{ph}^{EW}. These results are valid for all the Hamiltonians (143). For the electron wave amplitude, the index of refraction is defined as

$$n_{amp,l}^{EW} = m_{r,l}^{\beta+1/2}\,(\mathcal{E}-V)_{r,l}^{1/2} \tag{152}$$

These expressions are exactly the same as the anagous electromagnetic expressions for the reflection/refraction of an electromagnetic wave from an interface between two dielectrics [3].

The theory outlined in this section can be extended for 1D or 2D inhomogenous materials, but not for three dimensions [19].

Dragoman and Dragoman [22], [23], [24] obtained a quantum-mechanical - electromagnetic analogy, similar to [19], in the sense that the electronic wave function does not correspond to the fields, but to the vector potential:

$$m^{\alpha}\psi \rightarrow A\ (\text{TM wave})\ \text{or}\ (\varepsilon/\mu)\ (\text{TE wave}) \tag{153}$$

13. Optics experiments with ballistic electrons

In 2DESs, there exists a unique oportunity to control ballistic carriers via electrostatic gates, which can act as refractive elements for the electron path, in complete analogy to refractive elements in geometrical optics. [8], [25], [26]. The refraction of a beam of ballistic electrons can be simply described, using elementary considerations. If in the "left" half-space, the potential has a constant value V, and in the "right" one, a different (but also constant) one, $V + \Delta V$, an incident electron arriving with an incident angle θ and kinetic energy

$$\mathcal{E} = \frac{\hbar^2 k^2}{2m^*} \tag{154}$$

emerges in the "right" half-plane under the angle θ', with the kinetic energy

$$\mathcal{E}' = \mathcal{E} + e\Delta V = \frac{\hbar^2 k'^2}{2m^*} \tag{155}$$

Translational invariance along the interface preserves the parallel component of electron momentum and thus

$$k \sin\theta = k' \sin\theta' \tag{156}$$

or

$$\frac{\sin\theta}{\sin\theta'} = \frac{k'}{k} = \sqrt{\frac{\mathcal{E}'}{\mathcal{E}}} \tag{157}$$

Considering that the energies $\mathcal{E}$, $\mathcal{E}'$ are Fermi energies, proportional (in a 2D system) to the electron densities n_{el} (not to be confused with refractive index!), the Snell's law takes the form:

$$\frac{\sin\theta}{\sin\theta'} = \sqrt{\frac{n'_{el}}{n_{el}}} \tag{158}$$

An electrostatic lens for ballistic electrons was set up in [8] and its focusing action was demonstrated in the GaAs / AlGaAs heterostructure. In this way, the close analogy between the propagation of ballistic electrons and geometrical optics has been put in evidence.

Another nice experiment used a refractive electronic prism to switch a beam of ballistic electrons between different collectors in the same 2DES [25]

The quantum character of ballistic electrons is clearly present, even they are regarded as beams of particles. Transmission and reflection of electrons on a sharp (compared to λ_F), rectangular barrier, induced via a surface gate at $T = 0.5K$, follow the laws obtained in quantum mechanics, for instance the following expression for the transmission coefficient:

$$T = \frac{1}{1 + 0.25\left(\frac{k}{k_0} - \frac{k_0}{k}\right)^2 \sin^2(ka)}$$

with k, k_0 - the wavevectors inside and outside the barrier, and a - its width (compare the previous formula with the results of III.I.7 of [13]). [27]

14. Conclusions

Several analogies between electromagnetic and quantum-mechanical phenomena have been analyzed. They rely upon the fact that both wave equation for electromagnetic and electric field with well defined frequency, obtained from the Maxwell equations, and the time-independent Schrodinger equation, have the same form - which is a Helmholtz equation.

However, the description of these analogies is by no way a simple dictionary between two formalisms. On the contrary, their physical basis has been discussed in detail, and they have been developed for very modern domains of physics - optical fibers, 2DESs, electron waveguides, electronic transport in mesoscopic and nanoscopic regime. So, the analogies examined in this chapter offers the opportunity of reviewing some very exciting, new and rapidely developing fields of physics, interesting from both the applicative and fundamental perspective. It has been stressed that the analogies are not simple curiosities, but they bear a significant cognitive potential, which can stimultate both scientific understanding and technological progress in fields like waveguides, optical fibers, nanoscopic transport.

15. Acknowledgement

The author acknowledges the financial support of the PN 09 37 01 06 project, granted by ANCS, received during the elaboration of this chapter.

16. References

[1] R.J. Black, A. Ankiewicz: Fiber-optic analogies with mechnics, Am.J.Phys. 53, 554 (1985)
[2] J.W. Goethe, The Teory of Colours, MIT Press, 1982
[3] J.D. Jackson, Classical Electrodynamics, 3rd edition, John Wiley & Sons, 1999
[4] C.K. Kao, Nobel Lecture, December 8, 2009
[5] K. C. Kao, G. A. Hockham, Proc. IEEE, vol. 113, p. 1151 (1966)
[6] T. G. Brown, in: M. Bass, E. W. Van Stryland (eds.): Fiber Optics Handbook, McGraw-Hill, 2002
[7] P. K. Tien, Rev.Mod.Phys. 49, 361 (1977)
[8] J. Spector, H.L. Stormer, K.W. Baldwin, L.N. Pfeiffer, K.W. West, Appl. Phys. Lett. 56, 2433 (1990)
[9] Yu. V. Sharvin, Sov.Phys.JETP 21, 655 (1965)
[10] V. S. Tsoi, JETP Lett. 19, 70 (1974)
[11] S. Datta, Electronic transport in mesoscopic systems, Cambridge University Press, 1995
[12] H. J. Pain, The Physics of Vibrations and Waves, John Wiley & Sons, Ltd, 6th edition, 2005
[13] A. Messiah, Quantum mechanics, Dover (1999)
[14] A. Hondros, P. Debye, Ann. Phys. 32, 465 (1910)
[15] A. W. Snyder, J. D. Love: Optical Waveguide Theory, Chapman and Hall, London, 1983
[16] K. Kawano, T. Kioth: Introduction to optical waveguide analysis: solving Maxwell's equations and Schrodinger equation, John Wiley & Sons, 2001
[17] S. Flugge, Practical quantum mechanics, Springer, 1971
[18] A. Zajonc: Catching the light, Oxford University Press, 1993
[19] G.N. Henderson, T.K. Gaylord, E.N. Glytsis, Phys.Rev. B45, 8404 (1992)
[20] R.A. Morrow, K.R. Brownstein, Phys.Rev. B30, 687 (1984)
[21] O. von Ross, Phys.Rev. B27, 7547 (1983)

[22] D. Dragoman, M. Dragoman, Opt.Commun. 133, 129 (1997)
[23] D. Dragoman, M. Dragoman, Progr. Quantum Electron. 23, 131 (1999)
[24] D. Dragoman, M. Dragoman: Quantum-classical analogies (1999)
[25] J. Spector, H.L. Stormer, K.W. Baldwin, L.N. Pfeiffer, K.W. West, Appl. Phys. Lett. 56, 2433 (1990)
[26] J. Spector, H.L. Stormer, K.W. Baldwin, L.N. Pfeiffer, K.W. West, Appl. Phys. Lett. 56, 1290 (1990)
[27] X. Ying, J.P. Lu, J.J. Heremans, M.B. Santos, M.S. Shayegan, S.A. Lyon, M. Littman, P. Gross, H. Rabitz, Appl.Phys.Lett. 65, 1154 (1994)

Part 3

Numerical Methods in Electromagnetism

7

Fast Preconditioned Krylov Methods for Boundary Integral Equations in Electromagnetic Scattering

Bruno Carpentieri
University of Groningen, Institute of Mathematics and Computing Science, Groningen
The Netherlands

1. Introduction

An accurate numerical solution of Electromagnetic scattering problems is critically demanded in the simulation of many real-life applications, such as in the design of industrial processes and in the study of wave propagation phenomena. Electromagnetic (EM) scattering problems address the physical issue of computing the diffraction pattern of the EM radiation that is propagated by a complex body, illuminated by an incident wave. An explicit solution is possible only for simple targets, e.g. for spherical bodies; complicated geometries impose to use approximate numerical techniques.

Until the emergence of high-performance computing in the early eighties, the analysis of scattering problems was afforded by using approximate high frequency techniques such as the shooting and bouncing ray method (SBR) (Lee et al. (1988)). Ray-based asymptotic methods like SBR and the uniform theory of diffraction are based on the idea that EM scattering becomes a localized phenomenon as the size of the scatterer increases with respect to the wavelength. In the last decades, due to impressive advances in computer technology and the introduction of innovative algorithms with limited computational and memory requirement, a more rigorous numerical solution has become possible for many practical applications.

Finite-difference (FD) (Kunz & Luebbers (1993); Taflove (1995)), finite-element (FE) (Silvester & Ferrari (1990); Volakis et al. (1998)) and finite-volume (FV) methods (Bonnet et al. (1998); Botha (2006)) can be used to discretize the Maxwell's equations into a finite volume surrounding the scatterer, giving rise to sparse systems of linear equations. Upon inversion of the system, a solution is computed for all excitations. More recently, alternative approaches based on integral equations are becoming increasingly popular for solving high-frequency EM scattering problems. They reformulate the Maxwell's equations in the frequency domain and solve for the electric and the magnetic currents induced on the surface of the object. Thus integral methods require only a simple description of the surface of the target by means of triangular facets (see an example of discretization in Figure 1). This means that a 3D problem is reduced to solving a 2D surface problem, simplifying considerably the mesh generation especially in the case of moving objects. No artificial boundaries need to be imposed and boundary conditions are automatically satisfied in the case of perfectly conducting objects.

Another interest to use surface discretizations is that they noticeably reduce the effect of grid dispersion errors. Grid dispersion errors occur when a wave has a different phase velocity on

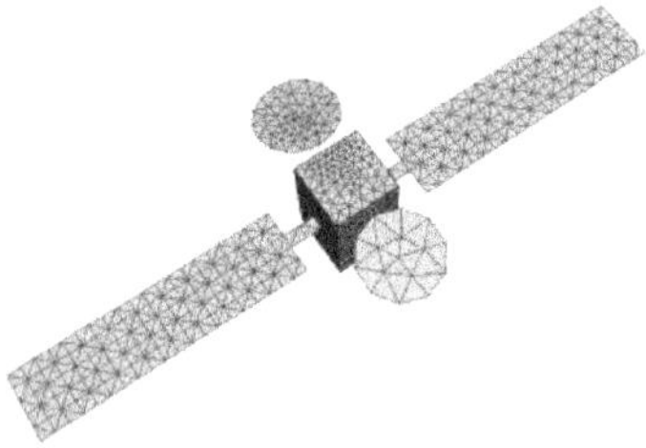

Fig. 1. Example of surface discretization in an integral equation context. Each unknown of the problem is associated to an edge in the mesh. Courtesy of the EMC-CERFACS Group in Toulouse.

the grid compared to the exact solution; they tend to accumulate in space and may introduce spurious solutions over large 3D simulation regions (Bayliss et al. (1985); Jr. (1994); Lee & Cangellaris (1992)). For second-order accurate differential schemes, to alleviate this problem the grid density may grow up to $\mathcal{O}((kd)^3)$ unknowns in 2D and of $\mathcal{O}((kd)^{4.5})$ in 3D, where k is the wavenumber and d is the approximate diameter of the simulation region. Therefore, the overall solution cost may increase considerably also for practical (*i.e.* finite) values of wavenumber (Chew et al. (1997)).

Boundary element discretizations are applied in many scientific and engineering areas beside electromagnetics and acoustics, e.g. in biomagnetic and bioelectric inverse modeling, magnetostatic and biomolecular problems, and many other applications (Forsman, Gropp, Kettunen & Levine (1995); Yokota, Bardhan, Knepley, Barba & Hamada (2011)). The potential drawback is that they lead, upon discretization, to large and dense linear systems to invert. Hence fast numerical linear algebra methods and efficient parallelization techniques are urged for solving large-scale boundary element equations efficiently on modern computers. In this chapter we overview some relevant techniques. In Section 2 we introduce the boundary integral formulation for EM scattering from perfectly conducting objects. In Section 4 we discuss fast iterative solution strategies based on preconditioned Krylov methods for solving the dense linear system arising from the discretization. In Section 5 we focus our attention on the design of the preconditioner, that is a critical component of Krylov methods in this context. We conclude our study in Section 5 with some final remarks.

2. The integral equation context

In an integral equation context, the standard EM scattering problem may be formulated in variational form as follows:

Find the surface current $\vec{j}$ such that for all tangential test functions $\vec{j}^t$, we have

$$\int_\Gamma \int_\Gamma G(|y-x|) \left(\vec{j}(x) \cdot \vec{j}^t(y) - \frac{1}{k^2} \mathrm{div}_\Gamma \vec{j}(x) \cdot \mathrm{div}_\Gamma \vec{j}^t(y) \right) dxdy =$$

$$= \frac{i}{kZ_0} \int_\Gamma \vec{E}_{inc}(x) \cdot \vec{j}^t(x) dx. \quad (1)$$

Eqn. (1) is called Electric Field Integral Equation (EFIE) (see Bilotti & Vegni (2003); Li et al. (2005)); we denote by $G(|y-x|) = \frac{e^{ik|y-x|}}{4\pi|y-x|}$ the Green's function of Helmholtz equation, Γ is the boundary of the object, k the wave number and $Z_0 = \sqrt{\mu_0/\varepsilon_0}$ the characteristic impedance of vacuum (ϵ is the electric permittivity and μ the magnetic permeability). Given a continuously differentiable vector field $\vec{j}(x)$ represented in Cartesian coordinates on a 3D Euclidean space as $\vec{j}(x_1,x_2,x_3) = j_{x_1}(x_1,x_2,x_3)\vec{e}_{x_1} + j_{x_2}(x_1,x_2,x_3)\vec{e}_{x_2} + j_{x_3}(x_1,x_2,x_3)\vec{e}_{x_3}$, where $\vec{e}_{x_1}, \vec{e}_{x_2}, \vec{e}_{x_3}$ are the unit basis vectors of the Euclidean space, we denote by $\mathrm{div}\vec{j}(x)$ the divergence operator defined as

$$\mathrm{div}\vec{j}(x) = \frac{\partial j_{x_1}}{\partial x_1} + \frac{\partial j_{x_2}}{\partial x_2} + \frac{\partial j_{x_3}}{\partial x_3}.$$

The EFIE formulation can be applied to arbitrary geometries such as those with cavities, disconnected parts, breaks on the surface; hence, it is very popular in industry.

For closed targets, the Magnetic Field Integral Equation (MFIE) can be used, which reads

$$\int_\Gamma (\vec{R}_{ext}\, j \wedge \vec{v}).\vec{j}^t + \frac{1}{2}\int_\Gamma \vec{j}.\vec{j}^t = -\int_\Gamma (\vec{H}_{inc} \wedge \vec{v}).\vec{j}^t.$$

The operator $\vec{R}_{ext}\, j$ is defined as

$$\vec{R}_{ext}\, j(y) = \int_\Gamma \vec{grad}_y G(|y-x|) \wedge \vec{j}(x)dx,$$

and is evaluated in the domain exterior to the object.

Both formulations suffer from interior resonances which make the numerical solution more problematic at some frequencies known as resonant frequencies, especially for large objects. The problem can be solved by combining linearly EFIE and MFIE. The resulting integral equation, known as Combined Field Integral Equation (CFIE), is the formulation of choice for closed targets. We point the reader to Gibson (2008) for a thorough presentation of integral equations in electromagnetism.

On discretizing Eqn. (1) in space by the Method of Moments (MoM) over a mesh containing n edges, the surface current $\vec{j}$ is expanded into a set of basis functions $\{\vec{\varphi}_i\}_{1\le i\le n}$ with compact support (the Rao-Wilton-Glisson basis, Rao et al. (1982), is a popular choice), then the integral equation is applied to a set of tangential test functions $\vec{j}^t$. Selecting $\vec{j}^t = \vec{\varphi}_j$, we are led to compute the set of coefficients $\{\lambda_i\}_{1\le i\le n}$ such that

$$\sum_{1\le i\le n} \lambda_i \left[\int_\Gamma\int_\Gamma G\left(|y-x|\right)\left(\vec{\varphi}_i(x)\cdot\vec{\varphi}_j(x) - \frac{1}{k^2} div_\Gamma \vec{\varphi}_i(x)\cdot div_\Gamma \vec{\varphi}_j(y)\right)dxdy\right] =$$

$$= \frac{i}{kZ_0}\int_\Gamma \vec{E}_{inc}(x)\cdot\vec{\varphi}_j(x)dx, \quad (2)$$

for each $1 \le i \le n$. The set of equations (2) can be recast in matrix form as

$$A\lambda = b, \quad (3)$$

where $A = \left[A_{ij}\right]$ and $b = [b_i]$ have elements, respectively,

$$A_{ij} = \int_\Gamma \int_\Gamma G(|y-x|)\left(\vec{\varphi}_i(x)\cdot\vec{\varphi}_j(y) - \frac{1}{k^2} div_\Gamma \vec{\varphi}_i(x)\cdot div_\Gamma \vec{\varphi}_j(y)\right) dxdy,$$

$$b_j = \frac{i}{kZ_0}\int_\Gamma \vec{E}_{inc}(x)\cdot\vec{\varphi}_j(y)dx.$$

The set of unknowns are associated with the vectorial flux across an edge in the mesh. The right-hand side varies with the frequency and the direction of the illuminating wave.

3. Fast matrix solvers for boundary element equations

Linear systems issued from boundary element discretizations may be very large in applications, although their size is typically much smaller compared to those arising from FE or FV formulations of the same problem. The number of unknowns grows linearly with the size of the scatterer and quadratically with the frequency of the incoming radiation (Bendali (1984)). A target with size of a few tens of wavelength, illuminated at $\mathcal{O}(1)$ GHz of frequency, may lead to meshes with several million points (Sylvand (2002)). Some efficient out-of-core dense direct solvers based on variants of Gaussian elimination have been proposed for solving blocks of right-hand sides, see e.g. Alléon, Amram, Durante, Homsi, Pogarieloff & Farhat (1997); Chew & Wang (1993). However, the memory requirements of direct methods are not affordable for solving such systems in realistic applications, even on modern parallel computers. Iterative methods can solve the problems of space of direct methods because they are based on matrix-vector (M-V) multiplications. In general terms, a modern integral equation solver is the mix of a robust iterative method, a fast algorithm for computing cheap approximate M-V products, and an efficient preconditioner to speed-up the convergence.

3.1 The choice of the iterative method

Krylov methods are among the most popular accelerators because of their ability to deliver good rates of convergence and to handle very large problems efficiently. They look for the solution of the system $Ax = b$ in the Krylov space $K_k(A,b) = span\{b, Ab, A^2b, ..., A^{k-1}b\}$. This is a good space from which to construct approximate solutions for a nonsingular linear system because it is intimately related to A^{-1}. In fact the inverse of any nonsingular matrix A can be written in terms of powers of A with the help of the minimal polynomial $q(t)$ of A, which is the unique monic polynomial of minimal degree such that $q(A) = 0$. If $\lambda_1, ..., \lambda_d$ are the distinct eigenvalues of A, m_j is the index of λ_j, and we define m as

$$m = \sum_{j=1}^{d} m_j,$$

then

$$q(t) = \prod_{j=1}^{d}(t-\lambda_j)^{m_j}. \tag{4}$$

Writing $q(t)$ in the form

$$q(t) = \sum_{j=0}^{m} \alpha_j t^j,$$

Solver	Iterations (CPU time)
CORS	601 (253*)
BiCOR	785 (334)
GMRES(50)	2191 (469)
QMR	878 (548)
BiCGSTAB	1065 (444)

Table 1. Number of iterations and CPU time (in seconds) required by Krylov methods to reduce the initial residual to $\mathcal{O}(10^{-8})$. An asterisk "*" indicates the fastest run. The problem is shown in Figure 2

we have

$$A^{-1} = -\frac{1}{\alpha_0}\sum_{j=0}^{m}\alpha_{j+1}A^j \quad , \quad \alpha_0 = \prod_{j=1}^{d}(-\lambda_j)^{m_j} \neq 0.$$

This shows that, if the minimal polynomial of A has degree m, then the solution of $Ax = b$ lies in the space $K_m(A, b)$. The smaller the degree of the minimal polynomial, the faster the expected rate of convergence of a Krylov method (see Ipsen & Meyer (1998)).

One issue is the choice of the suitable Krylov algorithm. Most integral formulations for surface and hybrid surface/volume scattering give rise to indefinite linear systems that cannot be solved using the Conjugate Gradient method (see discussions in Section 2). The GMRES method by Saad & Schultz (1986) is virtually always used for solving dense non-Hermitian linear systems as it is an optimal iterative solver, in the sense that it minimizes the 2-norm of the residual over the corresponding Krylov space. It generally requires the least number of iterations to converge. However, the optimality of GMRES comes at a price. The cost of applying the method increases with the iterations, and it may sometimes become prohibitively expensive for solving practical applications. As an attempt to limit the costs of GMRES, the algorithm is often restarted. After a given number of steps k, the approximate solution is computed from the generated Krylov subspace. Then the Krylov subspace is destroyed, and a new space is reconstructed using the latest residual.

On the other hand, non-optimal methods attempt to limit the costs of GMRES while preserving its favourable convergence properties. In Table 1, we show the number of iterations required by Krylov methods to reduce the initial residual to $\mathcal{O}(10^{-8})$ starting from the zero vector on the problem shown in Figure 2. For simplicity, the right-hand side of the linear system is set up so that the initial solution is the vector of all ones. We do not use preconditioning. In addition to restarted GMRES, we consider complex versions of iterative algorithms based on Lanczos biorthogonalization, such as BiCGSTAB (van der Vorst (1992)) and QMR (Freund & Nachtigal (1994)) and on the recently developed Lanczos biconjugate A-orthonormalization, such as BiCOR and CORS (Carpentieri et al. (2011); Jing, Huang, Zhang, Li, Cheng, Ren, Duan, Sogabe & Carpentieri (2009)). We clearly observe the importance of the choice of the iterative method. In our experiments, the CORS method is the fastest non-Hermitian solver with respect to CPU time on most selected examples except GMRES with large restart. Indeed, unrestarted GMRES may outperform all other Krylov methods and should be used when memory is not a concern. However, reorthogonalization costs may penalize the GMRES convergence in large-scale applications, so using high values of restart may not be convenient (or even not affordable for the memory) as shown in Carpentieri et al. (2005). In Table 1 we select a value of 50 for the restart parameter.

The BiCOR and CORS methods are introduced in Carpentieri et al. (2011); Jing, Huang, Zhang, Li, Cheng, Ren, Duan, Sogabe & Carpentieri (2009). They search for the approximate

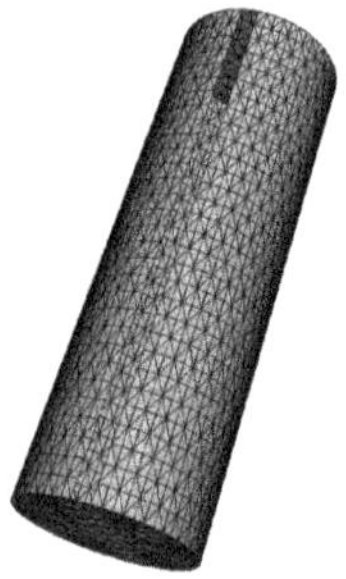

Fig. 2. Test problem: an open cylindric surface. Characteristics of the associated linear system: size=6268, frequency=362 MHz, $\kappa_1(A) = \mathcal{O}(10^5)$. Courtesy of the EMC-CERFACS Group in Toulouse.

solution in the Krylov subspace $K_m(A, r_0)$ by applying a Petrov-Galerkin approach and imposing the residual be orthogonal to the constraints subspace $A^H K_m(A^H, r_0^*)$; the shadow residual r_0^* is chosen to be equal to $r_0^* = Ar_0$. The basis vector representations for the subspaces $K_m(A, r_0)$ and $A^H K_m(A^H, r_0^*)$ are computed by means of the biconjugate A-Orthonormalization procedure. Starting from two vectors v_1 and w_1 chosen to satisfy the condition $\omega_1^H A v_1$, the method ideally builds up a pair of biconjugate A-orthonormal bases $v_j, j = 1, 2, \ldots, m$ and $w_i, i = 1, 2, \ldots, m$, respectively for the dual Krylov subspaces $K_m(A; v_1)$ and $K_m(A^H; w_1)$, satisfying the condition $\omega_i^H A v_j = \delta_{i,j}, 1 \leq i, j \leq m$. We point the reader to Jing, Carpentieri & Huang (2009) for further experiments with iterative Krylov methods for surface integral equations.

A significant amount of work has been devoted in the last years to design fast algorithms that can reduce the $\mathcal{O}(n^2)$ computational complexity for the M-V product operation required at each step of a Krylov method, such as the Fast Multipole Method (FMM) (Greengard & Rokhlin (1987); Rokhlin (1990)), the panel clustering method (Hackbush & Nowak (1989)), the $\mathcal{H}$-matrix approach (Hackbush (1999)), wavelet techniques (Alpert et al. (1993); Bond & Vavasis (1994)), the adaptive cross approximation method (Bebendorf (2000)), the impedance matrix localization method (Canning (1990)), the multilevel matrix decomposition algorithm (Michielssen & Boag (1996)) and others. In particular, the combination of iterative Krylov subspace solvers and FMM is a popular approach for solving integral equations. For Helmholtz and Maxwell problems, FMM algorithms enable to speedup M-V multiplications with boundary element matrices down to $\mathcal{O}(n \log n)$ algorithmic and memory complexity depending on the problem and on the specific implementation, see e.g. Cheng et al. (2006); Darrigrand (2002); Darve & Havé (2004); Dembart & Epton (1994); Engheta et al. (1992); Song & Chew (1995); Tausch (2004). Two-level implementations of FMM can reduce the cost of the M-V product operation from $\mathcal{O}(n^2)$ to $\mathcal{O}(n^{3/2})$, a three level algorithm down to $\mathcal{O}(n^{4/3})$ and the Multilevel Fast Multipole Algorithm (MLFMA) to $\mathcal{O}(n \log n)$.

Multipole techniques exploit the rapid decay of the Green's function and compute interactions amongst degrees of freedom in the mesh at different levels of accuracy depending on their physical distance. The 3D mesh of the object is partitioned recursively into boxes of roughly equal size until the size becomes small compared with the wavelength. The hierarchical partitioning of the object is typically represented using a tree-structured data called *oct-tree* (see Figure 3). Multipole coefficients are computed for all boxes starting from the smallest

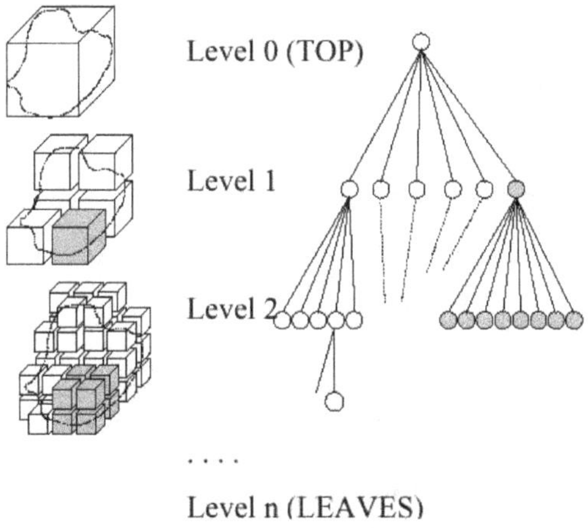

Fig. 3. The *oct-tree* data structure representation in the FMM algorithm. Each cube has up to eight children and one parent box except for the largest cube which encloses the whole domain.

ones, that are the leaves, and recursively for each parent cube in the tree by summing together multipole coefficients of its children. Interactions of degrees of freedom within one observation box and its close neighboring boxes are computed exactly using MoM; depending on the frequency, they generate between 1% and 2% of the entries of A. Interactions with boxes that are not neighbors of the observation box but whose parent in the oct-tree is a neighbor of the box parent are computed using FMM (see Figure 4). All other interactions are computed

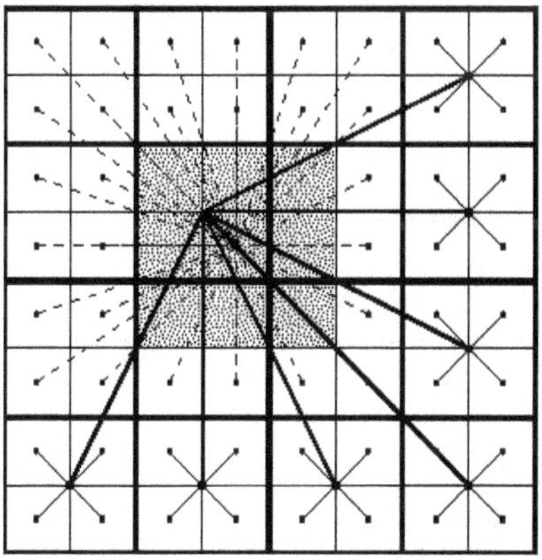

Fig. 4. Interactions in the multilevel FMM algorithm. Interactions for the gray boxes are computed directly. We denote by dashed lines cubes that are not neighbors of the cube itself but whose parent is a neighbor of the cube's parent. These interactions are computed using the FMM. All other interactions are computed hierarchically on a coarser level.

hierarchically on a coarser level by traversing the oct-tree. Multiple techniques have been efficiently implemented on distributed memory parallel computers proving to be scalable to several million discretization points, see for instance the FISC code developed at University of Illinois by Song & Chew (1998); Song et al. (1997; 1998), the INRIA/EADS integral equation code AS_ELFIP by Sylvand (2002; 2003), the Bilkent University code by Ergül & Gürel (2007; 2008) and others.

4. Algebraic preconditioning for boundary integral equations

Krylov methods may converge very slowly in practice, mainly due to bad spectral conditioning of the linear system. Relation (4) implies that the dimension of the solution space, and therefore the convergence properties, are mostly dictated by the eigenvalue distribution of A. The spectral properties may vary noticeably depending on the integral operator as well as on object shape and material. Problems with cavities or open surfaces typically require more iterations to converge than closed objects of the same physical size, and nonuniform meshes often produce ill-conditioned MoM matrices. On EFIE, the iteration count of Krylov solvers may increase as $\mathcal{O}(n^{0.5})$ when the number of unknowns n is related to the wavenumber, see for instance experiments reported in Song & Chew (1998), whereas on CFIE the number of iterations typically increases as $\mathcal{O}(n^{0.25})$.

On the other hand, if preconditioning A by a nonsingular matrix M the eigenvalues of $M^{-1}A$ fall into a few clusters, say t of them, whose diameters are small enough, then $M^{-1}A$ behaves numerically like a matrix with t distinct eigenvalues. As a result, we would expect t iterations of a Krylov method to produce reasonably accurate approximations. The matrix M is called the *preconditioner* matrix; preconditioning can be applied from the left as $M^{-1}Ax = M^{-1}b$ as well as from the right as $AM^{-1}y = b$ with $x = M^{-1}y$.

Optimal analytic preconditioners have been proposed for surface integral equations, see e.g. Antoine et al. (2004); Christiansen & Nédélec (2002); Steinbach & Wendland (1998). But they are problem-dependent. In this study, we consider purely algebraic techniques which compute the preconditioner only using information contained in the coefficient matrix of the linear system. Although far from optimal for any specific problem, algebraic methods can be applied to different operators and to changes in the geometry only by tuning a few parameters, and may often be developed from existent public-domain software implementations.

We are interested to develop techniques that have $\mathcal{O}(n \log n)$ algorithmic and memory complexity in the construction and in the application phase like FMM, and may be implemented efficiently within multipole codes. For memory concerns, we compute the preconditioner by initially decomposing the linear system in the form

$$(S + B)x = b \tag{5}$$

where S is a sparse matrix retaining the most relevant contributions to the singular integrals and is easy to invert, while B can be dense. If the continuous operator $\mathcal{S}$ underlying S is bounded and the operator $\mathcal{B}$ underlying B is compact, then $\mathcal{S}^{-1}\mathcal{B}$ is compact and

$$\mathcal{S}^{-1}(\mathcal{S} + \mathcal{B}) = \mathcal{I} + \mathcal{S}^{-1}\mathcal{B}.$$

We may expect that the preconditioned system $\left(I + S^{-1}B\right) x = S^{-1}b$ has a good clusterization of eigenvalues close to one, see e.g. Chen (1994) and (Chen, 2005, pp. 182-185).

The simplest approach to compute the local matrix S is to drop the small entries of A below a threshold (Cosnau (1996); Kolotilina (1988); Vavasis (1992)). When all the entries of A are not explicitly available, it may be necessary to use information extracted from the physical mesh of the problem. In an integral equation context, the surface of the object is discretized using a triangular mesh; each degree of freedom (DOF), or equivalently each unknown of the linear system, is associated to an edge of the mesh. Therefore, the sparsity pattern of S can be defined according to the concept of level k neighbours (see Figure 5(a)). Level 1 neighbours of a DOF are the DOF plus the four DOFs belonging to the two triangles that share the edge corresponding to the DOF itself. Level 2 neighbours are all the level 1 neighbours plus the DOFs in the triangles that are neighbours of the two triangles considered at level 1, and so forth. Due to the very localized nature of the Green's function, by retaining a few (two or three) levels of neighbours for each DOF an effective approximation may be constructed.

Comparative experiments show that there is little to choose. Both matrix- and mesh-based approaches can provide very good approximations S to the dense coefficient matrix for low sparsity ratio between 1% and 2% (Carpentieri et al. (2000)). The mesh-based approach is straightforward to implement in FMM codes as the object is typically partitioned using geometric information (see Figure 5(b)). Multipole algorithms yield a matrix decomposition

$$A = A_{diag} + A_{near} + A_{far}, \tag{6}$$

where A_{diag} is the block diagonal part of A associated with interactions of basis functions belonging to the same box, A_{near} is the block near-diagonal part of A associated with interactions within one level of neighboring boxes (they are 8 in 2D and 26 in 3D), and A_{far} is the far-field part of A. Therefore, in a multipole setting a suitable choice for the local matrix may be $S = A_{diag} + A_{near}$.

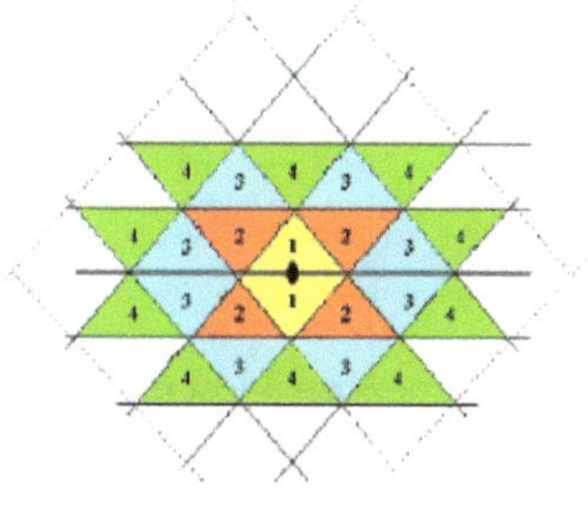

(a) Topological neighbours of a DOF in the discretization mesh.

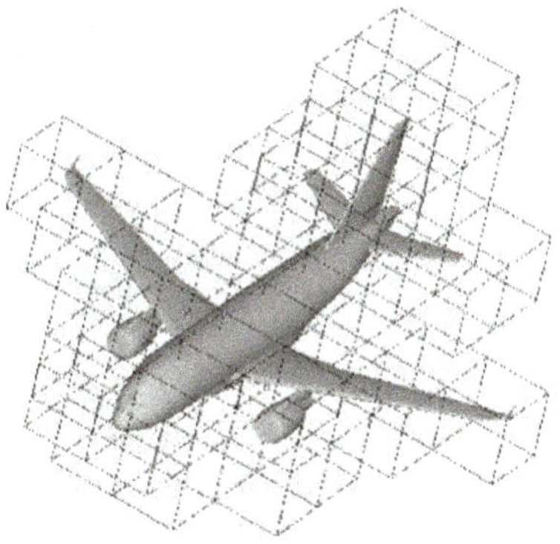

(b) Box-wise partitioning in the FMM context. Courtesy of EADS-CCR Toulouse.

Fig. 5. Mesh-based pattern selection strategies to compute local interactions in an integral equation context.

4.1 Comparison of standard preconditioners

To illustrate the difficulty of finding a good preconditioner for this problem class, in Table 2 we report one experiments with the GMRES solver and various algebraic preconditioners applied to a scattering problem from an open cylindric surface illuminated at 200 MHz of frequency and modeled using EFIE. The system has $n = 1299$ unknowns and is a low resolution testcase than the problem in Figure 2. In connection with GMRES, we consider preconditioners M of either implicit type (which approximately factorize S) or of explicit type (which approximately invert S) at roughly the same number of nonzero entries in M. We adopt the following acronyms:

- *None*, means that no preconditioner is used;
- *Diag*, a simple diagonal scaling, *i.e.* M is the diagonal of S;
- *SSOR*, the symmetric successive overrelaxation method $M = \frac{(D+\omega E)D^{-1}(D+\omega E^T)}{\omega(2-\omega)}$, where we denote by D the diagonal of S and E is the strict lower triangular part of S;
- *ILU*(0) by Saad (1996), the lower/upper incomplete LU factorization $M = \widetilde{L}\widetilde{U}$, $\widetilde{L} \approx L$, $\widetilde{U} \approx U$, $S = LU$, where the sparsity pattern of $\widetilde{L}$ (resp. $\widetilde{U}$) is equal to that of the lower (resp. upper) triangular part of S;
- *SPAI* by Grote & Huckle (1997), an approximate inverse preconditioner $M \approx S^{-1}$ computed by minimizing $\|I - SM\|_F$. The same pattern of S is imposed to M.
- *AINV* by Benzi et al. (1996), a sparse approximate inverse computed in factorized form by applying an incomplete biconjugation process to S, and dropping small entries below a threshold in the inverse factors.

Density of S = 3.18% - Density of M = 1.99%

Precond.	GMRES(30)	GMRES(80)	GMRES(∞)
None	-	-	302
Diag	-	-	272
SSOR	-	717	184
ILU(0)	-	454	135
SPAI	308	70	70
AINV	-	-	-

Table 2. Number of iterations using GMRES and various preconditioners on a test problem, a cylinder with an open surface, discretized with $n = 1299$ edges. The tolerance is set to 10^{-8}. The symbol '-' means that no convergence was achieved after 1000 iterations. The results are for right preconditioning.

We see that many standard methods fail. Simple preconditioners, like the diagonal of A, diagonal blocks, or a band, may be effective when the coefficient matrix has some degree of diagonal dominance (Song et al. (1997)). For ill-conditioned and indefinite matrices, more robust methods are needed. Techniques that are successful for solving partial differential equations may be successfully adopted for integral equations; in the next section, we analyse some of these methods.

4.2 Sparse approximate inverses preconditioner

Approximate inverse methods are very attractive for parallelism. They explicitly compute and store an approximation of the inverse of the coefficient matrix $M \approx S^{-1}$, which may be used as preconditioner by performing one or more sparse M-V products operations at each step of an iterative solver. As shown in Figure 6, due to the rapid decay of the Green's function the entries of A^{-1} may have a very similar structure to those of A, so that a very sparse preconditioner M may effectively capture the large contributions to the inverse.

(a) Pattern of large entries of A

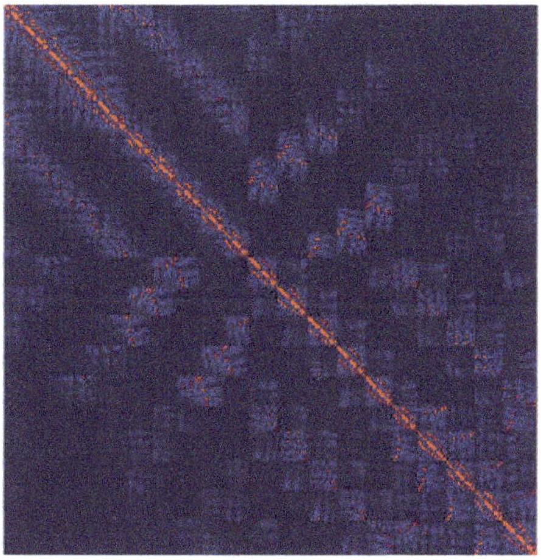

(b) Pattern of large entries of A^{-1}

Fig. 6. Structure of the large entries of A (on the left) and of A^{-1} (on the right). Large to small entries are depicted in different colors, from red to green, yellow and blue. The test problem is a small sphere.

The actual entries of M may be computed by minimizing the error matrix $\|I - SM\|_F$ for right preconditioning ($\|I - MS\|_F$ resp. left preconditioning). The Frobenius-norm allows to decouple the constrained minimization problem into n independent linear least-squares problems, one for each column (resp. row) of M when preconditioning from the right (resp. from the left). The independence of the least-squares problems can be immediately seen from the identity

$$\|I - SM\|_F^2 = \sum_{j=1}^{n} \|e_j - Sm_{\bullet j}\|_2^2, \tag{7}$$

where e_j is the jth canonical unit vector and $m_{\bullet j}$ is the column vector representing the jth column of M. In the case of right preconditioning, the analogous relation

$$\|I - MS\|_F^2 = \|I - S^T M^T\|_F^2 = \sum_{j=1}^{n} \|e_j - S^T m_{j\bullet}\|_2^2 \tag{8}$$

holds, where $m_{j\bullet}$ is the column vector representing the jth row of M. The preconditioner is not guaranteed to be nonsingular in general, and additionally it does not preserve any possible symmetry of A. The condition to ensure non-singularity of M may be derived from the following estimates of the accuracy of the approximate inverse (Grote & Huckle (1997)):

THEOREM **1.** *Let $r_j = Sm_j - e_j$ be the residual associated with column m_j for $j = 1,2,\ldots,n$, and $q = \max_{1\leq j\leq n}\left\{nnz\left(r_j\right)\right\} \ll n$. Suppose that $\left\|r_j\right\|_2 < t$ for $j = 1,2,\ldots,n$, then we have*

$$\begin{array}{ll} \|SM - I\|_F \leq \sqrt{n}t, & \|M - S^{-1}\|_F \leq \|S^{-1}\|_2 \sqrt{n}t, \\ \|SM - I\|_2 \leq \sqrt{n}t, & \|M - S^{-1}\|_2 \leq \|S^{-1}\|_2 \sqrt{n}t, \\ \|SM - I\|_1 \leq \sqrt{q}t, & \|M - S^{-1}\|_1 \leq \|S^{-1}\|_1 \sqrt{q}t. \end{array}$$

■

Owing to this result, all the eigenvalues of SM lie in the disk centered in 1 and of radius $\sqrt{q}t$; the value of q is not known a priori, though, so that one might enforce the condition $\sqrt{n}t < 1$ to prevent singularity or near-singularity of the preconditioned matrix. In practice it may be too costly to compute M with such a small t. For some problems, it may be observed a lack of robustness of the approximate inverse due to the clustering of small eigenvalues in the spectrum of the preconditioned matrix. Stabilization techniques based on eigenvalue deflation may be used to enhance the robustness of M, see e.g. Carpentieri et al. (2003).

The most critical component is the computation of the nonzero structure of M. From Figure 6, we see that the sparse pattern of S may be a suitable pattern for M. Denoting by

$$\mathcal{P} = \{\, (i,j) \in [1,n]^2 \text{ s.t. } m_{ij} \neq 0 \,\}$$

the nonzero structure of the approximate inverse, we may automatically determine the pattern of the nonzero entries of the jth column of M as

$$\mathcal{C}_j = \{i \in [1,n] \text{ s.t. } (i,j) \in \mathcal{P}\}.$$

and compute the associated entries by solving a small size dense least-squares problem. The least-squares solution involves only those columns of S indexed by $\mathcal{C}_j$; we indicate this subset by $S(:,\mathcal{C}_j)$. Because S is sparse, many rows in $S(:,\mathcal{C}_j)$ are usually null, not affecting the solution of the least-squares problems (7). Thus denoting by $\mathcal{R}_j$ the set of indices corresponding to the nonzero rows in $S(:,\mathcal{C}_j)$, by $\widehat{S} = S(\mathcal{R}_j,\mathcal{C}_j)$, by $\widehat{m}_j = m_j(\mathcal{C}_j)$, and by $\widehat{e}_j = e_j(\mathcal{C}_j)$, the actual "reduced" least-squares problems to solve are

$$min\|\widehat{e}_j - \widehat{S}\widehat{m}_j\|_2, \; j = 1,..,n. \tag{9}$$

Usually problems (9) have much smaller size than problems (7) and can be efficiently solved by dense QR factorization. The parallel implementation of the approximate inverse is highly scalable as shown in Table 3, while the numerical performance typically tend to deteriorate for increasing matrix size as can be seen in Table 4.

Approximate inverses may be also computed in factorized form as $M = \widetilde{G}\widetilde{Z}$, where $\widetilde{G} \approx U^{-1}$ and $\widetilde{Z} \approx U^{-1}$ are approximation of the inverse triangular factors of S, see for instance Alléon, Benzi & Giraud (1997); Chen (1998); Rahola (1998); Samant et al. (1996). One example of such preconditioner is the $AINV$ method by Benzi et al. (1996), a sparse approximate inverse computed in factorized form by applying an incomplete biconjugation process to S and dropping small entries below a threshold in the inverse factors. However, disappointing results with factorized approximate inverses have been reported on this problem class, see e.g. Carpentieri et al. (2004). The reason of failure is that for many integral formulations like EFIE and CFIE, the inverse factors may be totally unstructured. In this case, selecting

n (procs)	Construction time (sec)	Elapsed time precond (sec)
112908 (8)	513	0.39
221952 (16)	497	0.43
451632 (32)	509	0.48
900912 (64)	514	0.60

Table 3. Parallel scalability of the approximate inverse for solving large-scale boundary integral equations on a model problem.

$dof\,/\,freq$	FROB		GMRES(∞)		GMRES(120)	
	Density	Time	Iter	Time	Iter	Time
23676 / 1.3 Ghz	0.94	2m	438	20m	+2000	55m
104793 / 2.6 "	0.19	6m	234	20m	253	17m
419172 / 5.2 "	0.05	21m	413	2h 44m	571	2h 26m
943137 / 7.8 "	0.02	49m	454	3h 35m•	589	5h 55m

Table 4. Numerical scalability of the approximate inverse for solving large-scale boundary integral equations. The symbol • means run on 32 processors. Notation: m means minutes, h hours.

a priori the sparse pattern for the factors can be extremely hard and dynamic pattern selection strategies, that drop small entries below a user-defined threshold during the computation, may be very difficult to tune as they can easily discard relevant information and lead to a very poor preconditioner. For those problems, finding the appropriate threshold to enable a good trade-off between sparsity and numerical efficiency is challenging and very problem-dependent.

4.3 Incomplete LU factorization preconditioner

ILU-type methods compute an approximate triangular decomposition of S by means of an incomplete Gaussian elimination process. The ILU preconditioner writes as $M = \widetilde{L}\widetilde{U}$, $\widetilde{L} \approx L$, $\widetilde{U} \approx U$ where L and U denote respectively the lower and upper triangular factors of the standard LU factorization of S. This class of methods is virtually always used for solving sparse linear systems. However, mixed success is reported on dense matrix problems, due to the indefiniteness of the systems arising from the discretization. The root of the problem is that small pivots often appear during the factorization, leading to highly ill-conditioned triangular factors and unstable triangular solves (Carpentieri et al. (2004)).

In Table 5 we show an experiment with an ILU preconditioner computed from the sparse approximation S to A, using different values of density for S. The test case is a sphere of 1 meter length illuminated at 300 MHz; the problem is modeled using EFIE and the mesh is discretized with 2430 edges. The set $\mathcal{F}$ of fill-in entries to be kept for the approximate lower triangular factor $\widetilde{L}$ is defined by

$$\mathcal{F} = \{\, (k,i) \mid lev(l_{k,i}) \le \ell \,\} \,,$$

where the integer ℓ denotes a user specified maximal fill-in level. The level $lev(l_{k,i})$ of the coefficient $l_{k,i}$ of L is computed as follows:

Initialization

$$lev(l_{k,i}) = \begin{cases} 0 \ \text{ if } \ l_{k,i} \neq 0 \ \text{ or } \ k = i \\ \\ \infty \ \text{otherwise} \end{cases}$$

Factorization

$$lev(l_{k,i}) = \min\left\{ lev(l_{k,i}) \, , \, lev(l_{i,j}) + lev(l_{k,j}) + 1 \right\}.$$

Observe that the larger ℓ, the higher the density of the preconditioner. We denote the resulting preconditioner by $ILU(\ell)$ Saad (1996).

In our results, increasing the fill-in parameter may produce much more robust preconditioners than $ILU(0)$ applied to a denser sparse approximation of the original matrix; $ILU(1)$ may deliver a good rate of convergence provided the coefficient matrix is not too sparse. However, the factorization of a very sparse approximation (up to 2%) of the coefficient matrix can be stable and accelerate significantly the convergence, especially if at least one level of fill-in is retained. Then, for higher values of the density of S the factors may become progressively ill-conditioned, the triangular solves unstable and consequently the preconditioner is useless. The table also shows that ill-conditioning of the factors is not related to ill-conditioning of A.

Density of S = 2%				
IC(level)	Density of L	$\kappa_\infty(L)$	GMRES(30)	GMRES(50)
$IC(0)$	2.0%	$2 \cdot 10^3$	378	245
$IC(1)$	5.1%	$1 \cdot 10^3$	79	68
$IC(2)$	9.1%	$9 \cdot 10^2$	58	48
Density of S = 4%				
IC(level)	Density of L	$\kappa_\infty(L)$	GMRES(30)	GMRES(50)
$IC(0)$	4.0%	$6 \cdot 10^9$	-	-
$IC(1)$	11.7%	$2 \cdot 10^5$	-	-
$IC(2)$	19.0%	$7 \cdot 10^3$	40	38
Density of S = 6%				
IC(level)	Density of L	$\kappa_\infty(L)$	GMRES(30)	GMRES(50)
$IC(0)$	6.0%	$8 \cdot 10^{11}$	-	-
$IC(1)$	18.8%	$5 \cdot 10^{11}$	-	-
$IC(2)$	29.6%	$7 \cdot 10^4$	-	-

Table 5. Number of iterations of GMRES varying the sparsity level of S and the level of fill-in of the approximate factor L on a spherical model problem ($n = 2430, \kappa_\infty(A) = \|A\|_\infty \|A^{-1}\|_\infty \approx \mathcal{O}(10^2)$). The symbol '-' means that convergence was not obtained after 500 iterations.

A complex diagonal compensation can help to compute a more stable preconditioner, by shifting along the imaginary axis the eigenvalues close to zero in the spectrum of the coefficient matrix. However, the value of the shift is not easy to tune *a priori* and its effect on the convergence is difficult to predict (Carpentieri et al. (2004)). Pivoting may be a more robust approach to overcome the problem according to reported experiment by Malas & Gürel (2007); in this case, the ith row of the factor is computed as soon as $permtol \times |s_{ij}| > |s_{ii}|$, where $permtol$ is the permutation tolerance and s_{ij} are the entries of S.

We follow a different approach. We report on experiments with multilevel inverse-based ILU factorization methods to possibly remedy numerical instabilities. Following Bollhöfer & Saad (2006), we initially rescale and reorder the initial matrix A as

$$P^T D_l A D_r Q = \hat{A}, \tag{10}$$

which yields $\hat{A}\hat{x} = \hat{b}$ for appropriate $\hat{x}, \hat{b}$. The initial step may consist of an optional maximum weight matching (Duff & Koster (1999)). By rescaling and a one-sided permutation, it attempts to improve the diagonal dominance. After that, a symmetric reordering is applied to reduce the fill-/bandwidth. The latter can also be used without an a priori matching step, only rescaling the entries and symmetrically permuting the rows and the columns. This is of particular interest for (almost) symmetrically structured problems. Next, an inverse-based ILU with static diagonal pivoting is computed. I.e., during the approximate incomplete factorization $\hat{A} \approx LDU$ such that L, U^H are unit lower triangular factors and D is block diagonal, the norms $\|L^{-1}\|, \|U^{-1}\|$ are estimated. If at factorization step l a prescribed bound κ is exceeded, the current row l and column l are permuted to the lower right end of the matrix. Otherwise the approximate factorization is continued. One single pass leads to an approximate partial factorization

$$\Pi^T \hat{A} \Pi = \begin{pmatrix} B & F \\ E & C \end{pmatrix} \approx \begin{pmatrix} L_B & 0 \\ L_E & I \end{pmatrix} \begin{pmatrix} D_B & 0 \\ 0 & S_C \end{pmatrix} \begin{pmatrix} U_B & U_F \\ 0 & I \end{pmatrix} \equiv L_1 D_1 U_1, \tag{11}$$

with a suitable leading block B and a suitable permutation matrix Π, where $\|L_1^{-1}\| \leq \kappa$, $\|U_1^{-1}\| \leq \kappa$. The remaining system S_C approximates $C - EB^{-1}F$. From the relations

$$\begin{cases} B\hat{x}_1 + F\hat{x}_2 = \hat{b}_1 \\ E\hat{x}_1 + C\hat{x}_2 = \hat{b}_2 \end{cases} \Rightarrow \begin{cases} \hat{x}_1 = B^{-1}(\hat{b}_1 - F\hat{x}_2) \\ (C - EB^{-1}F)\hat{x}_2 = \hat{b}_2 - EB^{-1}\hat{b}_1 \end{cases},$$

at each step of an iterative solver we need to store and invert only blocks with B and $S_C \approx C - EB^{-1}F$ while for reasons of memory efficieny, L_E, U_F are discarded and implicitly represented via $L_E \approx EU_B^{-1}D_B^{-1}$ (resp. $U_F \approx D_B^{-1}L_B^{-1}F$). When the scaling, preordering and the factorization is successively applied to S_C, a multilevel variant of (10) is computed. E.g., after a one additional level we obtain

$$\tilde{P}\tilde{D}_l A \tilde{D}_r \tilde{Q} = \left[\begin{array}{cc|c} B & F_1 & F_2 \\ E_1 & C_{11} & C_{12} \\ \hline E_2 & C_{21} & C_{22} \end{array}\right] \approx \left[\begin{array}{cc|c} L_B & 0 & 0 \\ L_{E_1} & I & 0 \\ \hline L_{E_2} & L_{C_{21}} & I \end{array}\right] \left[\begin{array}{cc|c} D_B & 0 & 0 \\ 0 & D_{C_{11}} & 0 \\ \hline 0 & 0 & S_{22} \end{array}\right] \left[\begin{array}{cc|c} U_B & U_{F_1} & U_{F_2} \\ 0 & I & U_{C_{12}} \\ \hline 0 & 0 & I \end{array}\right].$$

The multilevel algorithm ends at some step m when either S_C is factored completely or it becomes considerably dense and switches to a dense LAPACK solver. After computing an m-step ILU decomposition, for preconditioning we have to apply $L_m^{-1}AU_m^{-1}$. From the error equation $E_m = A - L_m D_m U_m$, we see that $\|L_m^{-1}\|$ and $\|U_m^{-1}\|$ contribute to the inverse error $L_m^{-1}E_m U_m^{-1}$. Monitoring the growth of these two quantities during the partial factorization is essential to preserve the numerical stability of the solver, as can be observed comparing results in Table 5 and Table 6.

Density of S = 2%			
threshold	Density of L	GMRES(30)	GMRES(50)
1.0e-3	0.30	29	29
Density of S = 4%			
$MILU$	Density of L	GMRES(30)	GMRES(50)
1.0e-3	0.39	26	26
Density of S = 6%			
$MILU$	Density of L	GMRES(30)	GMRES(50)
1.0e-3	0.46	24	24

Table 6. Number of iterations of GMRES using a multilevel inverse-based ILU factorization as preconditioner. The model problem is the same as in Table 5.

5. Concluding remarks

We have discussed some fast iterative solution techniques for solving surface boundary integral equations. High-frequency simulations of large structures are extremely demanding for scalable solvers and large computing resources. We have reviewed recent advances for the class of Krylov subspace methods, sparse approximate inverses, incomplete LU factorizations.

Other approach have been applied in this area of research. Multigrid methods are provably optimal algorithms for solving various classes of partial differential equations. Attempts to apply these techniques to dense linear systems have obtained mixed success. Early experiments on boundary element equations are reported with geometric versions on simple model problems, typically the hypersingular and single-layer potential integral operators arising from the Laplace equation (Bramble et al. (1994); Petersdorff & Stephan (1992); Rjasanow (1987)). Multigrids require a hierarchy of nested meshes to setup the principal components of the algorithm, *i.e.* a coarsening strategy to decrease the number of unknowns, grid transfer operators to move from a grid to another one, coarse grid operators and smoothing procedure, see e.g. Hackbusch (1985). Thus they are difficult to implement. On the other hand, algebraic multigrid algorithms use only single grid information extracted from either the graph or the entries of the coefficient matrix and are nearly as effective as geometric algorithms in reducing the number of iterations, see e.g Braess (1995); Brandt (1999); Ruge & Stüben (1987); Vanek et al. (1996). Langer et al. propose to apply an auxiliary sparse matrix reflecting the local topology of the mesh on the fine grid to setup all the components of the multigrid algorithm in a purely algebraic setting (Langer et al. (2003)). This *gray-box* approach is fairly robust on model problems and maintains the algorithmic and memory complexity of the M-V product operation (Langer & Pusch (2005)), thus it is well suited to be combined with MLFMA. See also Carpentieri et al. (2007) for another multigrid-type solver.

Preconditioners based on wavelet techniques are also receiving interest. The wavelet compression of integral operators with smooth kernels yields nearly sparse matrices with at most $\mathcal{O}(n \log^a n)$ nonzero entries, where a is a small constant that depends on the operator and the wavelet used, see e.g. earlier work by Beylkin et al. (1991); Dahmen et al. (1993); Harbrecht & Schneider (2004); Hawkins et al. (2007); Lage & Schwab (1999). The compressed matrix is spectrally equivalent to the original matrix and preconditioning is often needed (Chan & Chen (2000; 2002); Chan et al. (1997); Chen (1999); Ford & Chen (2001); Hawkins & Chen (2005); Hawkins et al. (2005)). Some efficient wavelet preconditioning algorithms have been proposed, based on bordered block structure (Ford & Chen (2001); Hawkins et al. (2005)), multi-level preconditioners (Chan & Chen (2002)), and sparse approximate inverses. However, most experiments with wavelet preconditioners are reported for model problems, e.g. Calderon-Zygmund type matrix, single and double layer potentials, the hyper-singular operator. For oscillatory kernels the compressed matrix may be fairly dense and wavelet techniques are less useful. For Helmholtz problems, wavelet Galerkin schemes yield matrices with approximately $\mathcal{O}(kn)$ (k is the wavenumber) which becomes $O(n^2)$ when the number of unknowns is related to k.

Further investigations are necessary to identify the best class of methods for the given problem and the selected computer hardware. The use of more powerful (but also more complex) computing facilities should help in the search for additional speed, but it will also mean that there will be even more factors that need to be considered when attempting to identify the optimal approach in the future.

6. References

Alléon, G., Amram, S., Durante, N., Homsi, P., Pogarieloff, D. & Farhat, C. (1997). Massively parallel processing boosts the solution of industrial electromagnetic problems: High performance out-of-core solution of complex dense systems, *in* M. Heath, V. Torczon, G. Astfalk, P. E. Bjï£¡rstad, A. H. Karp, C. H. Koebel, V. Kumar, R. F. Lucas, L. T. Watson & E. D. E. Womble (eds), *Proceedings of the Eighth SIAM Conference on Parallel Computing*, SIAM Book, Philadelphia. Conference held in Minneapolis, Minnesota, USA.

Alléon, G., Benzi, M. & Giraud, L. (1997). Sparse approximate inverse preconditioning for dense linear systems arising in computational electromagnetics, *Numerical Algorithms* 16: 1–15.

Alpert, B., Beylkin, G., Coifman, R. & Rokhlin, V. (1993). Wavelet-like bases for the fast solution of second-kind integral equations, *SIAM J. Scientific and Statistical Computing* 14: 159–184.

Antoine, X., Bendali, A. & Darbas, M. (2004). Analytic preconditioners for the electric field integral equation, *International Journal for Numerical Methods in Engineering* 61(8): 1310–1331.

Bayliss, A., Goldstein, C. I. & Turkel, E. (1985). On accuracy conditions for the numerical computation of waves, *J. Comp. Phys.* 59: 396–404.

Bebendorf, M. (2000). Approximation of boundary element matrices, *Numerische Mathematik* 86(4): 565–589.

Bendali, A. (1984). *Approximation par elements finis de surface de problemes de diffraction des ondes electro-magnetiques*, PhD thesis, *Université Paris VI* .

Benzi, M., Meyer, C. & Tuma, M. (1996). A sparse approximate inverse preconditioner for the conjugate gradient method., *SIAM J. Scientific Computing* 17: 1135–1149.

Beylkin, G., Coifman, R. & Rokhlin, V. (1991). Fast wavelet transforms and numerical algorithms, *Comm. Pure Appl. Math.* 44: 141–183.

Bilotti, F. & Vegni, C. (2003). MoM entire domain basis functions for convex polygonal patches, *Journal of Electromagnetic Waves and Applications* 17(11): 1519–1538.

Bollhöfer, M. & Saad, Y. (2006). Multilevel preconditioners constructed from inverse–based ILUs, *SIAM J. Scientific Computing* 27(5): 1627–1650.

Bond, D. & Vavasis, S. (1994). Fast wavelet transforms for matrices arising from boundary element methods, *Technical Report 94-174*, Cornell University Ithaca, NY, USA.

Bonnet, P., Ferrieres, X., Grando, J., Alliot, J. & Fontaine, J. (1998). Frequency-domain finite volume method for electromagnetic scattering, *Antennas and Propagation Society International Symposium, 1998. IEEE* 1(21-26): 252–255.

Botha, M. (2006). Solving the volume integral equations of electromagnetic scattering, *J. Comput. Phys.* 218(1): 141–158.

Braess, D. (1995). Towards algebraic multigrid for elliptic problems of second order, *Computing* 55(4): 379–393.

Bramble, J., Leyk, Z. & Pasciak, J. (1994). The analysis of multigrid algorithms for pseudodifferential operators of order minus one, *Math. Comput.* 63(208): 461–478.

Brandt, A. (1999). General highly accurate algebraic coarsening, *Electronic Trans. Num. Anal* 10: 1–20.

Canning, F. (1990). The impedance matrix localization (IML) method for moment-method calculations, *IEEE Antennas and Propagation Magazine* .

Carpentieri, B., Duff, I. & Giraud, L. (2003). A class of spectral two-level preconditioners, *SIAM J. Scientific Computing* 25(2): 749–765.

Carpentieri, B., Duff, I., Giraud, L. & monga Made, M. M. (2004). Sparse symmetric preconditioners for dense linear systems in electromagnetism., *Numerical Linear Algebra with Applications* 11(8-9): 753–771.

Carpentieri, B., Duff, I., Giraud, L. & Sylvand, G. (2005). Combining fast multipole techniques and an approximate inverse preconditioner for large electromagnetism calculations, *SIAM J. Scientific Computing* 27(3): 774–792.

Carpentieri, B., Duff, I. S. & Giraud, L. (2000). Sparse pattern selection strategies for robust Frobenius-norm minimization preconditioners in electromagnetism, *Numerical Linear Algebra with Applications* 7(7-8): 667–685.

Carpentieri, B., Giraud, L. & Gratton, S. (2007). Additive and multiplicative two-level spectral preconditioning for general linear systems., *SIAM J. Scientific Computing* 29(4): 1593–1612.

Carpentieri, B., Jing, Y.-F. & Huang, T.-Z. (2011). The BiCOR and CORS algorithms for solving nonsymmetric linear systems, *SIAM J. Scientific Computing* . In press.

Chan, T. & Chen, K. (2000). Two-stage preconditioners using wavelet band splitting and sparse approximation, *UCLA CAM report CAM00-26*, Dept of Mathematics, UCLA, USA.

Chan, T. & Chen, K. (2002). On two variants of an algebraic wavelet preconditioner, *SIAM Journal on Scientific Computing* 24(1): 260–283.

Chan, T., Tang, W.-P. & Wan, W. (1997). Fast wavelet based sparse approximate inverse preconditioner, *BIT* 37(3): 644–660.

Chen, K. (1994). Efficient iterative solution of linear systems from discretizing singular integral equations, *Elec. Trans. Numer. Anal.* 2: 76ï£¡91.

Chen, K. (1998). On a class of preconditioning methods for dense linear systems from boundary elements., *SIAM J. Scientific Computing* 20(2): 684–698.

Chen, K. (1999). Discrete wavelet transforms accelerated sparse preconditioners for dense boundary element systems, *Electronic Transactions on Numerical Analysis* 8: 138–153.

Chen, K. (2005). *Matrix Preconditioning Techniques and Applications*, Cambridge University Press.

Cheng, H., Crutchfield, W., Gimbutas, Z., Greengard, L., Ethridge, J., Huang, J., Rokhlin, V., Yarvin, N. & Zhao, J. (2006). A wideband fast multipole method for the Helmholtz equation in three dimensions, *J. Comput. Phys.* 216(1): 300–325.

Chew, W. C., Jin, J. M., Lu, C. C., Michielssen, E. & Song, J. M. (1997). Fast solution methods in electromagnetics, *IEEE Transactions on Antennas and Propagation* 45(3): 533–543.

Chew, W. & Wang, Y. (1993). A recursive T-matrix approach for the solution of electromagnetic scattering by many spheres, *IEEE Transactions on Antennas and Propagation* 41(12): 1633–1639.

Christiansen, S. & Nédélec, J.-C. (2002). A preconditioner for the electric field integral equation based on Calderon formulas, *SIAM J. Numer. Anal.* 40(3): 1100–1135.

Cosnau, A. (1996). Etude d'un préconditionneur pour les matrices complexes dense symmetric issues des équations de Maxwell en formulation intégrale., *Note technique ONERA*, ONERA. 142328.96/DI/MT.

Dahmen, W., Prössdorf, S. & Schneider, R. (1993). Wavelet approximation methods for pseudodifferential equations ii: Matrix compression and fast solution, *Advances in Computational Mathematics* 1(3): 259–335.

Darrigrand, E. (2002). Coupling of fast multipole method and microlocal discretization for the 3D Helmholtz equation, *J. Comput. Phys.* 181(1): 126–154.

Darve, E. & Havé, P. (2004). Efficient fast multipole method for low-frequency scattering, *J. Comput. Phys.* 197(1): 341–363.

Dembart, B. & Epton, M. (1994). A 3D fast multipole method for electromagnetics with multiple levels, *Tech. Rep. ISSTECH-97-004*, The Boeing Company, Seattle, WA.

Duff, I. S. & Koster, J. (1999). The design and use of algorithms for permuting large entries to the diagonal of sparse matrices, *SIAM J. Matrix Analysis and Applications* 20(4): 889–901.

Engheta, N., Murphy, W., Rokhlin, V. & Vassiliou, M. (1992). The fast multipole method (FMM) for electromagnetic scattering problems, *IEEE Transactions on Antennas and Propagation* 40(6): 634–641.

Ergül, Ö. & Gürel, L. (2007). Fast and accurate solutions of extremely large integral-equation problems discretized with tens of millions of unknowns, *Electron. Lett.* 43(9): 499–500.

Ergül, Ö. & Gürel, L. (2008). Efficient parallelization of the multilevel fast multipole algorithm for the solution of large-scale scattering problems, *IEEE Transactions on Antennas and Propagation* 56(8): 2335–2345.

Ford, J. & Chen, K. (2001). Wavelet-based preconditioners for dense matrices with non-smooth local features, *BIT* 41(2): 282–307.

Forsman, K., Gropp, W., Kettunen, L. & Levine, D. (1995). Volume integral equations in nonlinear 3D magnetostatics, *International Journal of Numerical Methods in Engineering* 38: 2655–2675.

Yokota, R., Bardhan, J. P., Knepley, M. G., Barba, L.A. & Hamada, T. (2011). Biomolecular electrostatics using a fast multipole BEM on up to 512 gpus and a billion unknowns, *Computer Physics Communications* 182(6): 1272–1283.

Freund, R. W. & Nachtigal, N. M. (1994). An implementation of the QMR method based on coupled two-term recurrences., *SIAM J. Scientific Computing* 15(2): 313–337.

Gibson, W. (2008). *The method of moments in electromagnetics.*, Boca Raton, FL: Chapman and Hall/CRC.

Greengard, L. & Rokhlin, V. (1987). A fast algorithm for particle simulations, *Journal of Computational Physics* 73: 325–348.

Grote, M. & Huckle, T. (1997). Parallel preconditionings with sparse approximate inverses., *SIAM J. Scientific Computing* 18: 838–853.

Hackbusch, W. (1985). *Multigrid Methods and Applications*, Springer-Verlag, Berlin.

Hackbush, W. (1999). A sparse matrix arithmetic based on $\mathcal{H}$-matrices, *Computing* 62(2): 89–108.

Hackbush, W. & Nowak, Z. (1989). On the fast matrix multiplication in the boundary element method by panel clustering, *Numerische Mathematik* 54(4): 463–491.

Harbrecht, H. & Schneider, R. (2004). Wavelet based fast solution of boundary integral equations, *in* N. C. et al. (ed.), *Abstract and Applied Analysis*, World Scientific Publishing Company, pp. 139–162. Proceedings of the International Conference in Hanoi, 2002.

Hawkins, S. & Chen, K. (2005). An implicit wavelet sparse approximate inverse preconditioner, *SIAM J. Scientific Computing* 27(2): 667–686.

Hawkins, S., Chen, K. & Harris, P. (2005). An operator splitting preconditioner for matrices arising from a wavelet boundary element method for the Helmholtz equation, *International Journal of Wavelets, Multiresolution and Information Processing* 3 (4): 601–620.

Hawkins, S., Chen, K. & Harris, P. (2007). On the influence of the wavenumber on compression in a wavelet boundary element method for the Helmholtz equation, *International Journal of Numerical Analysis and Modeling* 4(1): 48–63.

Ipsen, I. C. F. & Meyer, C. D. (1998). The idea behind Krylov methods, *The American Mathematical Monthly* 105(10): 889–899.

Jing, Y.-F., Carpentieri, B. & Huang, T.-Z. (2009). Experiments with Lanczos biconjugate A-orthonormalization methods for MoM discretizations of Maxwell's equations, *Progress In Electromagnetics Research, PIER 99* pp. 427–451.

Jing, Y.-F., Huang, T.-Z., Zhang, Y., Li, L., Cheng, G.-H., Ren, Z.-G., Duan, Y., Sogabe, T. & Carpentieri, B. (2009). Lanczos-type variants of the COCR method for complex nonsymmetric linear systems, *Journal of Computational Physics* 228(17): 6376–6394.

Jr., W. S. (1994). Errors due to spatial discretization and numerical precision in the finite-element method, *IEEE Trans. Ant. Prop.* 42(11): 1565–1569.

Kolotilina, L. Y. (1988). Explicit preconditioning of systems of linear algebraic equations with dense matrices., *J. Sov. Math.* 43: 2566–2573. English translation of a paper first published in Zapisli Nauchnykh Seminarov Leningradskogo Otdeleniya Matematicheskogo im. V.A. Steklova AN SSSR 154 (1986) 90-100.

Kunz, K. & Luebbers, R. (1993). *The Finite Difference Time Domain Method for Electromagnetics*, CRC Press, Boca Raton.

Lage, C. & Schwab, C. (1999). Wavelet Galerkin algorithms for boundary integral equations, *SIAM J. Scientific Computing* 20(6): 2195–2222.

Langer, U. & Pusch, D. (2005). Data-sparse algebraic multigrid methods for large scale boundary element equations, *Applied Numerical Mathematics* 54(3-4): 406–424.

Langer, U., Pusch, D. & Reitzinger, S. (2003). Efficient preconditioners for boundary element matrices based on grey-box algebraic multigrid methods, *International Journal for Numerical Methods in Engineering* 58(13): 1937–1953.

Lee, R. & Cangellaris, A. (1992). A study of discretization error in the finite element approximation of wave solution, *IEEE Trans. Ant. Prop.* 40(5): 542–549.

Lee, S., Ling, H. & Chou, R. (1988). Ray tube integration in shooting and bouncing ray method, *Micro. Opt. Tech. Lett.* 1: 285–289.

Li, J., Li, L. & Gan, Y. (2005). Method of moments analysis of waveguide slot antennas using the EFIE, *Journal of Electromagnetic Waves and Applications* 19(13): 1729–1748.

Malas, T. & Gürel, L. (2007). Incomplete LU preconditioning with multilevel fast multipole algorithm for electromagnetic scattering, *SIAM J. Scientific Computing* 29(4): 1476–1494.

Michielssen, E. & Boag, A. (1996). A multilevel matrix decomposition algorithm for analyzing scattering from large structures, *IEEE Transactions on Antennas and Propagation* 44(8): 1086 –1093.

Petersdorff, T. v. & Stephan, E. (1992). Multigrid solvers and preconditioners for first kind integral equations, *Numer. Meth. for PDE* 8: 443–450.

Rahola, J. (1998). Experiments on iterative methods and the fast multipole method in electromagnetic scattering calculations, *Technical Report TR/PA/98/49*, CERFACS, Toulouse, France.

Rao, S., Wilton, D. & Glisson, A. (1982). Electromagnetic scattering by surfaces of arbitrary shape, *IEEE Trans. Antennas Propagat.* AP-30: 409–418.

Rjasanow, S. (1987). Zweigittermethode für eine modellaufgabe bei bem-diskretisierung, *Wiss. Z. Tech. Univ. Karl-Marx-Stadt* 29(2): 230–235.

Rokhlin, V. (1990). Rapid solution of integral equations of scattering theory in two dimensions, *J. Comp. Phys.* 86(2): 414–439.

Ruge, J. & Stüben, K. (1987). *Multigrid Methods, Frontiers in Applied Mathematics 3*, S.F. McCormick ed., SIAM, Philadelphia, PA, chapter Algebraic multigrid (AMG), pp. 73–130.

Saad, Y. (1996). *Iterative Methods for Sparse Linear Systems.*, PWS Publishing, New York.

Saad, Y. & Schultz, M. (1986). GMRES: A generalized minimal residual algorithm for solving nonsymmetric linear systems., *SIAM J. Scientific and Statistical Computing* 7: 856–869.

Samant, A., Michielssen, E. & Saylor, P. (1996). Approximate inverse based preconditioners for 2D dense matrix problems, *Technical Report CCEM-11-96*, University of Illinois.

Silvester, P. & Ferrari, R. (1990). *Finite Elements for Electrical Engineers*, Cambridge University Press, Cambridge.

Song, J. & Chew, W. (1995). Multilevel fast-multipole algorithm for solving combined field integral equations of electromagnetic scattering, *Mico. Opt. Tech. Lett.* 10(1).

Song, J. & Chew, W. (1998). The Fast Illinois Solver Code: Requirements and scaling properties, *IEEE Computational Science and Engineering* 5(3): 19–23.

Song, J., Lu, C.-C. & Chew, W. (1997). Multilevel fast multipole algorithm for electromagnetic scattering by large complex objects, *IEEE Transactions on Antennas and Propagation* 45(10): 1488–1493.

Song, J., Lu, C., Chew, W. & Lee, S. (1998). Fast illinois solver code (FISC), *IEEE Antennas and Propagation Magazine* 40(3): 27–34.

Steinbach, O. & Wendland, W. (1998). The construction of some efficient preconditioners in the boundary element method., *Adv. Comput. Math.* 9(1-2): 191–216.

Sylvand, G. (2002). *La Méthode Multipôle Rapide en Electromagnétisme : Performances, Parallélisation, Applications*, PhD thesis, Ecole Nationale des Ponts et Chaussées.

Sylvand, G. (2003). Complex industrial computations in electromagnetism using the fast multipole method, *in* G. Cohen, E. Heikkola, P. Joly & P. Neittaanmdki (eds), *Mathematical and Numerical Aspects of Wave Propagation*, Springer, pp. 657–662. Proceedings of Waves 2003.

Taflove, A. (1995). *Computational Electrodynamics: The Finite-Difference Time-Domain Method*, Artech House, Boston.

Tausch, J. (2004). The variable order fast multipole method for boundary integral equations of the second kind, *Computing* 72(3-4): 267–291.

van der Vorst, H. (1992). Bi-CGSTAB: a fast and smoothly converging variant of Bi-CG for the solution of nonsymmetric linear systems., *SIAM J. Scientific and Statistical Computing* 13: 631–644.

Vanek, P., Mandel, J. & Brezina, M. (1996). Algebraic multigrid by smoothed aggregation for second and fourth order elliptic problems, *Computing* 56: 179–196.

Vavasis, S. (1992). Preconditioning for boundary integral equations, *SIAM J. Matrix Analysis and Applications* 13: 905–925.

Volakis, J., Chatterjee, A. & Kempel, L. (1998). *Finite element methods for electromagnetics*, IEEE Press, Piscataway, NJ.

8

Three-Dimensional Numerical Analyses on Liquid-Metal Magnetohydrodynamic Flow Through Circular Pipe in Magnetic-Field Outlet-Region

Hiroshige Kumamaru, Kazuhiro Itoh and Yuji Shimogonya
University of Hyogo
Japan

1. Introduction

In conceptual design examples of a fusion reactor power plant, a lithium-bearing blanket in which a great amount of heat is produced is cooled mainly by helium gas, water or liquid-metal lithium (Asada et al. Ed., 2007). The liquid-metal lithium is an excellent coolant having high heat capacity and thermal conductivity and also can breed tritium that is used as fuel of a deuterium-tritium (D-T) fusion reactor. In cooling the blanket, however, the liquid-metal lithium needs to pass through a strong magnetic field that is used to magnetically confine high-temperature reacting plasma in a fusion reactor core. There exists a large magnetohydrodynamic (MHD) pressure drop arising from the interaction between the liquid-metal flow and the magnetic field. In particular, the MHD pressure drop becomes considerably larger in the inlet region or outlet region of the magnetic field than in the fully-developed region inside the magnetic field for the reason mentioned later in this chapter.

A three-dimensional calculation is indispensable for the exact calculation of MHD channel flow in the inlet region or outlet region of magnetic field, also as described later in this chapter. There exist a few three-dimensional numerical calculations on the MHD flows in rectangular channels with a rectangular obstacle (Kalis and Tsinober, 1973), with abrupt widening (Itov et al., 1983), or with turbulence promoter such as conducting strips (Leboucher, 1999). All these calculations, however, were carried out for low Hartmann numbers (corresponding to low strength of the applied magnetic field) and low Reynolds number, because of instability problems in numerical calculations.

As to the MHD channel flow in the magnetic-field inlet-region, three-dimensional numerical calculations were conducted for the cases of Hartmann number of ~10 and Reynolds number of ~100 (Khan and Davidson, 1979). The calculations were based on what is called the parabolic approximation, in which the flow and magnetic field effects are assumed to transfer only in the main flow direction. However, the calculations based on parabolic approximation cannot predict exactly the MHD flow in the magnetic-field inlet-region. Were performed full three-dimensional calculations (without any assumptions) on the MHD rectangular-channel flow in the magnetic-field inlet-region (Sterl, 1990). The calculations were conducted mainly for the ranges of Hartmann numbers from 50 to 70 and Reynolds numbers from 2.5 to 5, and for a smoothly-increasing applied magnetic field. However, these ranges of Reynolds numbers and Hartmann numbers are unrealistic as conditions that appear even in laboratory conditions. The laboratory conditions reach Reynolds numbers up to ~1000 and Hartmann numbers up to ~100 simultaneously.

In fusion reactor conditions, the Reynolds number and the Hartmann number reach $\sim 10^4$ and $\sim 10^4$, respectively, the channel walls are electrically-conducting, the magnetic field changes in steps at the inlet or the outlet, and the flow changes from non-MHD turbulent flow to MHD laminar flow. However, because of instability problems in numerical calculations, it is quite difficult to obtain three-dimensional numerical solutions on MHD flows in the magnetic-field inlet-region or outlet-region even in the laboratory conditions that reach Reynolds numbers up to ~1000 and Hartmann numbers up to ~100 simultaneously.

Within the present limit of computer performance, the authors have already performed full three-dimensional calculations on the MHD flow through a circular pipe in the magnetic-field inlet-region, in simulating typical laboratory conditions (Kumamaru et al., 2007). In the calculations, the Hartmann number and the Reynolds number are ~100 and ~1000, respectively, the channel walls are electrically-insulating, the applied magnetic field changes in steps, and a laminar non-MHD flow enters the calculation domain. In this study, full three-dimensional calculations are performed on the MHD flow through a circular pipe in the magnetic-field outlet-region for the same conditions as for the magnetic-field inlet-region.

Figure 1 shows schematically the coordinate system, the applied magnetic field and the induced electric currents, together with the directions of Lorentz force, in the outlet region of the magnetic field. The applied magnetic field is imposed in the y direction, having a constant value for $z=0\sim z1$, a linear decrease from $z=z1\sim z2$, and a value of zero for $z=z2\sim z0$, as shown in Fig. 1(a).

In the region of fully-developed MHD flow near $z=0$, the induced electric current which is produced by the vector product of flow velocity and applied magnetic field flows in the negative x direction as shown in Figs. 1(b) and 1(c1). The induced current returns by passing through regions very near the walls (in an x-y plane at the same z) where the flow velocity is nearly zero, in the case of insulating walls. (The induced current can also pass through the walls in the case of conducting walls.) The induced current loop has a relatively large electrical resistance, since the current needs to flow in the thin regions near the walls. The Lorentz force which is caused by the vector product of induced current and applied magnetic field acts in the negative z direction and produces a large pressure drop.

In the outlet region of magnetic field from $z \approx z1$ to $z \approx z2$, the induced electric current flows in the negative x direction, as was the case of fully-developed region, as shown in Fig. 1(b).

However, the induced current can pass through the large region downstream the magnetic field section (in an x-z plane with the same y) where no magnetic field or small magnetic field is applied. The electric resistance in this region is much smaller than the resistance in the thin region near the walls mentioned above. Hence, the induced current becomes larger in the outlet region than in the fully-developed region. The Lorentz force and thus the pressure drop also may become considerably larger in the outlet region than in the fully-developed region (Moreau, 1990).

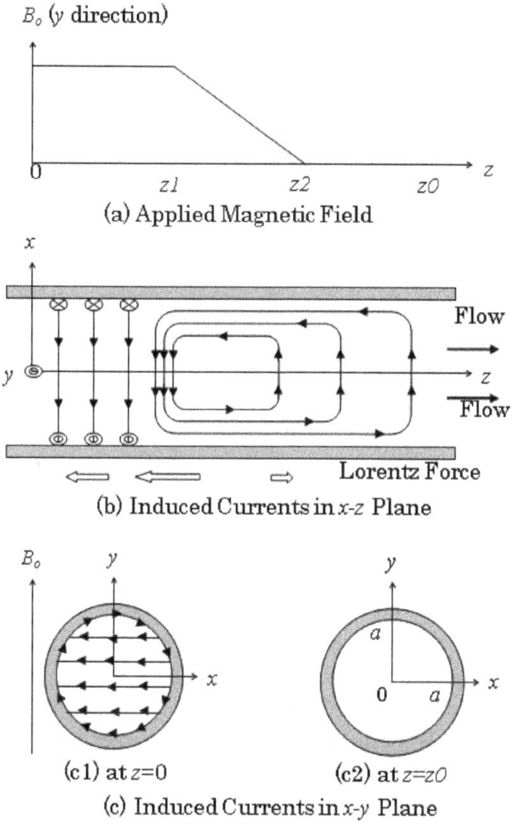

Fig. 1. Coordinate system for magnetic-field outlet-region.

On the other hand, the induced current in the last section of the outlet region near z=$z2$ can flow in the positive x direction as shown in Fig. 1(b). Thus, a smaller Lorentz force may act in the flow direction and thus a small pressure recovery may occur in this section of the outlet region (Moreau, 1990).

The induced electric currents in the outlet region, flowing in both x- and z-directions and in y-direction, cannot be calculated by a two-dimensional model. It is also important that a sufficiently large fluid region downstream the magnetic field section is included in a calculation domain. For these reasons, in this study, in order to obtaine mainly the pressure drop quantitatively, the authors have performed three-dimensional numerical calculations on the MHD flow through a circular pipe in the outlet region of the magnetic field, including the region of no magnetic field downstream the magnetic field region. To the authors' knowledge, there have been no numerical calculations or experimental studies on the MHD flow through a circular pipe in the magnetic-field outlet-region. In this study, calculation results on the magnetic-field outlet-region have also been compared with authors' calculation results on the magnetic-field inlet-region.

2. Numerical analyses

Numerical calculations are performed for an MHD flow in a circular pipe with an inner radius of a, shown in Fig. 1. A fully-developed MHD laminar flow enters the calculation domain at z=0, and a fully-developed non-MHD flow leaves the domain at z=$z0$. The applied magnetic field is imposed in the y direction as shown in Fig. 1(a), as was stated previously.

The basic equations which describe a liquid-metal MHD flow are the continuity equation, the momentum equation and the induction equation. The equations are expressed respectively by:

$$\nabla \cdot \boldsymbol{v} = 0 , \tag{1}$$

$$\rho\left[\frac{\partial \boldsymbol{v}}{\partial t} + (\boldsymbol{v} \cdot \nabla)\boldsymbol{v}\right] = -\nabla p + \eta \nabla^2 \boldsymbol{v} + \frac{1}{\mu}(\nabla \times \boldsymbol{B}) \times \boldsymbol{B_O} , \tag{2}$$

$$\frac{\partial \boldsymbol{B}}{\partial t} = \nabla \times \left(\boldsymbol{v} \times \boldsymbol{B_O}\right) + \frac{1}{\sigma\mu}\nabla^2 \boldsymbol{B} . \tag{3}$$

Here, $\boldsymbol{v}$ is velocity vector, p pressure, $\boldsymbol{B}$ induced magnetic field vector and t time; $\boldsymbol{B_0}$ is applied magnetic field vector, and ρ is density, η viscosity, μ magnetic permeability and σ electric conductivity. The vector $\boldsymbol{B}$ is an induced magnetic field produced by the induced electric current, and is treated as an unknown variable together with the velocity $\boldsymbol{v}$ and the pressure p. The induced electric current $\boldsymbol{j}$ can be calculated by the Ampere equation $\boldsymbol{j} = (1/\mu)(\nabla \times \boldsymbol{B})$ from $\boldsymbol{B}$. The third term in the right-hand side of Eq. (2) represents the Lorentz force. The induction equation, i.e. Eq. (3), is derived from Maxwell's equations and Ohm's law in electromagnetism.

The basic equations are expressed in nondimensional forms by introducing the following nondimensional variables (indicated by superscript *) and nondimentional numbers:

$$t^* = \frac{t}{a/\bar{v}_z}, \quad r^* = \frac{r}{a}, \quad z^* = \frac{z}{a},$$

$$v_r^* = \frac{v_r}{\bar{v}_z}, \quad v_\theta^* = \frac{v_\theta}{\bar{v}_z}, \quad v_z^* = \frac{v_z}{\bar{v}_z}, \quad p^* = \frac{p}{\rho \bar{v}_z^2},$$

$$B_r^* = \frac{B_r}{\bar{v}_z \mu \sqrt{\sigma\eta}}, \quad B_\theta^* = \frac{B_\theta}{\bar{v}_z \mu \sqrt{\sigma\eta}}, \quad B_z^* = \frac{B_z}{\bar{v}_z \mu \sqrt{\sigma\eta}}, \tag{4}$$

$$Re = \frac{\overline{v}_z a}{\nu}, \quad Ha = B_0 a \sqrt{\frac{\sigma}{\eta}}, \quad Rm = \frac{\overline{v}_z a}{\nu_m}. \tag{5}$$

Here, r, θ, z are coordinates in the cylindrical coordinate systen; $\overline{v}_z$ mean velocity in z-direction, ν kinematic viscosity and ν_m ($=1/\sigma\mu$) magnetic kinematic viscosity. Nondimensional numbers Re, Ha and Rm are Reynolds number, Hartmann number and magnetic Reynolds number, respectively. The final nondimensional basic equations become respectively:

$$\nabla \cdot \boldsymbol{v} = 0, \tag{6}$$

$$\frac{\partial \boldsymbol{v}}{\partial t} + (\boldsymbol{v} \cdot \nabla)\boldsymbol{v} = -\nabla p + \frac{1}{Re}\nabla^2 \boldsymbol{v} + \frac{1}{Re}(\nabla \times \boldsymbol{B}) \times \boldsymbol{Ha}, \tag{7}$$

$$\frac{\partial \boldsymbol{B}}{\partial t} = \frac{1}{Rm}\nabla \times (\boldsymbol{v} \times \boldsymbol{Ha}) + \frac{1}{Rm}\nabla^2 \boldsymbol{B}. \tag{8}$$

Superscript * is omitted to simplify the description in Eqs. (6) through (8) and in the following description. Note that the Hartmann number $\boldsymbol{Ha}$ is a given (known) vector having only y-component as a given function of z, i.e. $Ha(z)$.

The coordinate system is transformed from the Cartesian coordinate system (x, y, z) to the curvilinear coordinate system (ξ, η, ζ) in order to deal with a channel with an arbitrary flow cross-section in the future, and is thereafter transformed into the cylindrical coordinate system (r, θ, z) as a special case of the curvilinear coordinate system. Considering the symmetry, the numerical calculations are carried out for the region of $0<r<1$ and $0<\theta<\pi/2$. (Note that the inner surface of the wall corresponds to $r=1$ ($x=1$ or $y=1$) in the nondimensional coordinates.)

As the boundary condition on the flow velocities, the inflow boundary condition is adopted at the flow inlet, i.e. at $z=0$, by fixing a fully-developed MHD flow velocity (Kumamaru & Fujiwara, 1999). The outflow boundary condition is given at the flow outlet, i.e. at $z=z0$, by fixing the reference pressure. No-slip condition is given at the wall and the symmetry condition is adopted at $\theta=0$ and $\theta=\pi/2$. As the boundary condition on the induced magnetic fields, $\partial B / \partial \boldsymbol{z} = 0$ and $\boldsymbol{B}=0$ are specified at the flow inlet and the flow outlet, respectively. The former reflects the situation that the induced current does not change in the z-direction at the flow inlet in a fully-developed MHD flow region, and the latter represents that no induced current exists at the flow outlet in a fully-developed non-MHD flow region. At the wall, $\boldsymbol{B}=0$ is specified assuming that the walls are electrically insulating (nonconducting). The boundary conditions on the induced magnetic fields at the symmetry plane of $\theta=0$ and $\theta=\pi/2$ are not intuitively clear. Hence, by performing a calculation for the whole cross section in the case of small Hartmann numbers, it has been confirmed that the conditions are given by:

$$\theta = 0:\ A(-\theta) = -A(\theta), B(-\theta) = B(\theta), C(-\theta) = -C(\theta), \tag{9a}$$

$$\theta = \pi / 2: \\ A(\pi/2-\theta) = -A(\pi/2+\theta), B(\pi/2-\theta) = B(\pi/2+\theta), C(\pi/2-\theta) = C(\pi/2+\theta)' \tag{9b}$$

where *A*, *B* and *C* are the *x*, *y* and *z* components of ***B***, respectively.

The discretization of the equations is carried out by the finite difference method. The calculations are performed using a non-uniform expanding 15 x 15 x 30 grid with grid elements closely spaced near the channel wall of r=1 and the region between or around z=$z1$ and z=$z2$. The first-order accurate upwind differencing is adopted for the fluid convection terms in Eq. (7). The solution procedure follows the MAC method that is widely used in numerical calculations.

Even for the fully-developed region, it is difficult to obtain a stable numerical solution for large Hartmann numbers (Kumamaru & Fujiwara, 1999). In the present three-dimensional calculations, stable numerical solutions have been obtained for Hartmann numbers up to 100 and Reynolds numbers up to 1000 by applying the following means or procedures. (1) The grids are arranged closely near the wall of r=1, i.e. at r=0.0, … , 0.95, 0.97, 0.99, 0.995, 1.0, on referring to a velocity profile of the classical Hartmann flow, i.e. fully-developed MHD flow in infinite parallel plates (Kumamaru and Fujiwara, 1999). (2) Simultaneous linear equations on the pressure, i.e. Poisson equation, are solved not by the iterative method but by the elimination method. (3) First, a solution is obtained for *Re* (Reynolds number) of 0.01, and thereafter *Re* is increased gradually to a final value, i.e. 1000.

3. Analysis results

3.1 Pressure along flow axis

Numerical calculations have been performed for a circular pipe with insulating wall under the conditions of a Reynolds number (*Re*) of 1000, a Hartmann number (*Ha*) of 100 (for the fully-developed MHD region) and a magnetic Reynolds number (*Rm*) of 0.001. The Hartmann number (relating to the applied magnetic field) is 100 from z=0 to $z1$, decreases linearly from z=$z1$ to $z2$, and is zero from z=$z2$ to $z0$ (See Fig. 1(a)). The values of $z1/z2$ are changed from 10/20 to 10/10.05. (Note that both $z1$ and $z0$-$z2$ are fixed to 10 in all the cases.) These values for the nondimensional numbers and parameters are selected in order to simulate those typical to laboratory scales and conditions.

Figure 2 shows calculated pressures along the flow axis, i.e. the z-axis, for the cases of $z1/z2$ from 10/20 to 10/10.05. Figure 3 presents a calculated result only for the case of $z1/z2$=10/12 as a standard case, indicated by a solid line, together with a corresponding result for the magnetic-field inlet-region, indicated by a dotted line, which will be explained in Sec. 3.4. From z=0 to $z \approx z1$, the pressure decreases steeply following the pressure drop of a fully-developed MHD flow. From $z \approx z1$ to $z \approx z2$, the pressure decreases more sharply than in the region of $z<z1$, since a large Lorentz force is produced in the negative z direction as was mentioned in Chap. 1 and again will be explained in Sec. 3.2. In $z>z2$, the pressure decreases slowly, representing the frictional pressure drop as a non-MHD laminar flow.

The steeper the gradient of the applied magnetic field becomes, the more sharply the pressure decreases from $z \approx z1$ to $z \approx z2$. However, the pressure drop through the magnetic-field outlet-region becomes saturated for the steeper gradient of the magnetic field (See the cases of $z1/z2$=10/10.1 and 10/10.05). For the slower gradient of the magnetic field, the effect of the length along the flow axis (i.e. z-axis) contributes more to the pressure drop through the outlet region than the effect of the outlet region (Compare the cases of $z1/z2$=10/20 and 10/15).

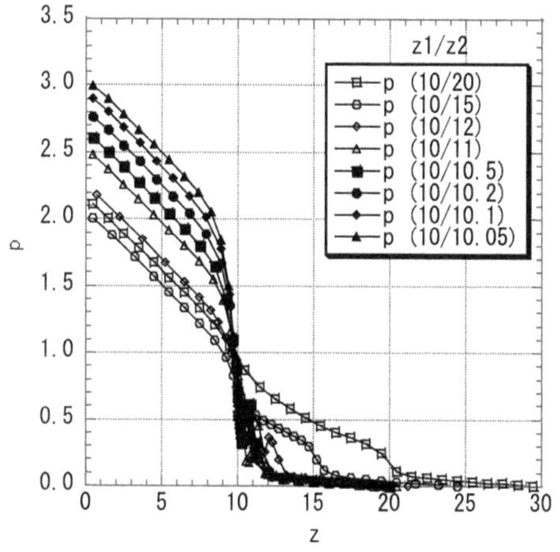

Fig. 2. Pressures along *z*-axis for *z1*/*z2*=10/20 to 10/10.05.

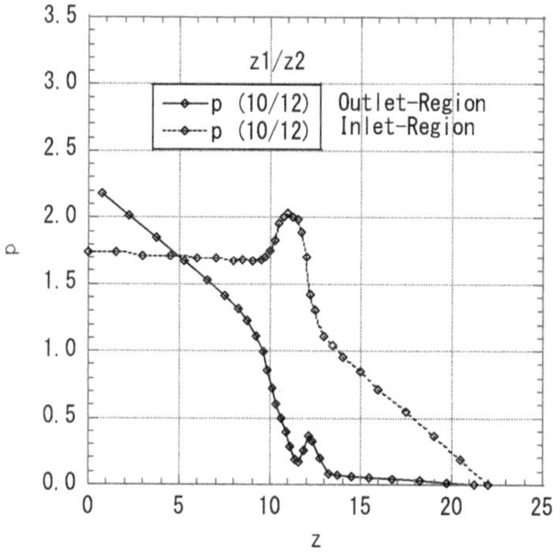

Fig. 3. Pressures along *z*-axis for *z1*/*z2*=10/12.

The small pressure recovery, which was also pointed out in Chap. 1 and again will be expained in Sec. 3.2, is observed in the region near $z \approx z2$ for the cases of *z1*/*z2*=10/12~10/10.05. The pressure drop appears again outside the magnetic-field region. This may be due to rapid change in velocity distibution in this region, which will be explained in Sec. 3.4.

The pressure drops in the fully-developed region of *z*<*z1*, $-\Delta p/\Delta z$, are almost the same for all the cases. The pressure drops agree with a value calculated numerically by the authors for the fully-developed MHD flow, $-\Delta p/\Delta z \approx 0.123$ (Kumamaru and Fujiwara, 1999), and also agree nearly with a value predicted by Schercliff's theoretical approximate equation, $-\Delta p/\Delta z \approx 0.118$ (Schercliff, 1956; Lielausis, 1975), for the case of *Ha*=100 and *Re*=1000. As mentioned in Chap. 1, no experimental data on the pressure drop through the magnetic-field outlet-region have been reported. However, pressure drops through the magnetic-field inlet-region calculated numerically by the authers agreed nearly with those estimated by an existing equation based on experimental data (Kumamaru 2007; Lielausis, 1975).

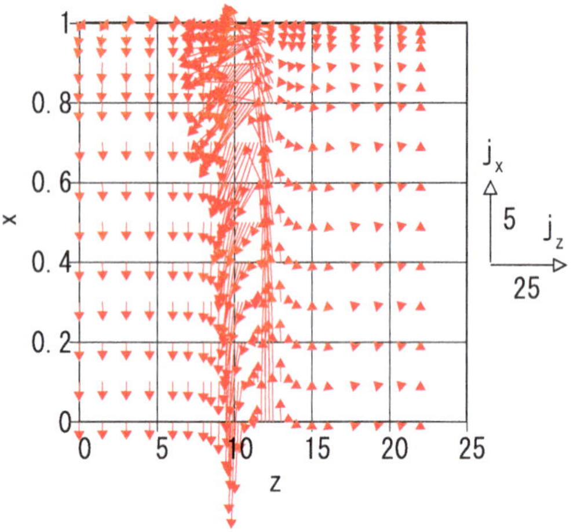

Fig. 4. Induced currents in *x*-*z* plane at *y*=0 for *z*1/*z*2=10/12.

3.2 Induced current distribution

Figure 4 illustrates induced electric current distribution in the *x*-*z* plane at *y*=0 for the case of *z*1/*z*2=10/12, i.e. the standard case. Figures 5(a), (b) and (c) give induced current distributions in the *x*-*y* planes at *z*=4.5, 10 and 12, respectively, for the same case. On the right side of each figure, is shown the magnitude of the (nondimensional) induced current vector in the each coordinate direction. In Figs. 5(a) through (c), the induced current vector is reduced by a factor of 4 for two vectors from the wall at each circumferential angle, in order to make the figures compact.

In the fully-developed region from *z*=0 to $z \approx 8$, the induced current, flowing mainly in the negative *x*-direction, does not change in the *z*-direction, as shown in Fig. 4. This constant induced current produces constant Lorentz force (acting in the negative *z* direction) and results in constant pressure drop along the *z*-axis as shown in Fig. 3. The induced current returns by passing in an extremely thin region very near the wall, as shown in Fig. 5(a). Almost no induced current (less than 10^{-2}) flows in the region from $z \approx 14$ to *z*=22, as shown in Fig. 4, since no magnetic field is applied.

In the magnetic-field outlet-region from $z \approx 8$ to $z \approx 14$, the induced current forms a loop mainly in the *x*-*z* plane, as shown in Fig. 4. The induced current is larger in the outlet region than in the fully-developed region. This is because the electric resistance of the induced current loop in the outlet region is much smaller than the resistance of the loop in the fully-developed region. The induced current can return in the large downstream region in the outlet region, although the current needs to return only in the extremely thin region near the wall in the fully-developed region.

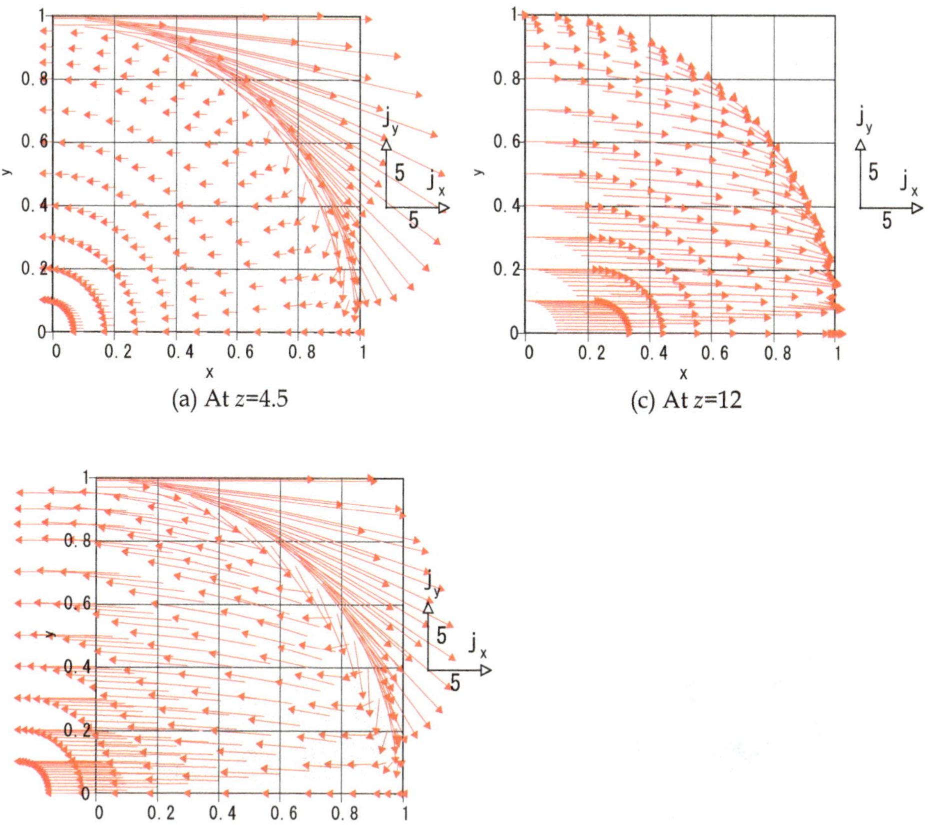

(a) At z=4.5

(c) At z=12

(b) At z=10

Fig. 5. Induced currents in x-y plane for $z1/z2$=10/12.

The induced current flows mainly in the negative x-direction from $z \approx 8$ to $z \approx 11$. Hence, in this region, a larger Lorentz force than in the fully-developed region acts in the negative z-direction, and a larger pressure drop is produced along the z-axis as shown in Fig. 3. On the other hand, the induced current flows mainly in the positive x-direction from $z \approx 11.5$ to $z \approx 13$. Thus, the Lorentz force is exerted in the positive z-direction, and a small pressure recovery along the z-axis happens from $z \approx 11.5$ to $z \approx 12$ as shown in Fig. 3. (No external magnetic field is applied from $z \approx 12$ to $z \approx 13$.) Also in the outlet region, there exists an induced current loop which returns in an extremely thin region near the wall, as shown in Fig. 5(b).

3.3 Velocity distribution

Figures 6(a), (b), (c), (d) and (e) show calculated velocity v_z distributions at z=4.5, 10, 11, 12 and 17.5, respectively, for the case of $z1/z2$=10/12, i.e. the standard case. There is no significant difference among velocity distributions from z=0 to z=8. The velocity profile is a

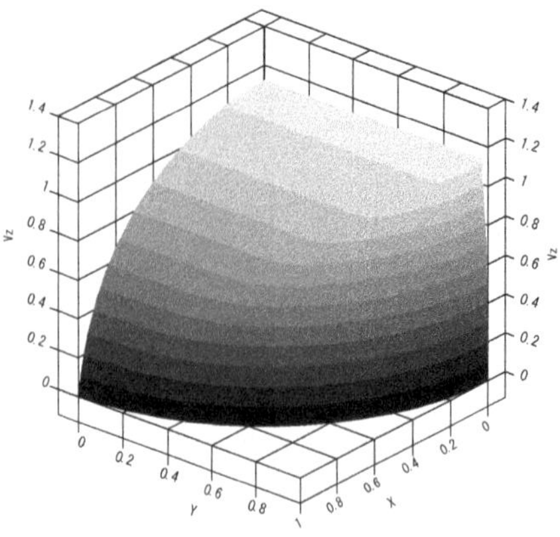

(a) At z=4.5

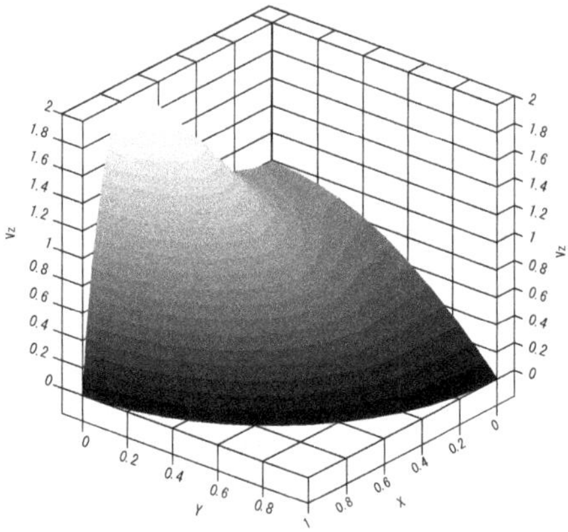

(d) At z=12

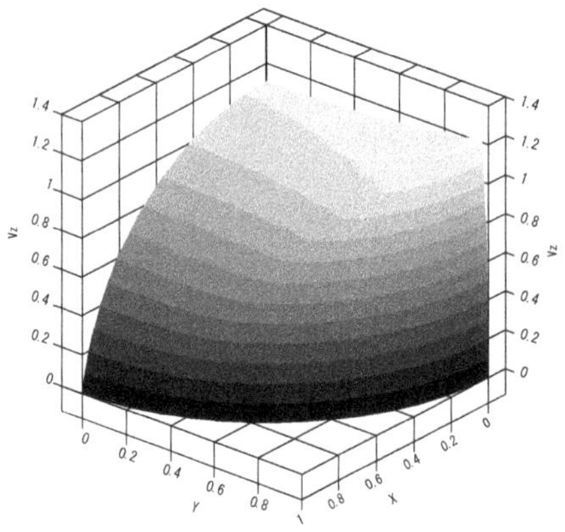

(b) At z=10

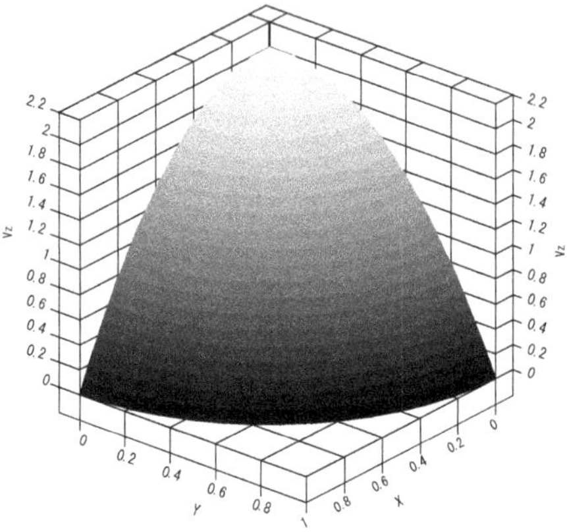

(e) At z=17.5

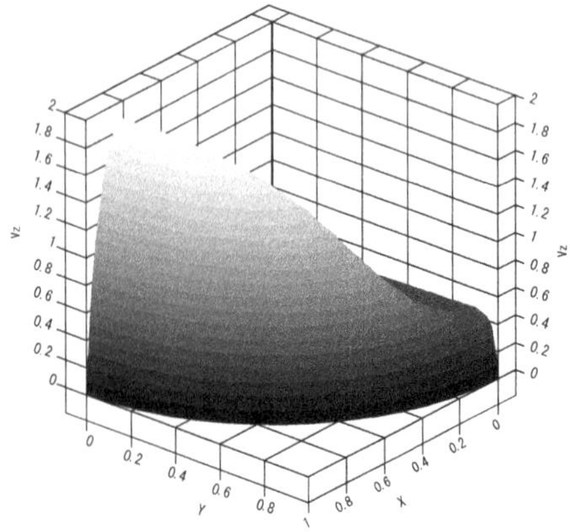

(c) At z=11

Fig. 6. Velocity distribution for $z1/z2$=10/12.

flat one, particularly in the direction of applied magnetic field, having a peak value of ~1.1, as shown in Fig. 6(a). The velocity distribution in this region agrees nearly with a profile calculated by the authors for the fully-developed MHD flow (Kumamaru, 1999).

The velocity distribution at z=10, shown in Fig. 6(b), is still nearly flat. The velocity profile changes sharply at $z \approx 11$ and shows what is called an M-shape distribution having a peak near the wall, as shown in Fig. 6(c). This is because the Lorentz force acting in the negative z-direction suppresses the flow in the z-direction in the fluid bulk region, though small Lorentz force acts in the negative z-direction in the region near the wall of x=1. The velocity distribution at z=12, at the outlet of applied magnetic field, shown in Fig. 6(d), is still nearly the same as that at z=11, shown in Fig. 6(c).

The velocity profile changes sharply, from z=12 to $z \approx 13$, from the M-shape distribution, shown in Fig. 6(d), to a parabolic distribution typical to a non-MHD flow, shown in Fig. 6(e). The pressure decrease from z=12 to $z \approx 13$, shown in Fig. 3, is attributable to this sharp change in velosity distribution. It is considered that the pressure decreases largely since the velocity increases quickly in the fluid bulk region. No significant difference exists among velocity profiles from $z \approx 13$ to z=22. The velocity profile is a parabolic one of a non-MHD laminar flow with a peak value of ~2.

3.4 Comparison with magnetic-field inlet-region

Figure 7 shows schematically the applied magnetic field in the y-direction, the induced currents in the x-z plane including the directions of Lorentz force and the pressure along the z-axis, in the inlet region and the outlet region of the magnetic field. The larger Lorentz force acts and thus the larger pressure drop occurs in the inlet and outlet regions than in the fully-developed MHD region for the reason mentioned in Chap. 1. On the other hand, a smaller Lorentz force may act in the flow direction and thus a small pressure recovery may occur in the first section of the inlet region and in the last section of the outlet region also for the reason mentioned in Chap. 1. The pressure drop behavior is not completely symmetric, since the fully- developed non-MHD flow enters the calculation domain in the inlet-region while the fully-developed MHD flow énters the domain in the outlet-region.

Figure 8 presents pressures along z-axis for the magnetic-field inlet-region, calculated by the authors and presnted in a previous paper (Kumamaru et al, 2007). The calculation parameters for Fig. 8 are the same as for Fig. 2, except for the Hartmann number change along z-axis. The Hartmann number (relating to the applied magnetic field) is 0 from z=0 to $z1$, increases linearly from z=$z1$ to $z2$, and is 100 from z=$z2$ to $z0$. The pressure change for the case of $z1/z2$=10/12, i.e. a standard case, in the inlet region, indicated by a dotted line, is also compared with the corresponding case in the outlet region in Fig. 3.

The pressure decreases slowly following the drop in a non-MHD laminar flow from z=0 to $z \approx z1$. The pressure recovery appears clearly in the region near $z \approx z1$. The pressure decreases more rapidly in the region from $z \approx z1$ to $z \approx z2$ than in the fully-developed MHD region of z>$z2$. The pressure decreases rapidly following the drop of a fully-developed MHD flow in the region of z>$z2$.

Figure 9 illustrates induced electric current distribution in the x-z plane at y=0 for the case of $z1/z2$=10/12, i.e. the standard case, in the magnetic-field inlet-region. The distribution in the

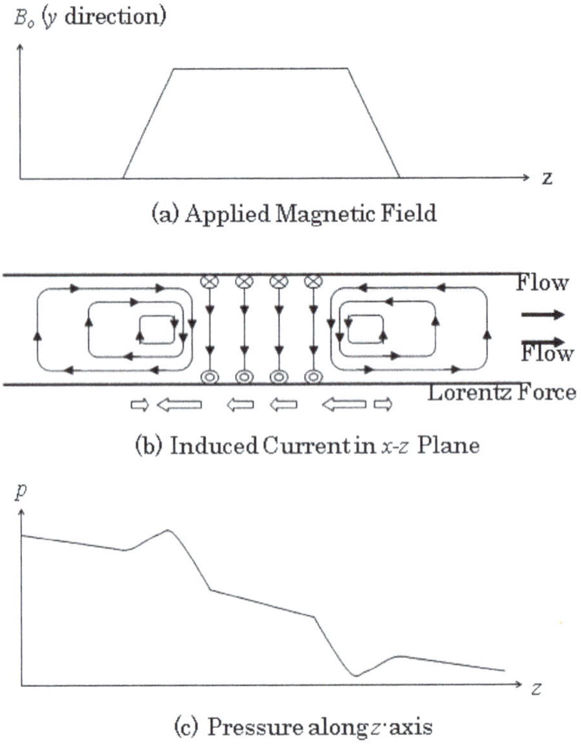

Fig. 7. Inlet and outlet regions of magnetic field.

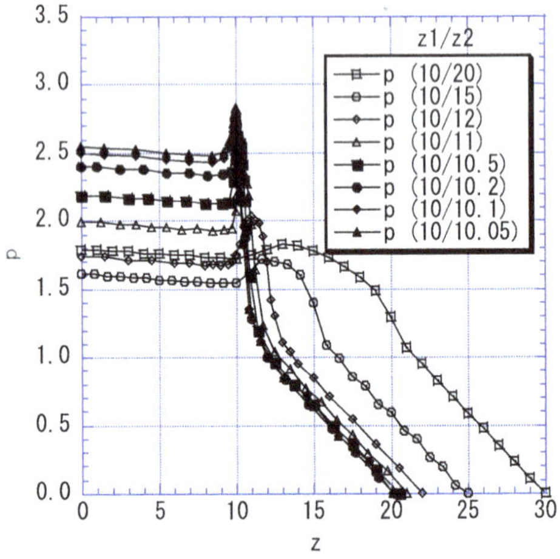

Fig. 8. Pressures along z-axis for magnetic-field inlet-region.

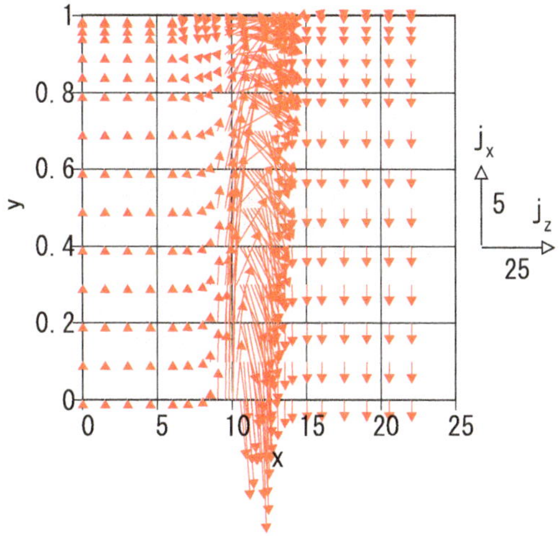

Fig. 9. Induced currents for magnetic-field inlet-region.

inlet region, Fig. 9, and that in the outlet region, Fig. 4, are nearly symmetric. For this reason, the sharp pressure drop in the inlet region from $z \approx 11$ to $z \approx 13$, i.e. $-\Delta p \approx 0.9$, agrees nearly with that in the outlet region from $z \approx 9$ to $z \approx 11.5$, i.e. $-\Delta p \approx 0.9$. However, the pressure recovery in the inlet region from $z \approx 9.5$ to $z \approx 11$, i.e. $-\Delta p \approx 0.4$ is larger than that in the outlet region from $z \approx 11.5$ to $z \approx 12$, i.e. $-\Delta p \approx 0.2$. The reason is examined later.

Figures 10(a), (b) and (c) show calculated velocity v_z distributions at z=10, 11 and 12, respectively, for the case of $z1/z2$=10/12, i.e. the standard case, in the magnetic-field inlet-region. The velocity profile at z=10, shown in Fig. 10(a), still keeps nearly a distribution typical to a non-MHD fully-developed laminar flow with a peak value of ~2. Hereafter, the velocity distribution becoms flatter along the channel axis, i.e. the z-axis, as shown in Figs. 10(b) (at z=11) and 10(c) (at z=12). However, the M-shape profile with extreme flow suppression in the fluid bulk region observed in the outlet region, as shown in Figs. 6(c) and (d), is not seen in the inlet region, as shown in Figs. 10(b) and (c). The reason may be that the non-MHD fully-develloped flow with the parabolic plofile enters the inlet region though the MHD fully-developed flow with the flat profile comes into the outlet region.

It is considered that, in addition to the pressure recovery due to the induced current in the positive x-direction, the velocity decrease in the fluid central region results in the pressure increase of $\Delta p \approx 0.4$ in $9.5<z<11$ of the magnetic-field inlet-region. On the other hand, it can be considered that after the pressure recovery of $\Delta p \approx 0.4$ due to the induced current in the positive x-direction, the pressure decreases by $-\Delta p \approx 0.3$ due to the velocity increase in $11.5<z<12$ of the magnetic-field outlet-region. From these differences, the pressure drop through the inlet region may become smaller than that through the outlet region.

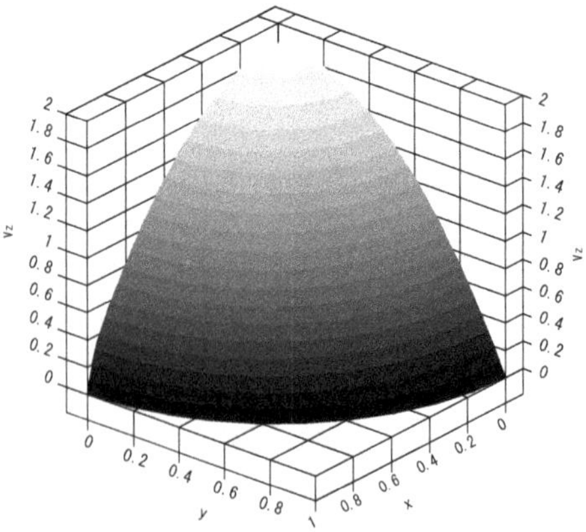

(a) At z=10

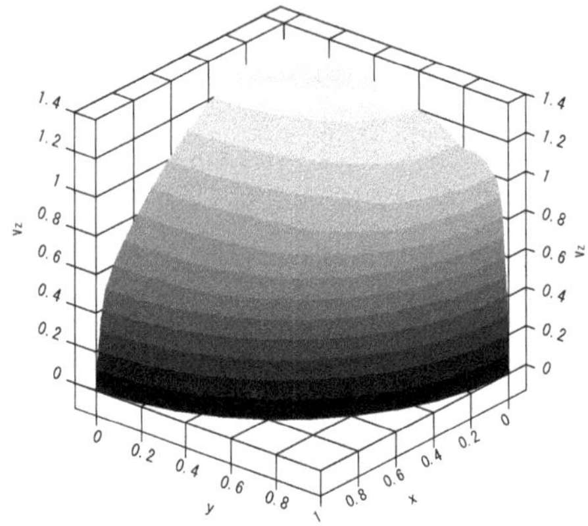

(b) At z=11

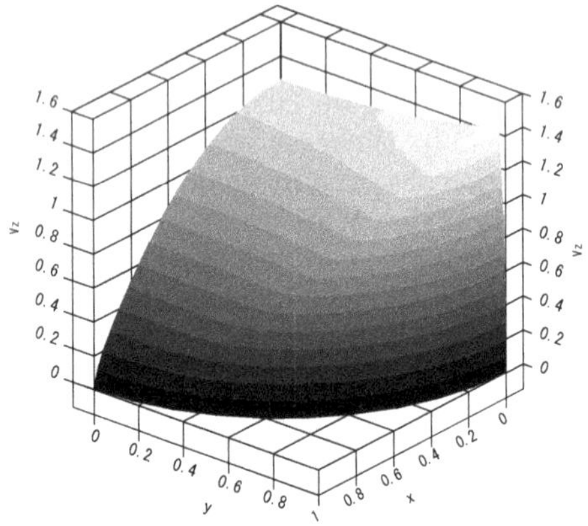

(c) At z=12

Fig. 10. Velocity distribution for magnetic-field inlet-region.

4. Conclusion

Three-dimensional numerical analyses have been performed on liquid-metal magnetohydrodynamic (MHD) flow through a circular pipe in the outlet region of magnetic field. The following conclusions have been obtained from the calculation results.

a. Along the flow axis, i.e. the circular pipe axis, the pressure decreases steeply as a fully-developed MHD flow, drops more sharply in the magnetic-field outlet-region, and finally decreases slowly as a normal fully-developed non-MHD flow.
b. If examined in detail, in the magnetic-field outlet-region, after the pressure drops most sharply, it recovers once and thereafter it drops sharply again outside the magnetic-field region.
c. The first sharp pressure drop and temporary pressure recovery are due to the formation of induced current loop which circulates in passing in the region downstream the magnetic-field region. The second sharp pressure drop is attributable to the change in velosity distribution outside the magnetic-field region.
d. The distribution of velocity in main flow direction changes from a flat profile of a fully-developed MHD flow, to an M-shaped profile and finally to a parabolic profile of a fully-developed non-MHD flow.
e. The total pressure drop through the magnetic-field outlet-region becomes larger than the corresponding drop through the magnetic-field inlet-region. The main reason may be that the difference in velocity profile change between the outlet region and the inlet region.

5. References

Aitov, T.N., Kalyutik, A.I. & Tananaev, A.V. (1983). Numerical Analysis of Three-Dimensional MHD-Flow in Channels with Abrupt Change of Cross Section, *Magnetohydrodynamics*, Vol. 19, pp. 223-229, ISSN 0024-998x

Asada, C. et al. Ed. (2007). *Handbook of Nuclear Engineering*, Ohmsha, Ltd., ISBN 978-4-274-20443-2, Tokyo, Japan [In Japanese]

Kalis, K.E. & Tsinober, A.B. (1973). Numerical Analysis of Three Dimensional MHD Flow Problems, *Magnetohydrodynamics*, Vol. 2, pp. 175-179, ISSN 0024-998x

Khan, S. and Davidson, J. N. (1979). Magneto-hydrodynamic Coolant Flows in Fusion Reactor Blankets, *Annals of Nuclear Energy*, Vol. 6, pp. 499-509, ISSN 0306-4549

Kumamaru, H. & Fujiwara, Y. (1999). Pressure Drops of Magnetohydrodynamic Flows in Rectangular Channel with Small Aspect Ratio and Circular Pipe for Very-Large Hartmann Numbers, *Proceedings of JSME/ ASME/SFEN 7th International Conference on Nuclear Engineering (ICONE-7)*, Tokyo, Japan, April 1999

Kumamaru, H., Shimoda, K. & Itoh, K. (2007). Three-Dimensional Numerical Calculations on Liquid-Metal Magnetohydrodynamic Flow through Circular Pipe in Magnetic-Field Inlet-Region, *J. of Nuclear Science and Technology*, Vol. 44, No. 5, pp. 714-722, ISSN 0022-3131 & 1881-1248

Leboucher, L. (1999). Monotone Scheme and Boundary Conditions for Finite Volume Simulation of Magnetohydrodynamic Internal Flows at High Hartmann Number, *J. of Computational Physics*, Vol. 150, pp. 181-198, ISSN 0021-9991

Lielausis, O. (1975). Liquid-Metal Magnetohydrodynamics, *Atomic Energy Review*, Vol. 13, pp. 527-581, ISSN 0004-7112

Moreau, R. (1990). *Magnetohydrodynamics*, Kluwer Academic Publishers, ISBN 978-90-481-4077-0, Dordrecht-Boston-London, Netherlands-USA-England

Schercliff, J.A. (1956). The Flow of Conducting Fluids in Circular Pipes under Transverse Magnetic Fields, *J. of Fluid Mechanics*, Vol. 1, pp. 644-666, ISSN 0022-1120

Sterl, A. (1990). Numerical Simulation of Liquid-Metal MHD Flows in Rectangular Ducts, *J. of Fluid Mechanics*, Vol. 216, pp. 161-191., ISSN 0022-1120

9

Time Reversal for Electromagnetism: Applications in Electromagnetic Compatibility

Ibrahim El Baba[1,2], Sébastien Lalléchère[1,2] and Pierre Bonnet[1,2]
[1]*Clermont University, Blaise Pascal University, BP 10448, F-63000, Clermont-Ferrand*
[2]*CNRS, UMR 6602, LASMEA, F-63177, Aubière*
France

1. Introduction

ElectroMagnetic Compatibility (EMC) is the branch of electromagnetism that studies generation, propagation and reception of involuntary electromagnetic energy in reference to the undesirable effect (electromagnetic interference) that this energy can induce. Since 1996, date of the directive 89/336/CEE (Directive 89/336/CEE, 1989) compulsory implementation concerning the electromagnetic compatibility (called CE) in Europe, and for much longer in United States, EMC has been playing an increasingly important role.

Most electrical and electronic equipment may be considered as sources of interference because it generates electromagnetic perturbations that pollute the environment and may disrupt the operation of other equipment (victims). The EMC is the ability of a device, equipment or system to operate satisfactorily within its electromagnetic environment and without producing itself an intolerable electromagnetic disturbance to anything in this environment. EMC hence controls the electromagnetic environment of the electronic equipment. To this end, EMC tackles several issues. Firstly, are the emission problems related to the generation of unwanted electromagnetic energy from a source and the measures that should be taken to reduce the generation of such disturbances and to prevent the escape of any remaining energy to the external environment. To verify that the perturbation level does not exceed a threshold value defined by standards, we measure the electric and/or magnetic fields radiated at a certain distance in the case of electromagnetic emissions, the voltage and/or current in the case of conducted disturbances. Secondly the susceptibility problems refer to the proper functioning of electrical equipment in presence of unplanned electromagnetic field. In the tests, we inject perturbation (conducting/radiating mode) on a device and check its good operation. Thirdly, for interference/noise disturbances, the EMC solutions are mainly obtained by addressing both the emissions and the vulnerability problems. This means minimizing the interference source levels and hardening the potential victims (shielding for example).

For measurements, EMC provides as test facilities different tools, the most popular are: the Anechoic Chamber (AC) (Emerson, 1973) and the Mode Stirred Reverberation Chamber (MSRC) (Corona et al., 2002; Hill, 1998). The AC is a cavity whose aim is to simulate the free space. Its walls are covered with ferrite tiles and/or polyurethane pyramids loaded with carbon absorbing electromagnetic waves and preventing their reflection. The second tool has grown in popularity over the past twenty years due to its ability to provide a statistically uniform and homogeneous electromagnetic field on a relatively large domain (called Working Volume: WV). In addition, high field's levels could be generated in the Reverberation Chamber (RC) for relatively low injected power. For mechanical stirring, the "statistical" uniformity is mainly based on the number N of the available independent configurations (i.e. the number N of the stirrer independent positions) for the RC and the studied frequency. When the number N tends to infinity, the intern electromagnetic field proprieties are statistically identical from one point to another in the WV. A statistically uniform and homogeneous distribution of the field in the MSRC signifies that the same energy attacks the Equipment Under Test (EUT) from all directions and with the same polarization, when averaged over the number N of the stirrer positions. A disadvantage of the AC is the high injected power needed, thus powerful amplifier are required, in addition to the high cost of the absorbers. In comparison with the AC case, low power in MSRC is needed.

Based on the principle of reciprocity, Time Reversal (TR) is a technique that allows focusing a field in time and space. Recently, it has been applied for EMC where better results have been reported in strongly reverberant or diffracting environments. Indeed, different studies (de Rosny, 2000; Moussa et al., 2009b) in acoustics and electromagnetics have verified how RC can provide an appropriate environment for TR. One of the main advantages of the MSRC is to provide the most critical illumination of the EUT. Paradoxically, this benefit may be considered as a disadvantage, since in this case it becomes impossible to know precisely the characteristics of the electromagnetic excitation. On the one hand, recent TR studies (Cozza & Moussa, 2009; Moussa et al., 2009a) have demonstrated how to make benefit of the re-focusing to control the wave incidence and polarization attacking the EUT. On the other hand, for the same input power, the TR enables to increase the achievable field levels in the MSRC. These promising applications of TR justify its characterization in MSRC.

During susceptibility tests of electronic equipments a problem may occur when the EUT is composed of several components with different field/current threshold values that cannot be exceeded. Indeed, various immunity levels can coexist on an electronic device (power supply, components, signal integrity, etc.) or on different zones of a complex structure (automobile, aircraft, etc.) since the expected reliability might be different from an area or device to another. But, in a classical susceptibility EMC MSRC test, the illumination is statistically the same for the whole EUT placed in the working volume and it may damage components that have a smaller threshold value than the incident field. A solution consists in performing the susceptibility test independently for each component. Unfortunately, "on table" tests are not always possible and also might not represent the reality. An alternative approach can be given via TR technique and selective focusing. As a matter of fact, at the focusing time, only one component can be illuminated by a desired field level while others parts of the system are aggressed by lower noise.

In this chapter, after presenting TR basis and theoretical principles, characteristic parameters of TR are numerically studied in free space and reverberating environment before introducing an original way for performing impulsive susceptibility testing.

2. Time reversal basis

2.1 Preamble

Originally developed in acoustics (Fink, 1992) by Mathias Fink team in the early 1990 at the ESPCI (Ecole Supérieure de Physique et de Chimie Industrielles) in Paris, TR is a physical process that is based on the principle of reciprocity. This technique allows a wave to propagate backward to its source. This retro-propagation is based on the reversibility of the wave equation in time. One of the results is to offer the possibility to focus a given wave both in time and space. Many studies have been led based on the acoustic wave equation, for applications concerning the detection and selective focusing (Prada & Fink, 1994), submarine telecommunications (Edelmann, 2005), and ultrasound and medical imaging (Quieffin, 2004) domains. More recently successful tests have been achieved in electromagnetics (de Rosny et al., 2007), mainly in telecommunications (Lerosey et al., 2004), detection and imaging (Liu et al., 2005; Maaref et al., 2008; Neyrat et al., 2008), and EMC (Davy et al., 2009; El Baba et al., 2009; 2010) fields.

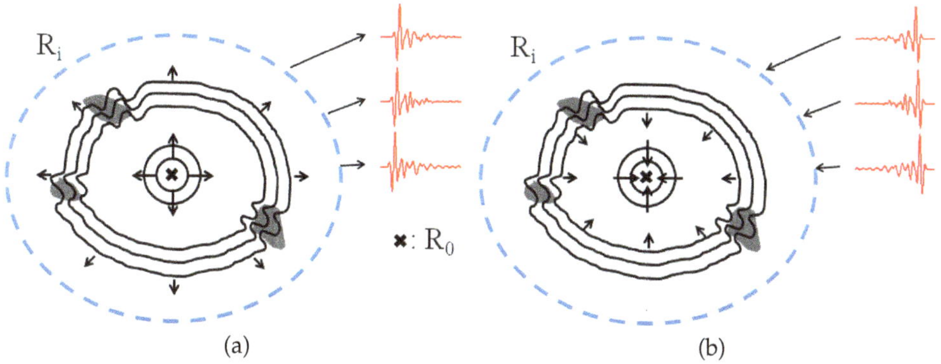

Fig. 1. (a) First and (b) second phase of time reversal process with a time reversal cavity.

In practice, the TR technique needs two phases. During the first (Fig. 1a), a source located at (R_0) emits an electromagnetic pulse that spreads in the medium. The source can be either active (transmission mode) or passive as a source of diffraction (in detection problems targets or diffusers act as passive sources). The electromagnetic radiation is recorded for a period Δt through an array of probes in reception (R_i) surrounding the source into a closed entity and forming a Time Reversal Cavity (TRC). Indeed, the data that arrive first in time travels a shorter distance than the data that arrive later. During the second phase (Fig. 1b), each probe retransmits its received signal in reversed time order, so the data that travel a longer distance are emitted earlier and the data that travel a shorter distance are emitted later (Last In First Out). Consequently, a returned wave propagates and acts as if it relives exactly its past life and this leads to a temporal and spatial focusing of the field at the original source location (R_0) where the focusing moment is considered as the time origin.

Unfortunately, from an experimental point of view and because of the large number of probes required for such operation, the TRC is not feasible. That's why classical TR experiments are conducted through a limited opening array forming the Time Reversal Mirror (TRM) (de Rosny & Fink, 2002).

In the case of a TRM, the focusing protocol by TR remains the same as in the TRC case (Fig. 2). The decrease in the angular aperture allows the convenient realization of such a mirror, but in the reemission step (Fig. 2b) only a part of the wave is time reversed leading to a loss of information reducing the focusing quality.

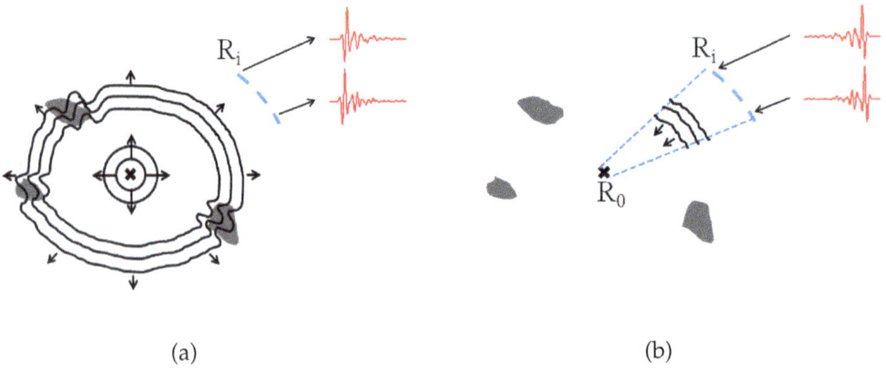

Fig. 2. (a) First and (b) second phase of time reversal process with a time reversal mirror.

This loss of information can be partially avoided if the scene takes place in a reverberant cavity. Different studies have shown that, in the case of a reverberant environment, the probe array can be replaced by a single probe (Fig. 3). Therefore, the first experiments in electromagnetism were realized in a RC. The properties of the cavity allow us to benefit from the different reflections suffered by the wave on the metal walls of the chamber, which ensures that a single probe collecting these echoes is sufficient to record necessary information for TR experiment.

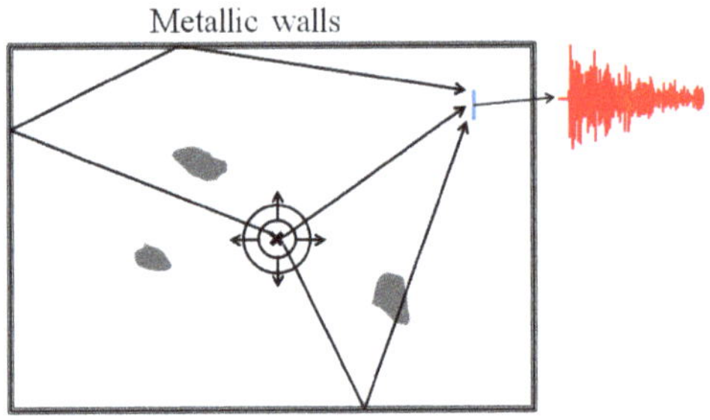

Fig. 3. TRM can be replaced by a single probe in a reverberant cavity.

2.2 Time reversal of electromagnetic waves

A propagation medium is called reversible if a field and its time reversed version can propagate in such an environment, i.e., if $\Phi(t)$ and $\Phi(-t)$ are solutions of the same propagation equation. In electromagnetism, the wave equation in a uniform and non-dissipative medium is given by

$$\frac{1}{c^2}\frac{\partial^2\Phi}{\partial t^2} = \Delta\Phi \tag{1}$$

where Φ stands for the electric E or magnetic H field, and c is the propagation celerity of the electromagnetic waves in the medium.

Assuming that $\Phi_0(t)$ is a solution of (1), the absence of first time derivative in the left-hand side leads to the existence of another solution chronologically reversed $\Phi_1(t) = \Phi_0(-t)$.

Therefore (1) is invariant under the TR action, and theoretically, an electromagnetic scene may be replayed in reverse from time $t = \Delta t$ to $t = 0s$.

Defining the electric field estimated at the position r (given by TRM or TRC devices) and time t by $E(r,t)$, the TR data are first recorded during the experiment time $T = \Delta t$. Then, fields are returned and reemitted following a reverse chronology, for example, an electric field $E(r,t)$ is retransmitted during the reversal phase as $E(r, T-t)$, $t \in [0;T]$.

An electromagnetic wave is described by four field vectors, the electric field E, the magnetic field H, the electric induction D, and the magnetic induction B. It has been shown in (Jackson, 1998) that E and D are even vectors; however H and B are odd pseudovectors under the time reversal action. Therefore if we consider T_{TR} the time inversion operator given by $T_{TR}\{\Phi(r,t)\} = \Phi(r,-t)$, we can write

$$\begin{aligned} T_{TR}\{E(r,t)\} = E(r,-t); \quad T_{TR}\{D(r,t)\} = D(r,-t); \\ T_{TR}\{H(r,t)\} = -H(r,-t); \; T_{TR}\{B(r,t)\} = -B(r,-t). \end{aligned} \tag{2}$$

3. Theoretical principles

An illustration of the electromagnetic TR in a reverberant environment can be obtained from Fig. 4, where R_0 is a point source representing the emission antenna, and R_i corresponds to the array of probes in reception (TRM).
The excitation pulse used is a Gaussian modulated sine pattern (Fig. 4a) emitted from the point R_0. Probes of the TRM (R_i) record the six components of the electromagnetic fields, these signals are returned by time reversal or by the phase conjugate of their Fourier transforms (method explained in section 3.1 below) and reemitted by R_i to obtain a time and space focusing at R_0 position.

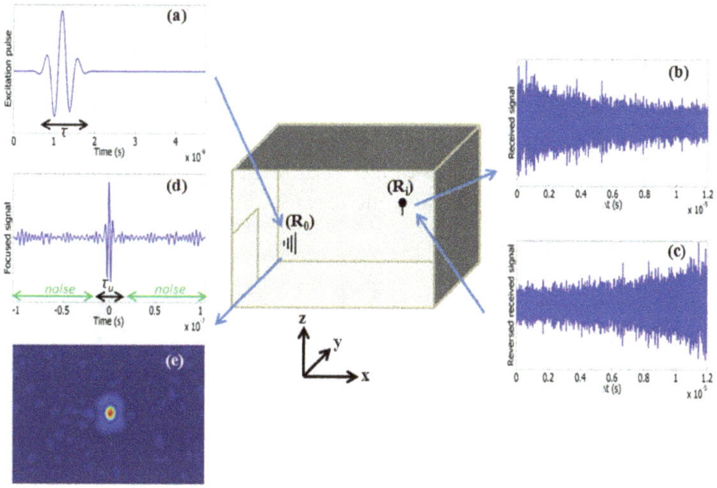

Fig. 4. TR set up, (a): excitation pulse, (b): received signal, (c): reversed received signal, (d) and (e): time and space focusing.

3.1 Time reversal and phase conjugate

Let $\tilde{\Phi}(r,\omega)$ the Fourier transform of the field $\Phi(r,t)$. It was verified that time reversing a signal corresponds to the inverse Fourier transform (FT_{inv}) of the phase conjugate of its Fourier transform

$$T_{TR}\{\Phi(r,t)\} = \Phi(r,-t) = FT_{inv}\{\tilde{\Phi}^*(r,\omega)\} \tag{3}$$

In Fig. 5a we plot the evolution of a signal with respect to time, and in Fig. 5b we verify that we can reverse a signal in time indifferently by the phase conjugate of its Fourier transform or by time reversal.

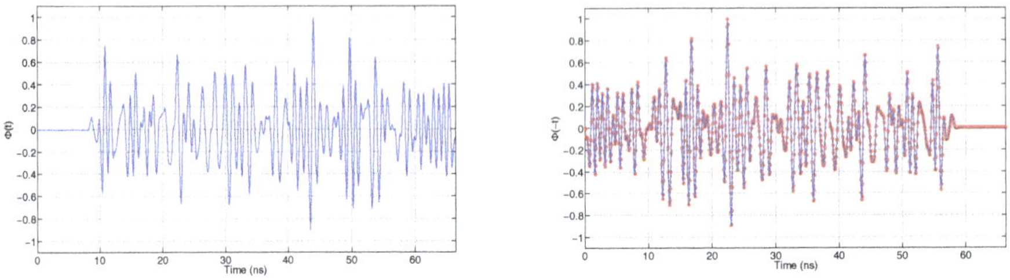

(a) (b)

Fig. 5. (a) Temporal signal $\Phi(t)$. (b) $\Phi(-t)$ by time reversal (plain blue curve) and by inverse Fourier transform of its phase conjugate (red markers).

3.2 Mathematical foundation

The received signal (Fig. 4b) by a TRM probe (following a pulse $x(t)$ (Fig. 4a) emitted from R_0) can be written

$$y_i(t) = k(t, R_0 \rightarrow R_i) \otimes x(t) \tag{4}$$

where $\otimes$ is the convolution product, and $1 \leq i \leq M$ with M the number of probes of the TRM, and $k(t, R_0 \rightarrow R_i)$ is the impulse response of the medium at a point R_i for a pulse emitted from R_0. After time reversal of $y_i(t)$ (Fig. 4c) and the reemission from R_i, the focused signal on R_0 (Fig. 4d) can be written as follows:

$$E_{TR}(t, R_0) = \sum_{i=1}^{M} k(t, R_i \rightarrow R_0) \otimes y_i(-t) = \sum_{i=1}^{M} k(t, R_i \rightarrow R_0) \otimes k(-t, R_0 \rightarrow R_i) \otimes x(-t) \tag{5}$$

The advantage of working in the frequency domain is to replace the convolution product by an ordinary product. Since reversing a signal versus time corresponds to the phase conjugate of its Fourier transform, the above equation (5) takes the following form in the frequency domain

$$E_{TR}(\omega, R_0) = \sum_{i=1}^{M} k(\omega, R_i \rightarrow R_0).k^*(\omega, R_0 \rightarrow R_i).x^*(\omega) \tag{6}$$

Switching into matrix notation, (6) takes the following form

$$E_{TR}(\omega, R_0) = K(\omega, R_i \rightarrow R_0).K^*(\omega, R_0 \rightarrow R_i).x^*(\omega) \tag{7}$$

Now, if the medium is reversible, thanks to the reciprocity theorem, the position of a point source and a probe can be reversed without altering the field. Consequently, the impulse response from R_0 to R_i is equal to the one from R_i to R_0, and therefore the matrix $K(\omega, R_0 \rightarrow R_i)$ is equal to the matrix $K(\omega, R_i \rightarrow R_0)$, in other words the matrix K is symmetric. In (7), the propagation matrix K is the Fourier transform of different impulse responses between one transmitter and M receivers.

3.3 Time reversal operator

One can build the so-called Time Reversal Operator (TRO) by considering the case where we have $M \times M$ transmitter receivers. The M transmitters emit successively M pulses $x_i(t)(i = 1, ..., M)$ that can be described in the frequency domain by a vector X containing M components for each frequency. The M components given by the receivers can be written by a matrix product KX. When signals are reversed (phase conjugate in the frequency domain) and retransmitted, the resulting vector is $K^t K^* X^*$, with K the $M \times M$ propagation matrix and K^t its matrix transpose. Therefore (7) can be written

$$Foc(\omega) = K^t(\omega)K^*(\omega)X^*(\omega) \tag{8}$$

with Foc the vector containing the M focused signals for each frequency. It is interesting to note that $T(\omega) = K^h(\omega)K(\omega)$, defined as the TRO (Derode et al., 2003), is a symmetrical square matrix where h is the Hermitian conjugate (conjugate-transpose). By performing a singular value decomposition of the propagation matrix we get $K(\omega) = U(\omega)\Lambda(\omega)V^t(\omega)$, where U and V are unitary matrices and Λ is a diagonal matrix whose elements are the singular values Λ_i. On the other hand, the eigenvalue decomposition of the TRO gives $T(\omega) = V(\omega)S(\omega)V^t(\omega)$, with $S(\omega) = \Lambda^t(\omega)\Lambda(\omega)$ the diagonal matrix of eigenvalues that are the propagation matrix singular values square, and V the unit matrix of eigenvectors. This decomposition of the TRO gives us information on the propagation medium. In the detection field, Decomposition of the Time Reversal Operator (DORT) (Yavuz & Teixeira, 2006) provides information on the diffraction strength of the target via the eigenvalues and information on the position via the eigenvector of the TRO.

4. Definitions, numerical methodologies and outputs

In this study, TR is applied in a numerical way in order to facilitate its characterization in different configurations. From a practical point of view, it is easier to carry out a parametric study numerically than experimentally. For instance, a numerical study offers the flexibility to choose between a TRC and a TRM, to vary the number of probes, their positions in many test cases, etc. The proposed methodology needs to gather from the TR principles and MSRC studies. This is why the chosen method must take into account the characteristics of each domain. Indeed, it is important to consider all elements present in the experimental RC device: cavity, stirrer, equipment (Corona et al., 2002). From a numerical point of view, the influence of the metallic elements must be considered in time simulation. Consequently, the fields temporal distribution must be numerically implemented with special care given to metal facets (i.e., considered as Perfect Electric Conductor, PEC). For all the above reasons, the numerical simulations were carried out using an own-made Finite Difference Time Domain (FDTD) electromagnetic code (Bonnet et al., 2005) with E / H formulation and later for more complex cases the commercial software CST MICROWAVE STUDIO® without neglecting the fact that we can use any numerical tool that solves Maxwell's equations.

4.1 Focusing quality

To characterize time and space focusing after the TR process, we will define multiple criteria and parameters.

4.1.1 Maximum magnitude of focusing

The first idea about the quality of focusing is obtained by considering the useful part of the reconstructed signal (see duration τ_u, Fig. 4d) and implementing the absolute maximum of the focused signal.

$$Max\,(R_0) = max_{t \in \tau_u}(|E_{TR}(t, R_0)|) \tag{9}$$

4.1.2 Focal spot

The second criterion characterizing spatial focusing around R_0 is the focal spot dimension (δ) which is described in two dimensions by distance along the x and y directions for which the total electric field focused at time $t = 0$ (which is the focusing time) is between $E_{TR}(R_0)$ and $E_{TR}(R_0)/2$ (in other words where $E_{TR}(R_0)/Max\,(R_0)$ belongs to $[-6\ dB;\ 0\ dB]$). The Fig. 6 illustrates this criterion in two dimensions (2-D), the principle can be extended to three dimensions (3-D). According to Fig. 6, we may write

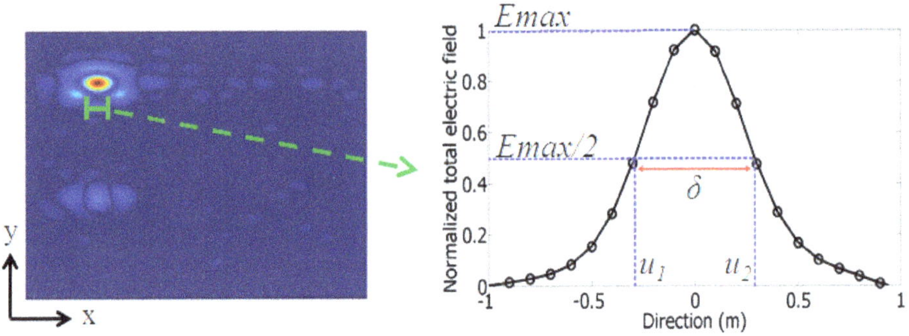

Fig. 6. Definition of the focal spot around the focusing point.

$$\begin{aligned} \delta_x &= u_{x2} - u_{x1} \\ \delta_y &= u_{y2} - u_{y1} \end{aligned} \tag{10}$$

where u_{x1}, u_{x2}, u_{y1}, and u_{y2} are the positions of both sides of R_0 at the focusing time ($t = 0$) along x and y directions. Note that the width of the focal spot in a reverberating environment, depending on the diffraction limit, is about $\lambda/2$ where λ is the wavelength corresponding to the frequency of the excitation pulse. On the other hand, in free space the focal spot is given by the following formula

$$\delta = \frac{\lambda F}{D} \tag{11}$$

with F the distance between the focusing point (R_0) and the TRM (R_i), and D the size of the TRM.

4.1.3 Signal to noise ratio

An important criterion to characterize focusing in a reverberation chamber is the Signal To Noise (STN) ratio, which was theoretically introduced in (de Rosny, 2000) as follows:

$$STN \cong \frac{4\sqrt{\pi}\Delta H \Delta\Omega}{\frac{\langle\alpha\rangle^4}{\langle\alpha^2\rangle^2} + \frac{\Delta H}{\Delta t}} \tag{12}$$

where we have: $\Delta\Omega$, the frequency bandwidth of the excitation pulse; α, the ensemble average of the eigenmodes magnitude of the chamber; ΔH, the Heisenberg's time given by the following formula

$$\Delta H = 2\pi n(\omega) \tag{13}$$

with $n(\omega)$, the average modal density of the reverberation chamber assumed constant over the entire bandwidth $\Delta\Omega$.

Numerically, this ratio can be calculated from the temporally focused signal in R_0, and it is known as the temporal STN ratio (STN_t) which is the ratio between the squared magnitude of the focused signal peak (Fig. 4d, $t = 0$) and the temporal noise around the peak. It is defined as the square of the focused field RMS on a part of the simulation time apart the useful signal (Fig. 4d, $t \notin \tau_u$).

The ratio is given by

$$STN_t = \frac{\langle E_{TR}(r = R_0, t = 0)\rangle^2}{\left\langle \overline{E_{TR}^2(r = R_0, t \notin \tau_u)} \right\rangle} \tag{14}$$

where $E_{TR}(R_0, t)$ represents the focused total electric field in R_0.

Similarly to (14), we can also calculate the spatial STN ratio (STN_s) which is the ratio of the squared magnitude of the focused signal peak in R_0 on the square of the field RMS value calculated over the rest of the studied domain at the focusing time ($t = 0$), this one can be considered as "spatial noise". So we have

$$STN_s = \frac{\langle E_{TR}(r = R_0, t = 0)\rangle^2}{\left\langle \overline{E_{TR}^2(r \neq R_0, t = 0)} \right\rangle} \tag{15}$$

It has been proved in (Moussa et al., 2009a) that for whole averages and as RC are ergodic systems, temporal STN ratio is equivalent to spatial one. In what follows, for the sake of simplicity, both temporal and spatial signal to noise ratios are denoted STN.

4.1.4 Delay spread

To characterize the temporal focusing (and linked to spatial aspects) in reverberation chamber, the delay spread parameter is defined as in (Ziadé et al., 2008). Indeed, the impulse response shown in Fig. 4b shows that a pulse emitted from a source will be received as a series of pulses (with different arrival times). This parameter stands for time separating last echo and straightforward way. The root mean square of the delay spread parameter (linking the standard deviation of time with the mean value) can be written for the E fields by

$$\tau_{RMS} = \sqrt{\frac{\int (\tau - \tau_m)^2 \left|E_{TR}(\mathbf{r}, \tau)\right|^2 d\tau}{\int \left|E_{TR}(\mathbf{r}, \tau)\right|^2 d\tau}} \tag{16}$$

with τ_m: mean value of E delays. Electric fields E_{TR} are given at location r and time τ. Thus, the average delay τ_m is given by

$$\tau_m = \frac{\int \tau \left|E_{TR}(\boldsymbol{r},\tau)\right|^2 d\tau}{\int \left|E_{TR}(\boldsymbol{r},\tau)\right|^2 d\tau}. \tag{17}$$

4.2 FDTD method for time reversal

In the FDTD method, Maxwell's equations are discretized following the Yee algorithm (Yee, 1966), these equations are invariant to time reversal transformation (Jackson, 1998). For more simplicity, we will consider here a 2-D formulation; the 3-D case can be straightforward extended by simple modifications. In the FDTD method, electric and magnetic fields are calculated by an explicit "leapfrog" scheme for time intervals separated by a half time step, in other words from the electric field at time $t = n - 1/2$ and the magnetic field at time $t = n$, the electric field at time $t = n + 1/2$ is calculated as we can see in the discretized Maxwell's equation below (18) (2-D TM mode)

$$E_z^{n+\frac{1}{2}}\left(i+\frac{1}{2},j+\frac{1}{2}\right) = \frac{2\epsilon - \sigma dt}{2\epsilon + \sigma dt} E_z^{n-\frac{1}{2}}\left(i+\frac{1}{2},j+\frac{1}{2}\right) + \frac{2dt}{2\epsilon + \sigma dt} \times \left[\frac{H_y^n\left(i+1,j+\frac{1}{2}\right) - H_y^n\left(i,j+\frac{1}{2}\right)}{dx} - \frac{H_x^n\left(i+\frac{1}{2},j+1\right) - H_x^n\left(i+\frac{1}{2},j\right)}{dy}\right] \tag{18}$$

where dx, dy and dz are the space steps in Cartesian directions (Ox), (Oy), and (Oz), while dt is the time step. σ and ϵ represent conductivity and permittivity of the medium.

Regarding time reversal, we need to reverse the calculation sequence. In fact, the electric field at time $t = n - 1/2$ is calculated from the electric field at time $t = n + 1/2$ and the magnetic field at time $t = n$. To do that, we only need to take the discretized Maxwell's equations and rewrite them under the desired shape (Sorrentino et al., 1993)

$$E_z^{n-\frac{1}{2}}\left(i+\frac{1}{2},j+\frac{1}{2}\right) = \frac{2\epsilon + \sigma dt}{2\epsilon - \sigma dt} E_z^{n+\frac{1}{2}}\left(i+\frac{1}{2},j+\frac{1}{2}\right) - \frac{2dt}{2\epsilon - \sigma dt} \times \left[\frac{H_y^n\left(i+1,j+\frac{1}{2}\right) - H_y^n\left(i,j+\frac{1}{2}\right)}{dx} - \frac{H_x^n\left(i+\frac{1}{2},j+1\right) - H_x^n\left(i+\frac{1}{2},j\right)}{dy}\right] \tag{19}$$

The above relation (19) should be applied to calculate fields at earlier moments from the later instants. To check the validity of this TR algorithm, consider a 2-D domain whose boundaries are simulated by PEC. An excitation point source emitting a Gaussian pulse is located in the middle of the Computational Domain (CD). The field propagates in the domain for a time $t = t_1$. Fig. 7a shows the distribution of the electric field E_z at time $t = t_1$. For $t > t_1$, time is reversed and we consider the field distribution E_z at time $t = t_1$ (Fig. 7a) as initial condition. After this time, and from the modified FDTD equations, we come to rebuild the field distribution of the source and found its position (Fig. 7b). This TR algorithm was effectively applied in (Neyrat, 2009) for buried object detection.

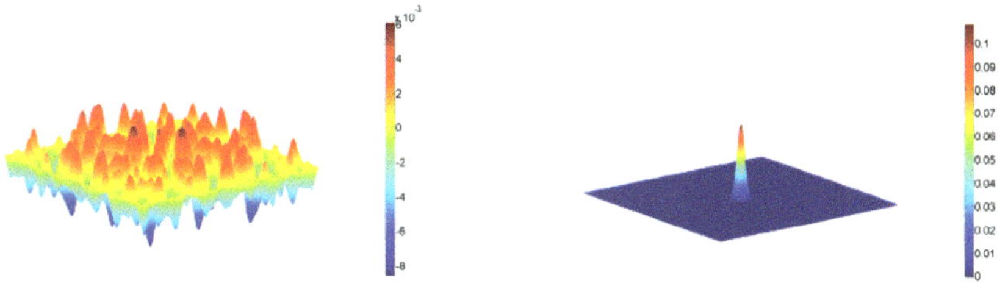

(a) (b)

Fig. 7. (a) Electric field E_z distribution at time $t = t_1$. (b) Reconstructed pulse obtained by inverse FDTD simulation.

The technique described above may only have numerical applications since fields have to be recorded for each discretisation point. In our case (applying TR in EMC), our objective is not detection, but focusing field numerically in a given place and time, and for instance, extending to experimental developments in further works. To avoid field registration throughout the whole domain which should be impossible experimentally, we will focus on the technique briefly described on Fig. 4. In this case, the field is recorded on probes during the first phase using the FDTD discretization (18), and in the second it is time reversed and retransmitted without changing the Maxwell's equations in the FDTD code. This alternative is the one used in most of TR experiments and TR numerical simulations.

4.3 Numerical configurations

Already mentioned, simulations were performed using an own-made code based on the FDTD method. Two *CDs* were considered: the first one is a 2-D TM mode given by $CD_1 = 3.3 \times 3.3\ m^2$, the second is a 3-D $CD_2 = 2.2 \times 1.5 \times 1\ m^3$ volume. The excitation signal used for the first phase of the TR process is a Gaussian modulated sine pattern (Fig. 8a)

$$x(t) = E_0 e^{-\left(\frac{t+\frac{z-z_0}{c}-t_0}{\ell}\right)^2} sin(2\pi f_c t) \tag{20}$$

where E_0 is the Gaussian magnitude, z_0 and t_0 are respectively the delays with respect to the origins of space and time, ℓ is the mid-height width of the pulse, and f_c is the central frequency.

The bandwidth $\Delta\Omega$ of this pulse is the frequency distance (Fig. 8b) of both sides of the central frequency with respect to the attenuation (Att) of the maximum amplitude (the amplitude corresponding to f_c). To calculate $\Delta\Omega$, we consider successively different attenuation levels. For instance, in the case of $Att = 2$ (corresponding to a $-6\ dB$ decrease) we divide the amplitude corresponding to f_c by 2 and calculate f_2 and f_1, and the bandwidth is given by $\Delta\Omega = f_2 - f_1$.

For all FDTD simulations in this chapter, we used a Gaussian modulated at a central frequency $f_c = 600\ MHz$ and bandwidth $\Delta\Omega = 350\ MHz$ calculated at $-6\ dB$. Simulations are performed with an uniform spatial discretization $dx = dy = 3.3\ cm$ for the 2-D domain, and $dx = dy = dz = 3.3\ cm$ for the 3-D domain (corresponding to $\lambda_{f_c}/15$, λ_{f_c}: wavelength corresponding to the central frequency f_c).

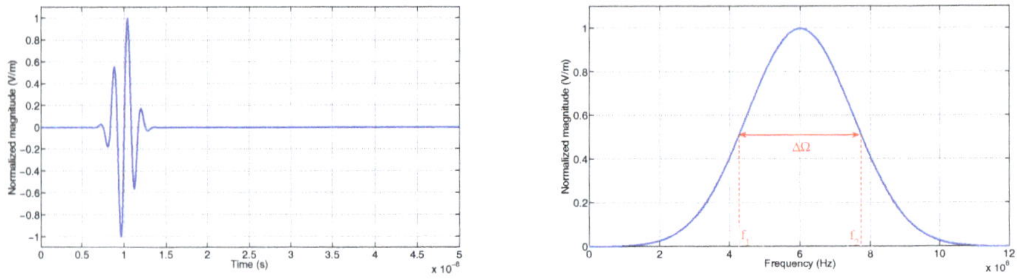

(a) (b)

Fig. 8. (a) Time response of the excitation signal used for the first phase of TR and (b) its spectrum.

5. Numerical results

In this section, we investigate the impact of various parameters on the TR process. These will be studied initially in free space, then we will see how the complexity of the environment can improve the focusing quality, and finally we will check how reverberant media are ideal environments to work with TR.

5.1 Preliminary study in free space

The first numerical example treated helps to qualify focusing relatively to the number of probes in the TRC. For this, we consider the 2-D CD_1 domain (Fig. 9) where the excitation source is located in the middle of the area and a TRC composed of 320 probes completely surrounding the point source. Free space is simulated by Mur absorbing boundary conditions (Mur, 1981).

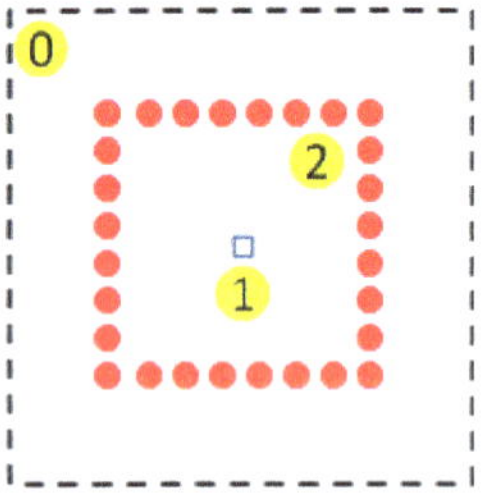

Fig. 9. CD_1 domain: (0) absorbing conditions, (1) source R_0, (2) TRC probes R_i.

A 7 *ns* pulse (Fig. 8a) is emitted from the point source and the TRC probes record the evolution of the electric field component E_z and the magnetic field components H_x and H_y (TM mode). After time reversal and reemission of the recorded signals by the TRC probes, we can find the position of the excitation source as shown in the spatio-temporal evolution of the absolute value of the electric field E_z on Fig. 10.

The returned excitation signals $x(-t)$ and the normalized temporal focusing signal E_{TR} at the point source are plotted on Fig. 11a. The original shape of the excitation signal and the position of the emission point are observed (here it is an active source but it can also be a diffracting object).

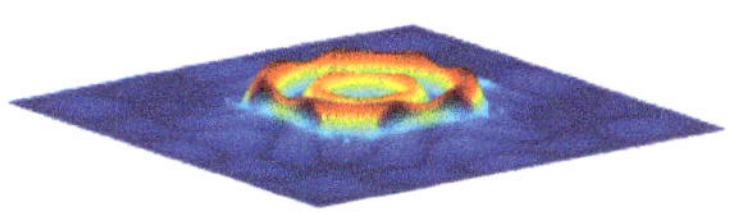

(a) $t = -1.47ns$

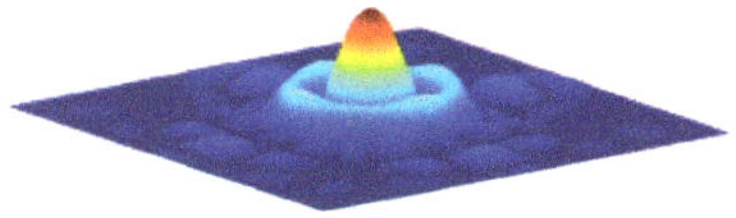

(b) $t = -0.29ns$

(c) $t = 0$

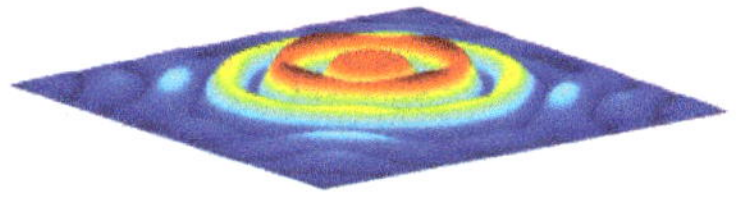

(d) $t = +2.12ns$

Fig. 10. Electric field spatio-temporal evolution around focusing point (focusing time set as time origin).

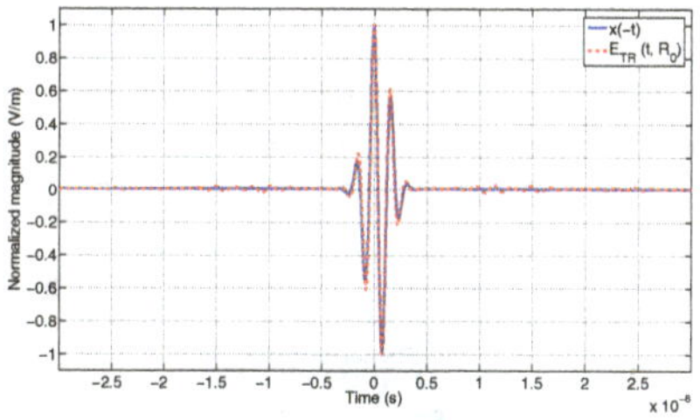

(a)

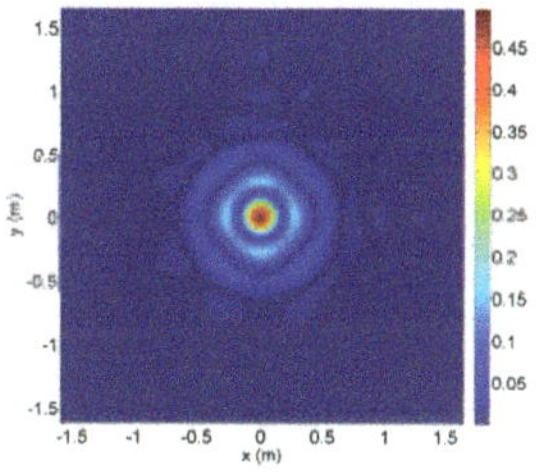

(b)

Fig. 11. (a) Temporal and (b) spatial focus on the point source at focusing time ($t = 0$).

The number of probes used in the previous example (320 probes) is the maximum number allowed by the used FDTD discretization. In Fig. 12a, curves (1), (2) and (3) demonstrate the importance of the probes number in the TRC on the maximum magnitude of focusing criterion. In addition, we note (Fig. 12b) that this criterion increases linearly with the number of probes.

For a 3-D domain (CD_2), two cases were treated. The first one deals with an excitation emitted by the point source along the three components of the electric field E_x, E_y and E_z, and the second one only E_x component is considered. The Fig. 13 shows the treated numerical configuration where the excitation point is in the middle of the CD_2 (Cartesian coordinates (0, 0, 0) which corresponds to the mesh (34, 23, 15), and the TRC is composed of 6114 probes corresponding to the maximum number allowed by the used FDTD discretization.

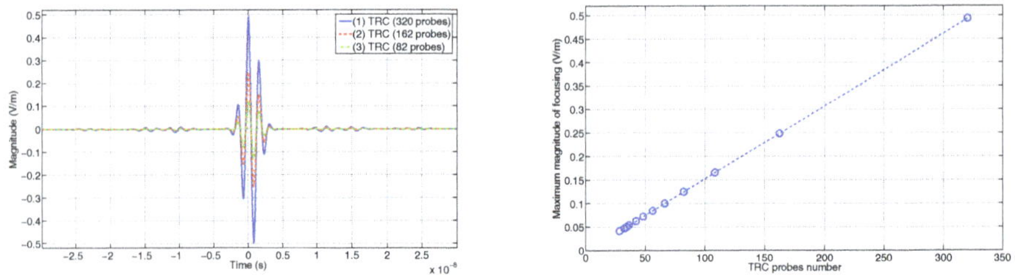

(a) (b)

Fig. 12. (a) Focused signals ($E_{TR}(t, R_0)$) for different number of probes uniformly distributed on the TRC. (b) Maximum magnitude of focusing criterion with respect to the TRC probes number.

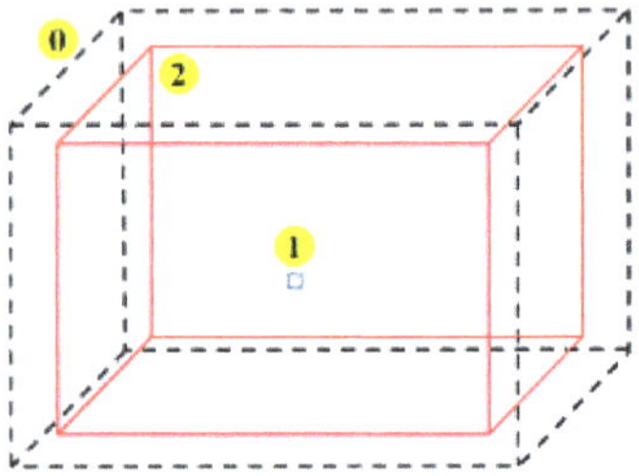

Fig. 13. CD_2 domain: (0) absorbing conditions, (1) source R_0, (2) TRC.

In the first case, we can see that the focused signal after TR is along the three polarizations x, y and z (Fig. 14a), and we can focus on the spatial distribution of the total electric field at focusing time (Fig. 14b) where an energy concentration appears around the point source.

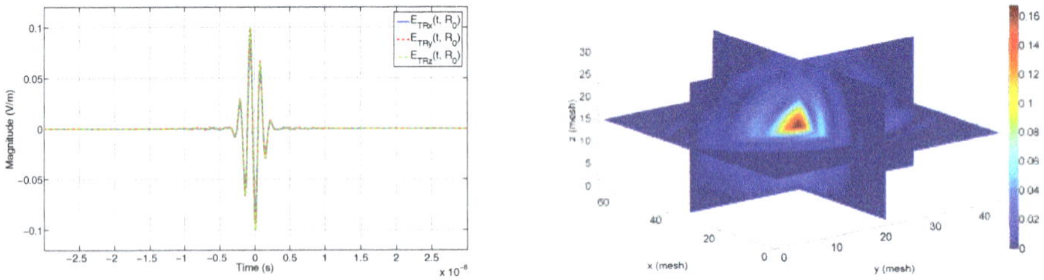

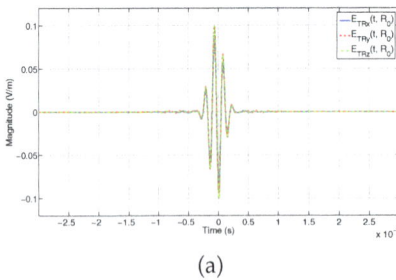

(a) (b)

Fig. 14. (a) Time focusing for emitted excitation along E_x, E_y, and E_z. (b) Total electric field cartography at focusing time ($t = 0$).

In the second case (where the excitation is along E_x), we see that the electric field is focused only along the x component (Fig. 15) and this can be verified if we extract E-field over a plan corresponding to $z = 0$ and we look to the field cartography at focusing time for all polarizations (Fig. 16). We clearly note that the electric field corresponding to E_y and E_z is almost zero compared to E_x.

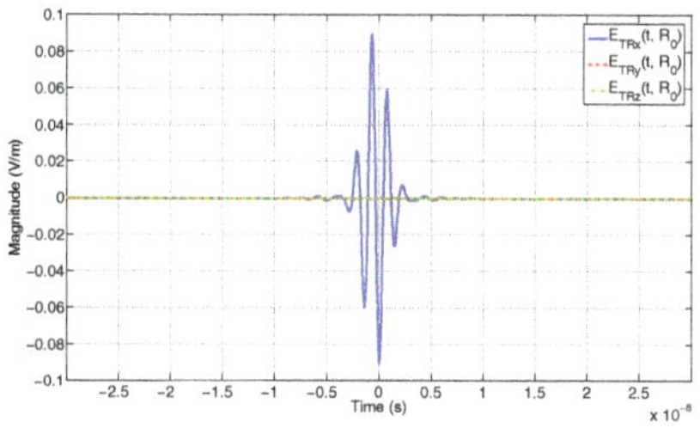

Fig. 15. Time focusing for excitation emitted along E_x.

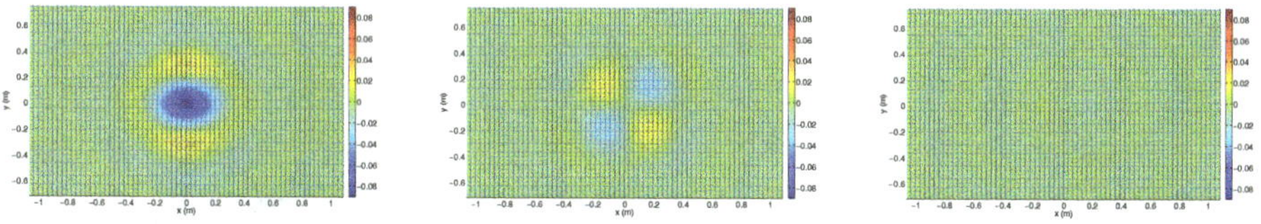

(a) (b) (c)

Fig. 16. E-field slice plan corresponding to (a) E_x, (b) E_y, and (c) E_z components at the focusing time ($t = 0$).

We deduce that it is theoretically possible to control the polarization of the wave attacking the EUT without changing the antenna polarization. This application can be very interesting especially in a reverberant environment, as we shall see later in this chapter.

Given the huge number of probes needed for the TRC and the inability to achieve such an experimental configuration, in the following simulations TRC is replaced by a TRM with limited opening (Fig. 17).

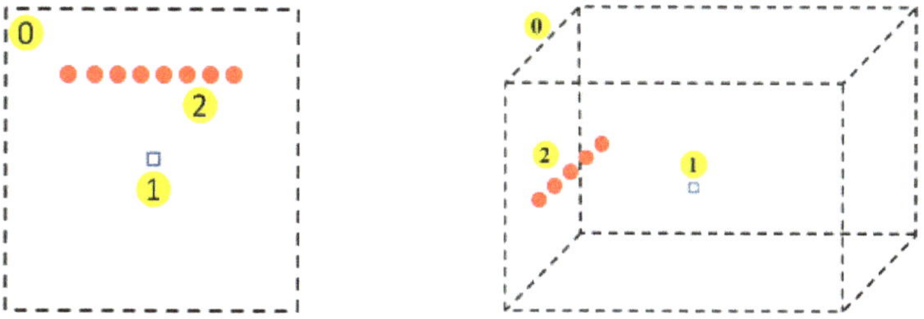

(a) (b)

Fig. 17. (a) CD_1 and (b) CD_2 domains: (0) absorbing conditions, (1) source R_0, (2) TRM probes R_i.

The previous simulations are repeated with a TRM of 41 probes for 2-D domain and 54 probes for 3-D domain, comparing temporal focusing (Figs. 18a, 19a) with those obtained with a TRC (Figs. 12a, 14a), we note that the maximum magnitude of focusing is greatly reduced. Moreover we note a spatial focusing damage. The Figs. 18b and 19b show that focusing is of weaker quality comparatively to TRC cases (Figs.11b, 14b).

So, unlike the previous case, recording fields on one side of the domain can not reconstruct the exact propagation of the wave as it spread, from the fact that the information is reduced

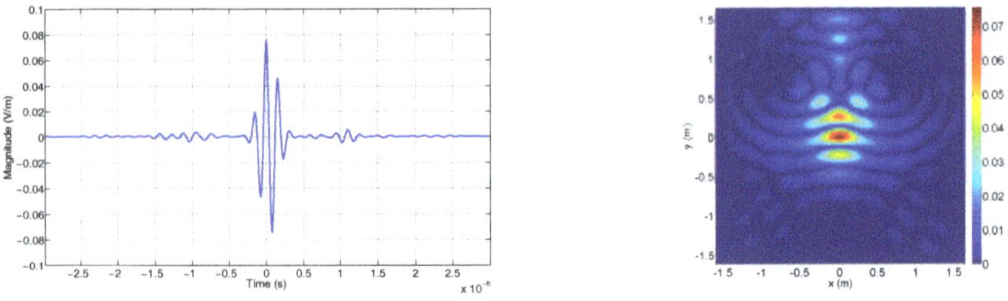

(a) (b)

Fig. 18. CD_1: (a) Temporal and (b) spatial focusing on the point source at the focusing time ($t = 0$) using a TRM.

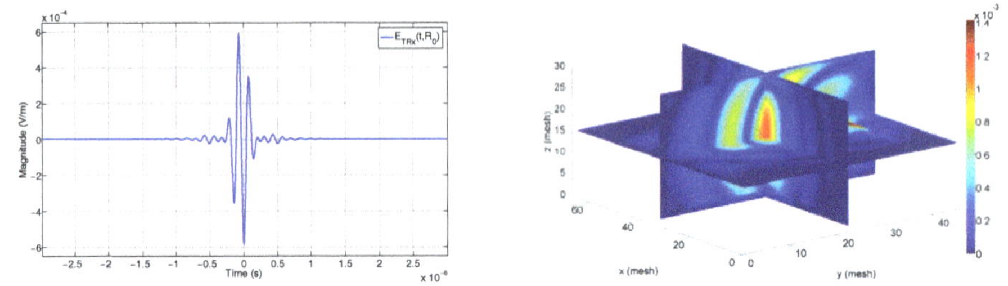

(a) (b)

Fig. 19. CD_2: (a) Temporal and (b) spatial focusing on the point source at the focusing time ($t = 0$) for an excitation along E_x, E_y, and E_z using a TRM.

especially for 3-D. This loss of information can be solved by making the domain more complex, which will allow recording more information without increasing the number of TRM probes. In the following, we will only consider the 3-D domain (CD_2).

5.2 Introducing multiple reflections

To collect more information on the wave propagation in the first phase of TR, it is better to increase the TRM angular opening or make the environment more complex. To achieve this, a metal plate modeled by PEC is added in the domain (Fig. 20).

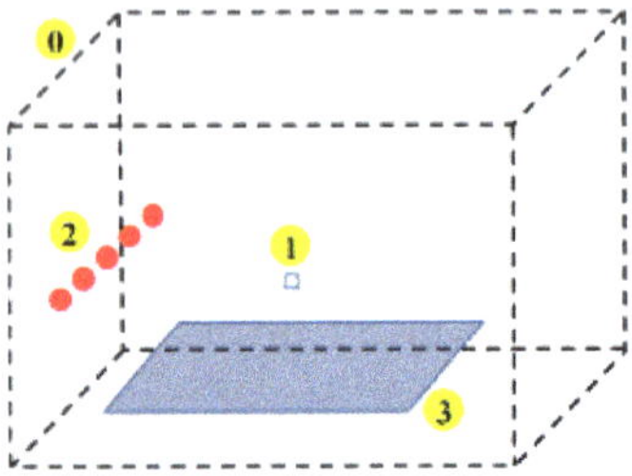

Fig. 20. CD_2 domain: (0) absorbing conditions, (1) source R_0, (2) TRM probes R_i, (3) metallic plate.

The goal here is to take advantage of reflections due to the presence of the metal plate. The Fig. 21a confirms the expectation about the presence of a diffracting object: we see that the signal received by a probe of the TRM (component x of the electric field E_x) in a complex environment contains more information. Indeed, the waves due to the PEC reflections improve the maximum magnitude of focusing (Fig. 21b).

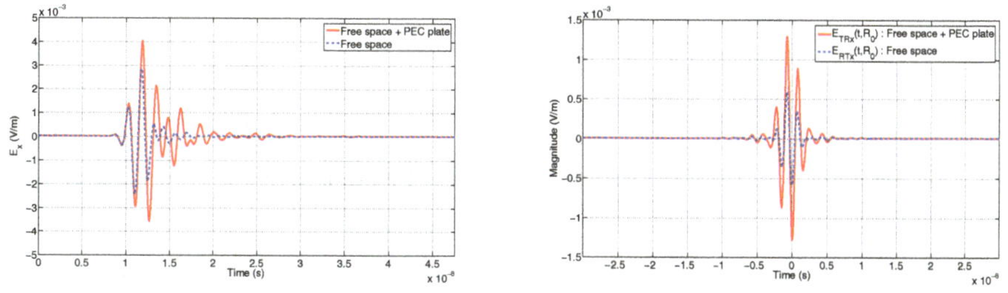

(a) (b)

Fig. 21. (a) Electric field E_x received by a TRM. (b) The temporal focusing on the source point in both cases (free space + metallic plate and free space) for an excitation along E_x, E_y, and E_z using a TRM.

Following this idea, we can easily imagine that suitable media to apply the TR are the reverberating environments, supporting the interests of its application in the MSRC. Indeed, the multiple reflections suffered by the wave on the metal walls of the chamber will allow us to replace the TRM by a limited number of probes.

5.3 Time reversal in a reverberant environment (reverberation chamber)

In this section, the previous configuration CD_2 is preserved with a 8-probes TRM and excitations along E_x, E_y and E_z. The purpose of this part is to show the benefits of TR application in the MSRC across different test cases:

- the "free space" data will be compared to a reverberant environment,
- the duration of the TR window, in other words duration of the reversed signal, is studied looking the STN ratio and the link with the modal density of the chamber,
- intrinsic properties of the wave propagation in the cavity will be treated by focusing on the randomness of the probes location,
- and finally, we will study the influence of the excitation source parameters in terms of the focal spot size.

5.3.1 Comparison with free space

The presence of perfectly metallic boundary conditions replacing absorbing conditions implies that the impulse response (Fig. 22a) received by one of the 8 probes of the TRM is composed of several reflections that never decrease, unlike free space case (Fig. 21a). It is important to note that real losses are not included in this section. As a result, the numerical energy injected after time reversal appears comparatively higher in RC than with absorbing conditions. This improves the focusing quality in terms of maximum focusing magnitude: 6.10^{-4} V/m with 51 probes as TRM in free space (Fig. 19a) and 0.04 V/m with a TRM of 8 probes in RC (Fig. 22b).

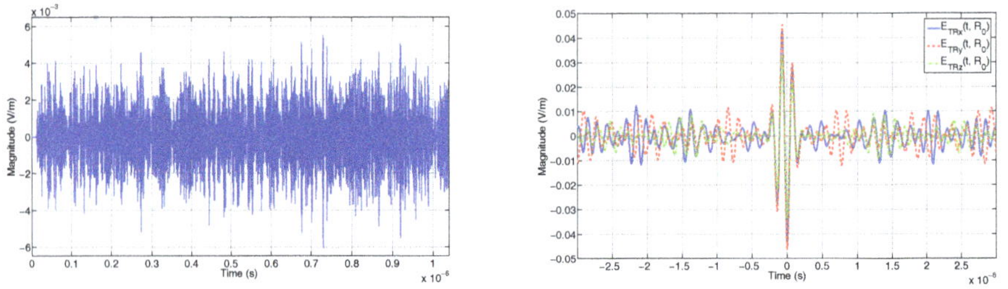

(a) (b)

Fig. 22. (a) Electric field E_x received by a probe of the TRM in RC. (b) Temporal focusing by TR in RC.

To investigate the spatial focusing by studying the focal spot size in all directions and check the temporal focusing using the delay spread criterion (τ_{RMS}), we recorded the total electric field around the focusing point along x, y and z axis. On the one hand, results given in Fig. 23a show that focusing is symmetrical and is $\lambda_{f_c}/2$ order (dimension of the focal spot $= 0.25\ m = \lambda_{f_c}/2$).

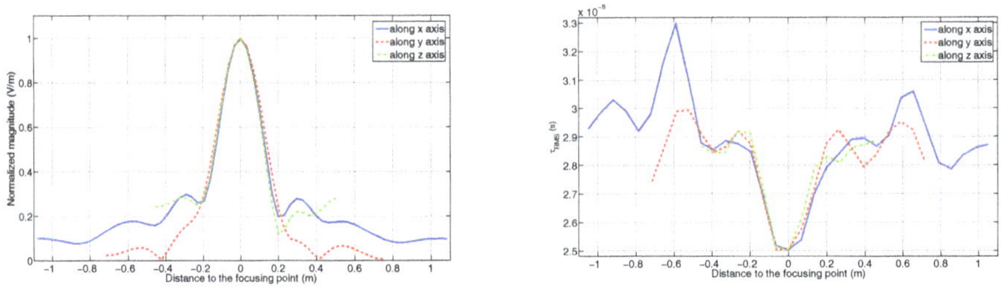

(a) (b)

Fig. 23. (a) Total electric field ($E_{TR}(r, t = 0)$) and (b) delay spread (τ_{RMS}) with respect to the distance to the focusing point.

On the other hand in Fig. 23b, the delay spread criterion is implemented. Indeed, the signals are sent back on the principle of first wave arrived last broadcast, so that all waves arrive simultaneously at the focusing point. The criterion τ_{RMS} measures the arrival time between first and last wave. Therefore, the smaller this parameter is over a given point in space, better the focusing is (in terms of agreement between time and space). For this we can see in Fig. 23b the smallest value of τ_{RMS} (along x, y, and z axis around R_0) corresponds to the point source. This means that TR has reduced the echoes and the excitation pulse was reconstructed even if we are in a reverberant environment.

The last studied parameter in this section is the propagation matrix K (section 3.3). This matrix can be constructed numerically in a simple way. To do this, an array of 24 point sources ($i = 1$ to 24) separated by 23 cm from each other is placed on one side of the domain and the same number of probes ($j = 1$ to 24) is used on the other side. We measure the 576 inter-element impulse responses ($k_{ij}(t)$) in both free space and RC. After a Fourier transform of each $k_{ij}(t)$, the propagation matrix K is known to the whole spectrum of the excitation pulse. For each frequency a singular value decomposition is applied. The singular values of K in both cases are shown in Fig. 24. Note that in the case of RC the number of singular values is much more representative than the free space case so the matrix K has a higher rank. For the central frequency $f_c = 600\ MHz$, we see that we have 20 singular values for the RC case and only 5 in free space (with a $-32\ dB$ threshold relatively to the first singular value). Physically the number of significant singular values is approximately the number of independent probes whose recorded impulse responses are not correlated, which is a crucial point in the application of TR in a reverberant environment, as we will see later in this chapter.

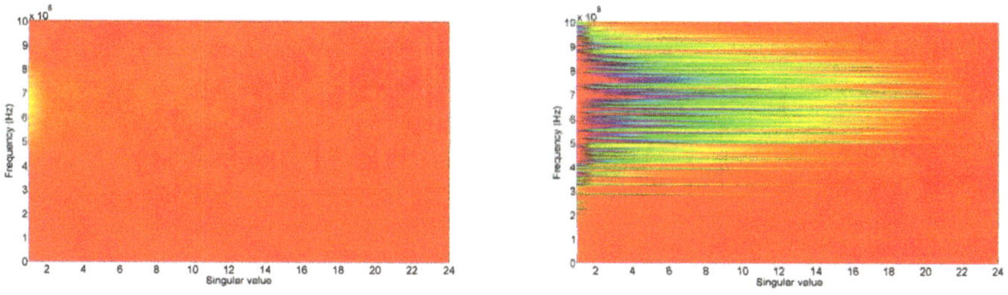

Fig. 24. Singular value decomposition of the propagation matrix K in (a) free space and (b) RC.

5.3.2 Influence of simulation duration

To study the influence of the reversed signal duration, in other words, the influence of the simulation duration on the STN ratio, different numerical tests were processed by varying the duration Δt and conserving one probe as TRM. To see more representative average data, each simulation of the second phase of TR is repeated nine times and at each time the receiving probe is placed in a different position of the chamber.

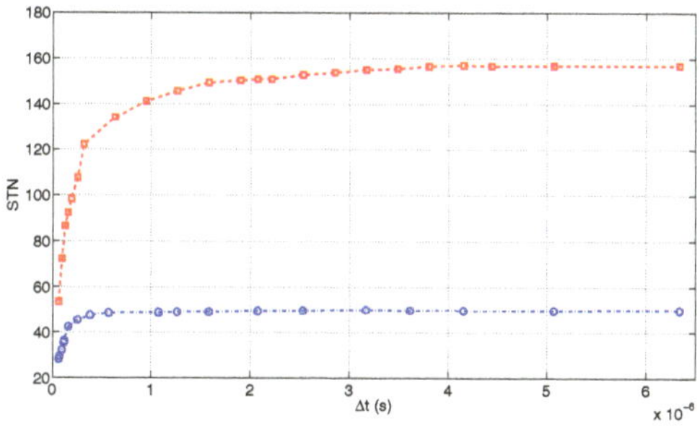

Fig. 25. STN ratio with respect to simulation duration (case 1: $f_c = 400\ MHz/\Delta\Omega = 260\ MHz$: blue circles markers, case 2: $f_c = 800\ MHz/\Delta\Omega = 260\ MHz$: red squares markers).

On Fig. 25, we plotted for each studied case the STN ratio, numerically calculated from (14), averaged over the nine positions of the TRM probe, with respect to the simulation duration Δt. We note that the STN ratio increases with the central frequency of the excitation signal, also this ratio becomes stable after a given time called Heisenberg time (ΔH). From (13), where $n(\omega)$ is deduced numerically by counting the resonant modes in the bandwidth (see later in this section), we obtain the Heisenberg time value. These results are verified numerically in Fig. 25 (e.g. in case 1, the Heisenberg time value $\Delta H = 0.3\ \mu s$ given by (13) is verified numerically).

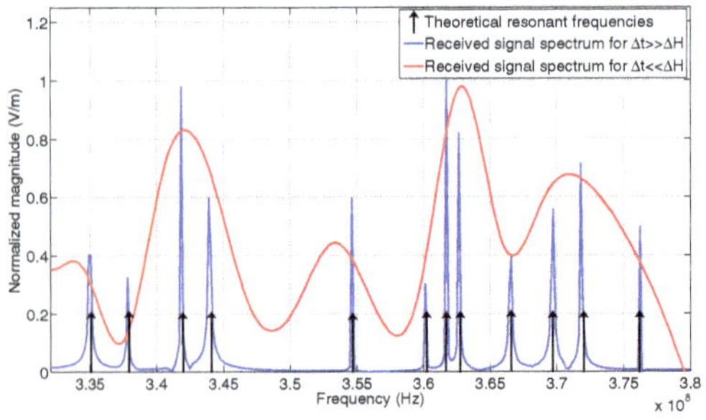

Fig. 26. Impulse response spectrum.

This behavior of STN saturation has been experimentally explained in terms of "information grains" and used in (de Rosny, 2000). In this model the impulse response of the system with $1/\tau$ as frequency width (τ: time duration of the modulated Gaussian), can be linked to a succession of uncorrelated information grains whose frequency width is around $1/\Delta t$. From Fig. 26, we see that the STN saturation seems to be a consequence of the existence of a finite number of resonant frequencies in the impulse response spectrum. Thus, for short time simulation durations, one information grain covers several frequencies (eigenmodes of the chamber). In this case, the number of information grains is equal to $\Delta t/\tau$, and the STN ratio increases with time. However, for longer durations, the number of information grains that can not be set only on the resonance frequencies stabilizes (number of information grains is equal to $1/(\tau\delta f)$ with δf: average distance between two successive eigenmodes); all the frequencies of the chamber are being resolved. The STN ratio becomes independent of the simulation duration (Fig. 25). This was predicted by the theoretical formula of STN ratio (12).

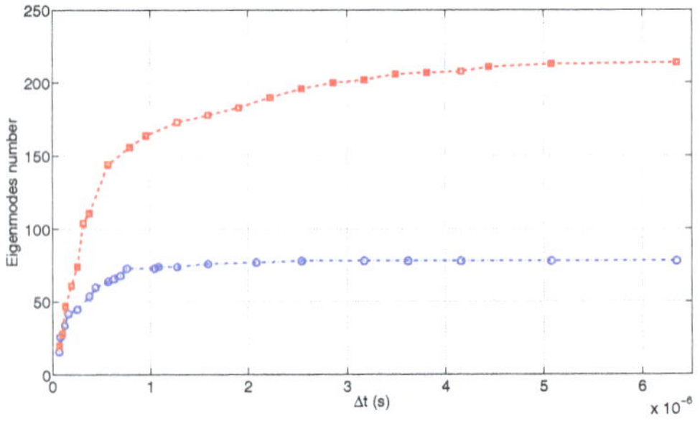

Fig. 27. Eigenmodes number with respect to simulation duration (case 1: $f_c = 400\ MHz/\Delta\Omega = 260\ MHz$: blue circles markers, case 2: $f_c = 800\ MHz/\Delta\Omega = 260\ MHz$: red squares markers).

This ratio depends on the product $\Delta H \Delta\Omega$ which is simply the number of eigenmodes. Thus Fig. 27 illustrates this saturation phenomenon: the evolution of the eigenmodes number of the RC in the bandwidth $\Delta\Omega$ is plotted as a function of simulation duration. The direct estimation of the number of resonance modes from Weyl theoretical formula (Liu et al., 1983) does not take into account the numerical characteristics of temporal simulation. As such, the eigenmodes number is defined from the mean total electric field spectrum recorded by the nine positions of the receiving probe: numerical calculation is achieved by counting the resonance peaks in the spectrum (Fig. 27). Note that the eigenmodes number stabilizes for duration greater than the corresponding Heisenberg time, which verifies the saturation of the STN ratio (Fig. 25).

5.3.3 Contribution of random receivers locations and number

Arnaud Derode has proved in (Derode et al., 1999) that the STN ratio increases with the root of the number of used probes, and this is caused by the fact that each supplementary probe brings an additional information uncorrelated with known information. In the RC case, uncorrelated information are the eigenmodes of the chamber. To study the influence of probe number in the TRM on the STN ratio, Various FDTD simulations are achieved for the case where $f_c = 400\ MHz$ and $\Delta\Omega = 260\ MHz$ using different numbers of probes (from 1 to 20) located randomly inside the CD (except on source and PEC boundaries). In order to compute mean values of outputs, each previous experiment is repeated 50 times for each number of probes (i.e. 50 random draws following a uniform law for 1 probe, then 50 for 2 probes, ...). We chose a TR window $\Delta t = 65\ ns$ much smaller than the corresponding Heisenberg time ($\Delta H = 300\ ns$).

The Fig. 28a shows that the maximum magnitude of focusing increases linearly with the number of probes. Contrary to maximum output, the use of STN criterion (more representative for a RC studies according to multiple reflections on PEC and thus to multiple sources) seems more relevant to characterize the quality of focusing. From the Fig. 28b (results obtained from the temporal STN ratio) and assuming a given number np of receiving probes ($np \cong 8$ or 9), focusing may appear independent from the probes number and the STN ratio stabilizes and does not follow the root law observed in (Derode et al., 1999). Indeed, after an increase, the STN ratio shows a level of saturation and the mean trend seems to reach a limit as a function of the probes number (from 10 to 20 probes in this case). This is due to the fact that the new probes do not provide any additional information since the eigenmodes of the RC are already resolved. To a weaker extent, the positions of the receiving probes need particular care. This result may provide great interests for MSRC studies since it is far more convenient to repeat some measurements using less probes for a given time than multiplying the number of field sensors for a shorter duration. Obviously, taking advantage of multiple scattering in RC, the use of a single TR probe needs a sufficient time of experiment to provide enough information (in comparison with a multi-probes setup).

5.3.4 Spatial resolution

In EMC tests, sometimes we need to obtain a field distribution following a focal spot as small as possible. In order to study the influence of the excitation pulse parameters in the focal spot, we computed the normalized total electric field as a function of the distance to R_0 along x axis by varying the central frequency and the bandwidth of the modulated Gaussian. We note (Fig. 29) that the size of the focal spot (as defined in section 4.1.2) decreases by increasing the central frequency and the bandwidth of the excitation pulse. Our interest is, for TR experiences in RC, to increase f_c and $\Delta\Omega$ to excite more resonant modes in the RC and influence the STN ratio quality.

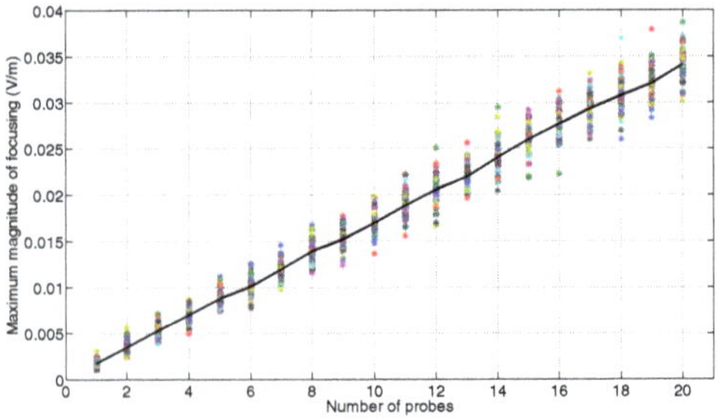

(a)

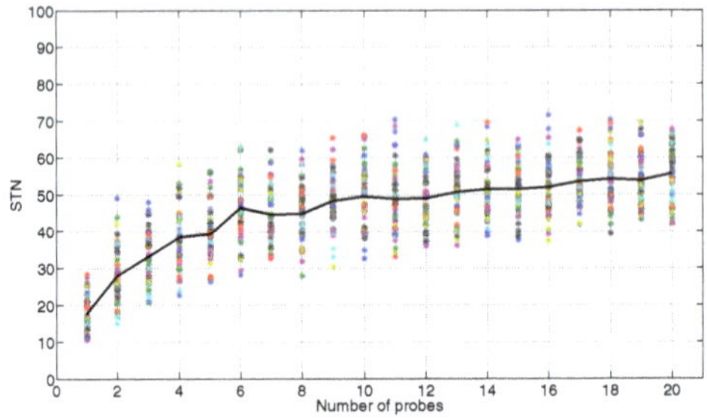

(b)

Fig. 28. (a) Maximum magnitude of focusing. (b) STN ratio, importance of focusing from 1 to 20 probes: regarding 50 random draws of probes locations each time (stars markers) and mean trend (plain line).

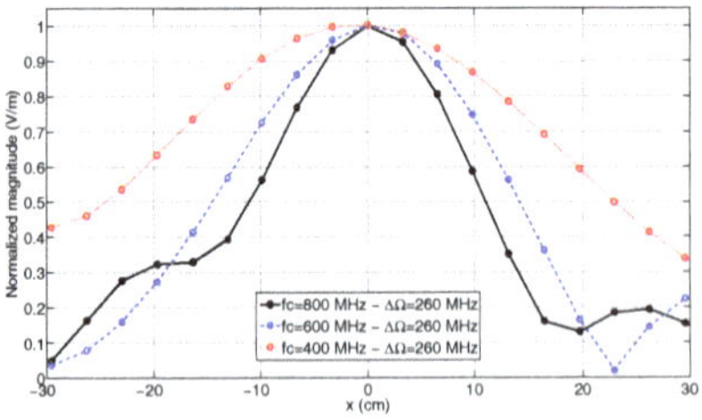

(a)

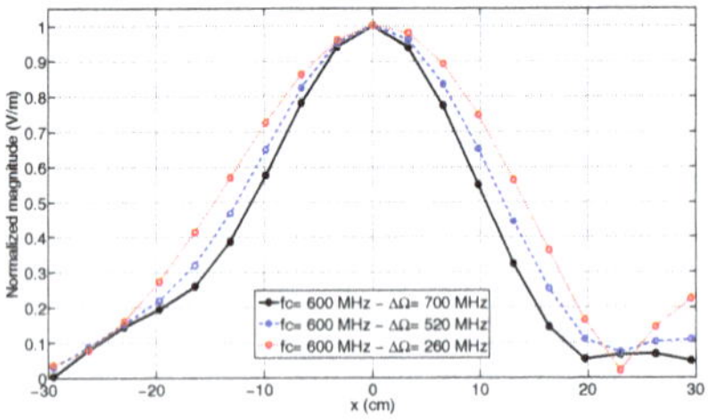

(b)

Fig. 29. Normalized total electric field as a function of the distance to R_0 along x: variations around (a) f_c and (b) $\Delta\Omega$.

In the final part of this chapter, the TR is numerically applied in the LASMEA (LAboratoire des Sciences et Matériaux pour l'Électronique et d'Automatique, Clermont Université) MSRC (for impulsive susceptibility test and selective focusing).

6. EMC application and selective focusing

Since 2001, a MSRC has been available for the EMC research & applications of LASMEA. Its dimensions and an internal view are given on Fig. 30. Historically, studies and tests in e MSRC were made in the frequency domain to provide an internal volume (WV) where the characteristics of the electromagnetic field illuminating the EUT are given with the same probability. A statistically uniform and homogeneous distribution of the electromagnetic field in MSRC means that the same part of energy attacks the EUT from all directions and with all polarizations (when averaged over the number of stirrer positions, i.e. over a full rotation of it). Nevertheless, this means that the illumination is statistically the same for the whole EUT, which can be a disadvantage, especially if the reliability is not the same for all components of the EUT.

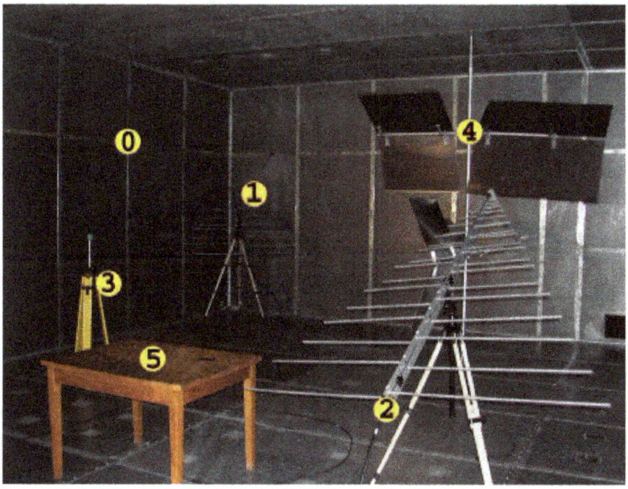

Fig. 30. Inner view of LASMEA MSRC ($6.7 \times 8.4 \times 3.5\ m^3$). Characteristics: (0) walls, (1) emitting and (2) receiving antennas, (3) field's probe, (4) mechanical stirrer, (5) working volume.

The application of the TR technique allows us to bring time techniques for EMC studies in MSRC. One of these intended applications is the impulsive susceptibility test and selective focusing that will be considered as a solution of the problem described above. To do this, we will use the commercial software CST MICROWAVE STUDIO® to numerically treat this case. In the next section, we will point to this software and the numerical set up used.

6.1 Numerical configuration

CST MICROWAVE STUDIO® is a specialist tool for the 3-D electromagnetic simulation. Inspired by the characteristics of LASMEA MSRC, the Fig. 31 shows the configuration of our example. Simulations are performed with a spatial discretization of 0.65 cm and 4 cm respectively corresponding to the smallest and largest mesh, and a time step of 24.4 ps.

The walls of the MSRC are modeled with a conductivity of $S_c = 1.1\ 10^6\ S/m$, furthermore the stirrer has a conductivity of $2.74\ 10^7\ S/m$. The support table and the EUT are made respectively of wood and aluminum ($S_c = 3.56\ 10^7\ S/m$). In this application, we wish to focus separately on the three components of the EUT modeled by dipoles (1, 2, and 3 in Fig. 31). The excitation signal is a Gaussian modulated sine pattern with a central frequency of 250 MHz and a bandwidth of 300 MHz calculated at $-20\ dB$. The TRM is composed of two 60 cm half-wavelength antennas.

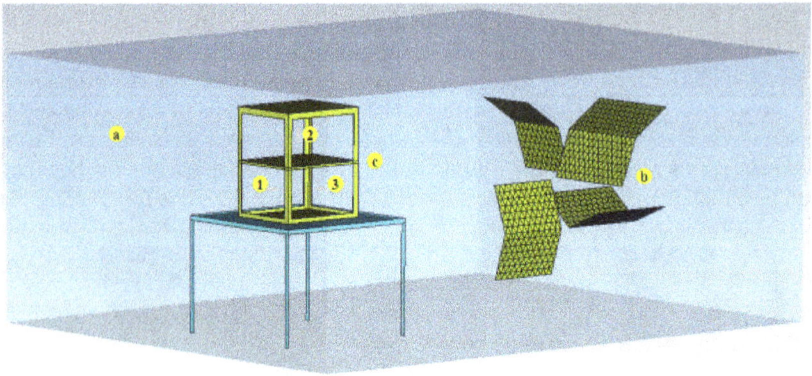

Fig. 31. LASMEA MSRC modeled by CST MICROWAVE STUDIO® : (a) walls, (b) mechanical stirrer, (c) EUT, (1)(2)(3) EUT three components.

6.2 Preliminary study

To justify the choice of the TRM antenna number, the duration of the TR window (Δt), and the ability to choose the polarization of the focused signal, a preliminary study is carried out maintaining the configuration of the Fig. 31 without the table and the EUT; but, here, an isotropic probe will be used to check focusing properties. Based upon TR principles, and because the simulator does not allow exciting probes and there are not designated to broadcast, necessary signals for the second phase of TR may be obtained directly by injecting excitation signal straightforward on the TRM antennas one by one. The three Cartesians components of these impulse responses are recorded by an isotropic field probe, and then back propagated from the TRM. Indeed, during the first phase, the isotropic probe records the electric field E_α (with $\alpha = x, y$ or z: Cartesian component of the field). We can choose the polarization α of the focused signal ($E_{TR\alpha}$) from back-propagating by the TRM recorded signal corresponding either to x, y or z without changing the polarization of the TRM antennas (Fig. 32).

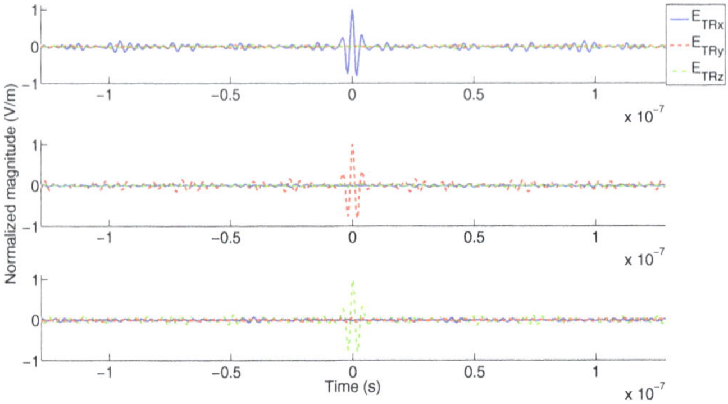

Fig. 32. Polarization control of the focused signal.

To determine the TR window duration (Δt), we need to calculate the Heisenberg time. Thus, we plotted on Fig. 33a the evolution of the STN ratio as a function of Δt for a TRM composed of a single antenna. Already mentioned above, we note that the STN ratio stabilizes after a certain duration, we can deduce that the Heisenberg time is about 8 μs. However an 8 μs simulation in comparison with the large dimensions of the LASMEA MSRC is disadvantageous in terms of computing time, so we chose to reduce simulation time by increasing the TRM antenna number. Fig. 33b shows that for $\Delta t = 0.75 \mu s$ the STN ratio stabilizes for a number of antenna greater than 8.

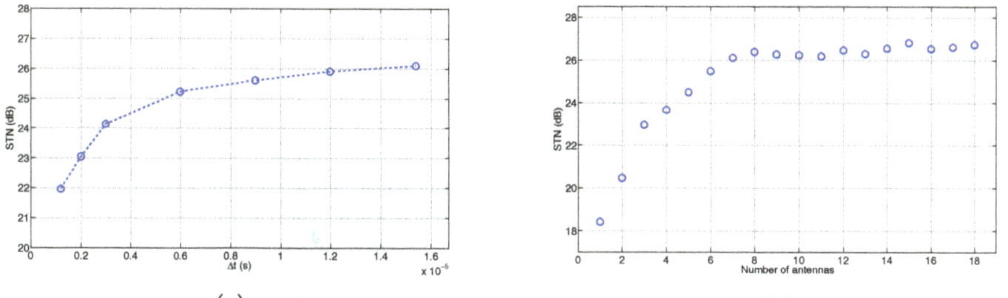

Fig. 33. STN ratio evolution: (a) as a function of the TR window for a TRM composed of 1 antenna, (b) as a function of the TRM antenna number with a TR windows of $\Delta t = 0.75$ μs.

The number of antennas needed for a TR experience is given by the ratio $\Delta H / \Delta t$, hence we chose the duration $\Delta t = 4$ μs with 2 antennas as TRM.

6.3 Selective focusing

In this section, we will check the possibility to focus the electric field on one of the three components of the EUT, while others are aggressed by lower levels (noise). To do this, we consider the example where the values 15 V/m, 70 V/m and 40 V/m correspond respectively to the three components threshold that should not be exceeded by the electric field. After recording the impulse responses $k_{ij}(t)$ with $1 \leq i \leq 2$: number of TRM antennas and $1 \leq j \leq 3$: number of components of the EUT, and given the linearity of the system, we can focus on any component with any desired focusing magnitude by a simple post-processing. Indeed, if for example we want to focus on the component number 2 (Figs. 34c, 34d), we will back-propagate through the first antenna of the TRM the signal $pk_{12}(-t)$ and the signal $pk_{22}(-t)$ by the second antenna, where p is the weight corresponding to the needed amplification. The p coefficient stands for the focusing magnitude control offered by TR (the focusing peak may be increased or decreased throughout the number of TRM antennas, the TR window duration, or an external amplification weight). We plotted in Fig. 34 temporal and spatial focusing corresponding to the "on demand" desired peak magnitude separately on each of the three components. The spatial focusing of the field corresponds to the absolute maximum value recorded over the entire simulation for each cell of the slice plan).

We note that for the different cases, the maximum of the field corresponds to the desired spatial location of each component. In addition, we note that for the second case, for example, we have focused on the component 2 (Figs. 34c, 34d) while respecting the threshold of the first component (component 1 was aggressed by a field whose numerical value is smaller than 15 V/m), same for the third case. To achieve this desired focusing magnitude on component 2

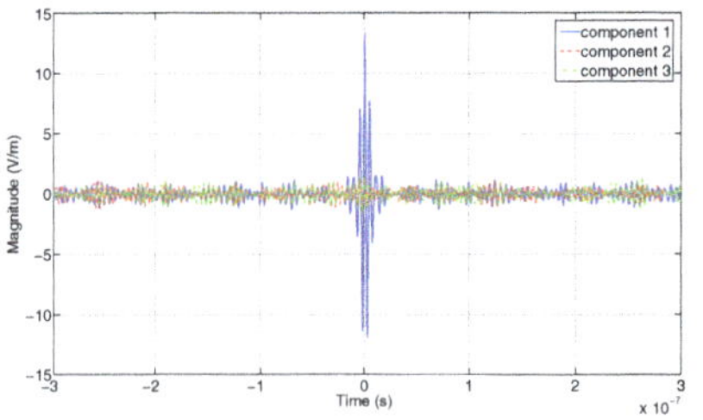

(a)

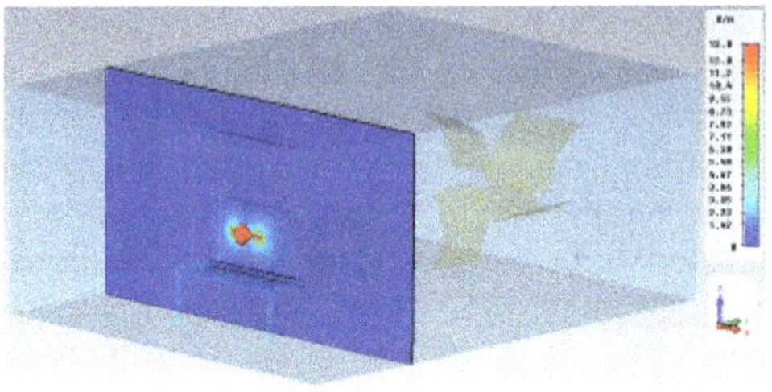

(b)

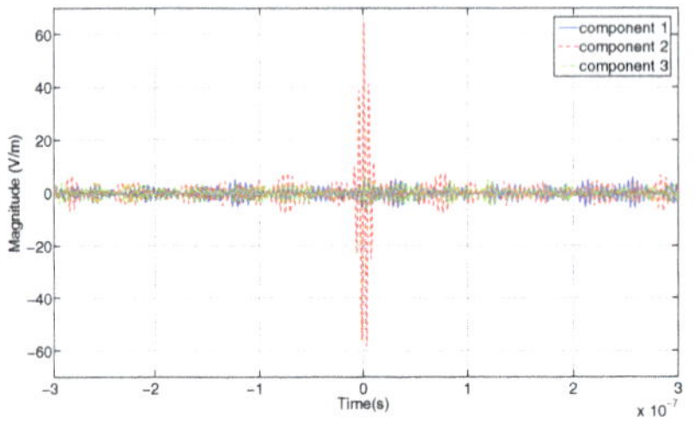

(c)

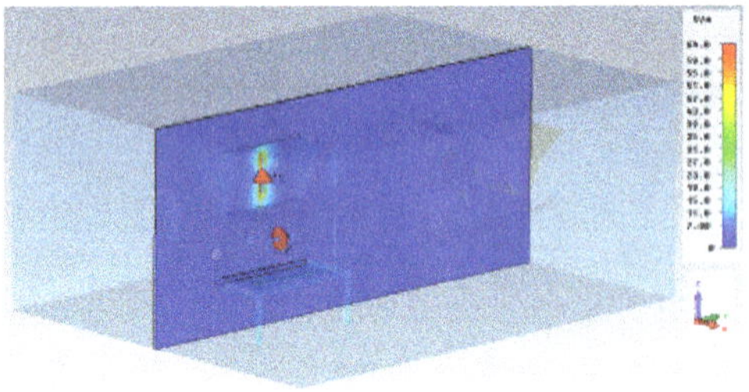

(d)

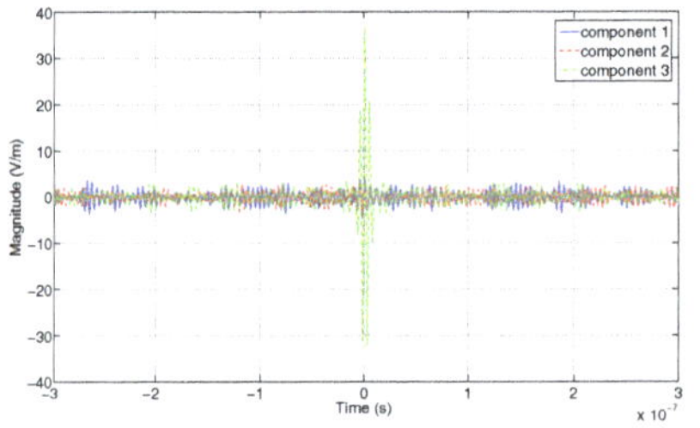

(e)

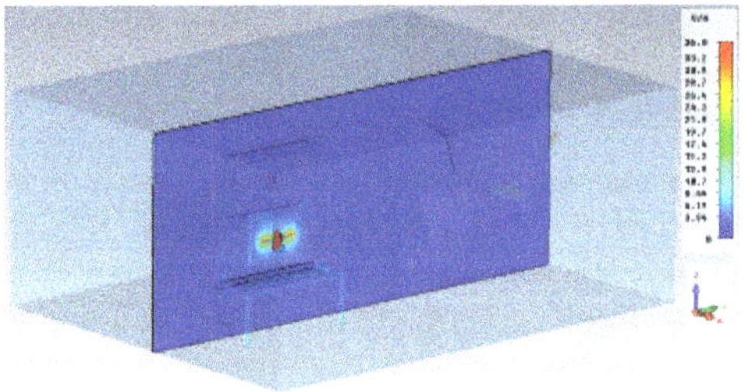

(f)

Fig. 34. Temporal focusing of the electric field on: (a) component 1 with $p = 1$, (c) component 2 with $p = 5$, (e) component 3 with $p = 3$. Spatial focusing corresponding to the absolute maximum value of the electric field: (b) component 1 with $p = 1$, (d) component 2 with $p = 5$, (f) component 3 with $p = 3$.

($64\ V/m$) smaller than the corresponding threshold ($70\ V/m$), the impulse responses $k_{12}(-t)$ and $k_{22}(-t)$ were multiplied by the weight $p = 5$; so we notice that following this way we can control the time, location, and magnitude of focusing (by the weight p).
Finally, if we wish, for example, to focus on the first and third components with respective magnitude of $13\ V/m$ and $35\ V/m$, we sum and back-propagate the needed impulse responses (on the first TRM antenna we back-propagate the signal $p_1k_{11}(-t) + p_3k_{13}(-t)$ with $p_1 = 1$ and $p_3 = 3$, on the second TRM antenna we back-propagate the signal $p_1k_{21}(-t) + p_3k_{23}(-t)$). The Fig. 35 justifies this approach and shows the ability of selective focusing by TR.

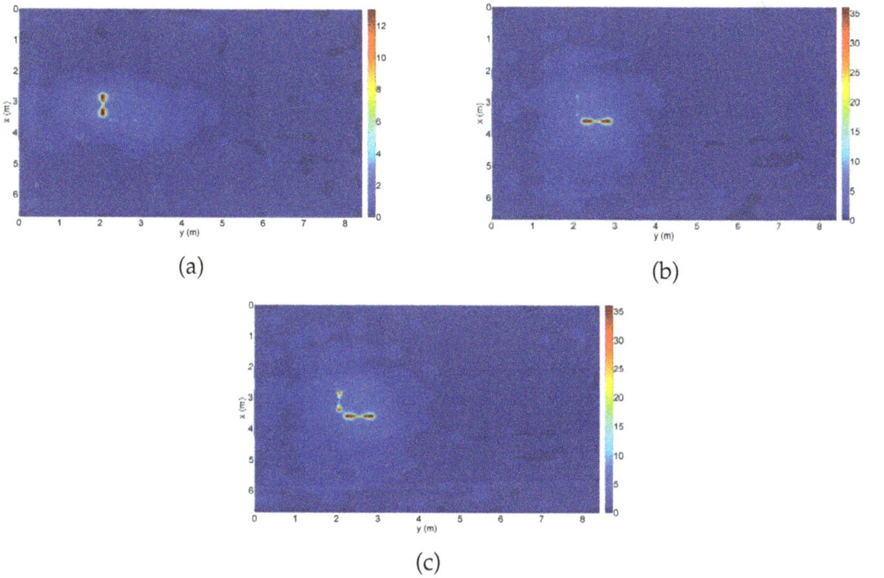

Fig. 35. Cutting plan ($z = 1.4\ m$) of the absolute maximum value of the electric field obtained on the: (a) component 1 with $p = 1$, (b) component 3 with $p = 3$, (c) two components togother, 1 with $p = 1$ and 3 with $p = 3$.

7. Conclusions

In this chapter, the TR method was presented in electromagnetism for applications concerning the EMC domain in a reverberating environment. Based upon the equivalence between backward propagation and reversibility of the wave equation, many TR experiments were led successfully in acoustics. In this chapter, after an introduction explaining the physical context, the theoretical principles of TR were described and illustrated numerically using the FDTD method. The use of the CST MICROWAVE STUDIO® commercial software laid emphasis on the industrial interest of TR for EMC test devices. First, the TR technique was applied in free space using a TRC and a TRM, and then the importance of the complexity of the medium was demonstrated. Relying on intrinsic RC behavior and due to multiple reflections, the results obtained by applying TR in a reverberating cavity were clearly improved; the aim was to accurately describe the influence of various parameters above focusing. Thus, a link between the modal density in a cavity and the TR focusing quality was clearly established through the STN ratio. A particular interest relies on the number and locations of TR probes and the excitation pulse parameters impact. Finally we introduced an original way to perform an impulsive susceptibility test study based on the MSRC use. We presented the possibility to choose the polarization of the wave aggressing the EUT, and to perform an "on demand" selective focusing. In further works, it would be interesting to experimentally confirm our numerical results, so one may expect to proceed to experimental analysis in LASMEA MSRC. At last, considering the characteristics of EMC applications in MSRC, a closer look might be set to the advantages of TR numerical tools for innovating studies in reverberation chambers.

8. References

Bonnet, P.; Vernet, R.; Girard, S. & Paladian, F. (2005). FDTD modelling of reverberation chamber, *Electronics Letters*, Vol. 41, No. 20, 2005, 1101-1102, ISSN 0013-5194.

Corona, P.; Ladbury, J. & Latmiral, G. (2002). Reverberation-chamber research - then and now: A review of early work and comparison with current understanding, *IEEE Transactions on Electromagnetic Compatibility*, Vol. 44, No. 1, 2002, 87-94, ISSN 0018-9375.

Cozza, A. & Moussa, H. (2009). Enforcing deterministic polarization in a reverberating environment, *Electronics Letters*, Vol. 45, No. 25, 2009, ISSN 0013-5194.

Davy, M.; de Rosny, J. & Fink, M. (2009). Focalisation et amplification d'ondes électromagnétiques par retournement temporel dans une chambre réverbérante, *Journées scientifiques d'URSI, Propagation et Télédétection*, pp. 13-20, France, 2009.

de Rosny, J. (2000). *Milieux réverbérants et réversibilité*, PhD thesis, Paris VI - Pierre et Marie CURIE University, 2000.

de Rosny, J. & Fink, M. (2002). Overcoming the diffraction limit in wave physics using a time-reversal mirror and a novel acoustic sink, *Physical Review Letters*, Vol. 89, No. 12, 2002, ISSN 1079-7114.

de Rosny, J.; Lerosey, G.; Tourin, A. & Fink, M. (2007). Time reversal of electromagnetic waves, *Modeling and computations in electromagnetics, Springer Berlin Heidelberg*, Vol. 59, 2007, 187-202, ISBN 3540737774.

Derode, A.; Tourin, A. & Fink, M. (1999). Ultrasonic pulse compression with one-bit time reversal through multiple scattering, *Journal of Applied Physics*, Vol. 85, No. 9, 1999, 6343-6352, ISSN 0021-8979.

Derode, A.; Tourin, A.; de Rosny, J.; Tanter, M.; Yon, S. & Fink, M. (2003). Taking advantage of multiple scattering to communicate with time-reversal antennas, *Physical Review Letters*, Vol. 90, No. 1, 2003, ISSN 1079-7114.

Directive 89/336/CEE (1989). Guide d'application de la Directive 89/336/CEE du Conseil du 3 mai 1989 concernant le rapprochement des législations des États membres relatives à la compatibilité électromagnétique. URL: $http\ :\ //cmrt.centrale-marseille.fr/electromagnetisme/veille/guide_89_336.pdf$

Edelmann, G. F. (2005). An overview of time-reversal acoustic communications, *Proceedings of TICA 2005*, 2005.

El Baba, I.; Lalléchère, S. & Bonnet, P. (2009). Electromagnetic time-reversal for reverberation chamber applications using FDTD, *Proceedings of ACTEA 2009, International Conference on Advances in Computational Tools for Engineering Applications*, pp. 157-167, ISBN: 978-1-4244-3833-4, Lebanon, 2009.

El Baba, I.; Patier, L., Lalléchère, S. & Bonnet, P. (2010). Numerical contribution for time reversal process in reverberation chamber, *Proceedings of APS-URSI 2010, IEEE Antennas and Propagation Society International Symposium*, ISBN: 978-1-4244-4967-5, Canada, 2010.

Emerson, W. (1973). Electromagnetic wave absorbers and anechoic chambers through the years, *IEEE Transactions on Antennas and Propagation*, Vol. 21, No. 4, 1973, 484-490, ISSN 0018-926X.

Fink, M. (1992). Time reversal of ultrasonic fields - Part I: Basic principles, *IEEE Transactions on Ultrasonics, Ferroelectrics, and Frequency Control*, Vol. 39, No. 5, 1992, 555-566, ISSN 0885-3010.

Hill, D.A. (1998). Plane wave integral representation for fields in reverberation chambers, *IEEE Transactions on Electromagnetic Compatibility*, Vol. 40, No. 3, 1998, 209-217, ISSN 0018-9375.

Jackson, J.D. (1998). *Classical Electrodynamics Third Edition*, John Wiley and Sons Inc, 1998.

Lerosey, G.; de Rosny, J.; Tourin, A.; Derode, A.; Montaldo, G. & Fink, M. (2004). Time reversal of electromagnetic waves, *Physical Review Letters*, Vol. 92, No. 19, 2004, ISSN 1079-7114.

Liu, B.H & Chang, D.C (1983). Eigenmodes and the composite quality factor of a reverberating chamber, *National Bureau of Standards*, USA, Technical Note, 1983.

Liu, D.; Kang, G.; Li, L.; Chen, Y.; Vasudevan, S.; Joines, W.; Liu, Q.H.; Krolik, J. & Carin, L. (2005). Electromagnetic time-reversal imaging of a target in a cluttered environment, *IEEE Transactions on Antennas and Propagation*, Vol. 53, No. 9, 2005, 3058-3066, ISSN 0018-926X.

Maaref, N.; Millot, P. & Ferrières, X. (2008). Electromagnetic imaging method based on time reversal processing applied to through-the-wall target localization, *Progress In Electromagnetics Research M*, Vol. 1, 2008, 59-67, ISSN 1937-8726.

Moussa, H.; Cozza, A. & Cauterman, M. (2009a). A novel way of using reverberation chambers through time-reversal, *Proceedings of ESA 2009, ESA Workshop on Aerospace EMC*, Italy, 2009.

Moussa, H.; Cozza, A. & Cauterman, M. (2009b). Directive wavefronts inside a time reversal electromagnetic chamber, *Proceedings of EMC 2009, IEEE International Symposium on Electromagnetic Compatibility*, pp. 159-164, ISBN: 978-1-4244-4266-9, USA, 2009.

Mur, G. (1981). Absorbing boundary conditions for the finite-difference approximation of the time-domain electromagnetic-field equations, *IEEE Transactions on Electromagnetic Compatibility*, Vol. EMC-23, No. 4, 1981, 377-382, ISSN 0018-9375.

Neyrat, M.; Guiffaut, C. & Reineix, A. (2008). Reverse time migration algorithm for detection of buried objects in time domain, *Proceedings of APS 2008, IEEE Antennas and Propagation Society International symposium*, ISBN: 978-1-4244-2041-4, USA, 2008.

Neyrat, M. (2009). *Contribution à l'étude de G.P.R. (Ground Penetrating Radar) multicapteurs. Méthodes directes et inverses en temporel*, PhD thesis, Limoges University, 2009.

Prada, C. & Fink, M. (1994). Eigenmodes of the time reversal operator: A solution to selective focusing in multiple-target media, *Wave Motion, Elsevier, Kidlington*, Vol. 20, No. 2, 1994, 151-163, ISSN 0165-2125.

Quieffin, N. (2004). *Etude du rayonnement acoustique de structures solides : vers un système d'imagerie haute résolution*, PhD thesis, Paris VI - Pierre et Marie CURIE University, 2004.

Sorrentino, R.; Roselli, L. & Mezzanotte, P. (1993). Time reversal in finite difference time domain method, *IEEE Microwave and Guided Wave Letters*, Vol. 3, No. 11, 1993, 402-404, ISSN 1051-8207.

Yavuz, M. & Teixeira, F. (2006). Full time-domain DORT for ultrawideband electromagnetic fields in dispersive, random inhomogeneous media, *IEEE Transactions on Antennas and Propagations*, Vol. 54, No. 8, 2006, 2305-2315, ISSN 0018-926X.

Yee, K. S. (1966). Numerical solution of initial boundary value problems involving Maxwell's equations in isotropic media, *IEEE Transactions on Antennas and Propagation*, Vol. Ap-14, No. 3, 1966, 302-307, ISSN 0018-926X.

Ziadé, Y.; Wong, M. & Wiart, J. (2008). Reverberation chamber and indoor measurements for time reversal application, *Proceedings of APS 2008, IEEE Antennas and Propagation Society International Symposium*, ISBN: 978-1-4244-2041-4, USA, 2008.

Part 4

Technological and Engineering Applications

10

Coupled-Line Couplers Based on the Composite Right/Left-Handed (CRLH) Transmission Lines

Masoud Movahhedi and Rasool Keshavarz
Electrical Engineering Department, Shahid Bahonar University of Kerman, Kerman, Iran

1. Introduction

Recently, the idea of complex materials in which both the permittivity and the permeability possess negative real values at certain frequencies has received considerable attention. In 1967, Veselago theoretically investigated plane-wave propagation in a material whose permittivity and permeability were assumed to be simultaneously negative (Veselago, 1968). For materials with negative permittivity and permeability, several names and terminologies have been suggested, such as "left-handed" media, media with "negative refractive index" (NIR), "backward-wave" (BW) media and "double-negative"(DNG) material (Caloz & Itoh, 2005). In this book chapter, materials with negative permittivity and permeability, and hence negative index of refraction, will be referred indistinctly as left-handed metamaterials (LHMs) or metamaterials (MTMs) (Caloz & Itoh, 2005).

Metamaterials have found many applications in electromagnetic problems. For instance, numerous novel MTM-based microwave components have been proposed to control amplitudes, frequencies, and wave numbers of propagating and non-propagating electromagnetic modes (Caloz & Itoh, 2005). Advances in MTMs have also stimulated the development of new couplers with unique coupling mechanisms. Recently, coupled-line couplers (CLCs) using composite right/left-handed transmission lines (CRLH TLs), which are the special realization of transmission lines based on the metamaterial concept, with broad bandwidth and arbitrary loose/tight coupling levels have been developed. But usually these couplers occupy large length and also, because of using stubs in their structures, width of them would be large. For eliminating this drawback, we have proposed some new backward and forward coupled line couplers with high coupling levels, broad bandwidths and compact sizes, base on the CRLH TLs.

Organization of this chapter is as follows. In Section 2 theory of CRLH TLs, interdigital capacitor and their equivalent circuit models and parameters, have been explained. Section 3, at first, reviews some conventional CRLH- based CLCs and in continues presents our proposed couplers. In this section, three CLCs based on the concepts of CRLH CLCs are presented; a symmetrical backward CLC (Section 3.3.1), an asymmetrical backward CLC (Section 3.3.2) and a symmetrical forward CLC (Section 3.3.3).

2. Composite Right/Left Handed Transmission Lines (CRLH TLs)

A conventional transmission line (right-handed TL) is represented by a series inductance (L_R) and a shunt capacitance (C_R), implying the use of a low pass topology (Pozar, 2004). By interchanging the position of the inductor and capacitor, the resulting structure is referred to as left-handed TL with a high pass configuration (Caloz & Itoh, 2005). In these purely left-handed transmission lines (PLH TLs), the phase and group velocities are opposite to each other (Pozar, 2004). PLH TLs cannot exist physically because, even if we intentionally provide only series capacitance and shunt inductance, parasitic series inductance (L_L) and shunt capacitance (C_L) effects, increasing with increasing frequency, will unavoidably occur due to currents flowing in the metallization and voltage gradients developing between the metal patterns of the trace and the ground plane (Caloz & Itoh, 2002). Thus, the composite right and left handed (CRLH) model represents the most general MTM structure possible. Equivalent circuit model of a CRLH TL for one cell is shown in Fig. 1(a) (Caloz & Itoh, 2004a). In this figure, L_R and L_L are right and left handed inductances, respectively, also C_R and C_L are right and left handed capacitances, respectively. Many lumped (using SMT chip components) or distributed implementations (microstrip, stripline, CPW, etc.) are possible for CRLH TLs. Interdigital/stub configuration is one of widely used of these implementations (Caloz & Itoh, 2005). A layout of this configuration for one cell is shown in Fig. 1(b). A CRLH TL is constructed of these unit cells connected in series as shown in Fig. 1(c). This structure consists of series interdigital capacitor of capacitanceC_L and parallel short-ended stub working as inductor of inductance L_L. Moreover, L_R and C_R are parasitic elements of interdigital capacitor.

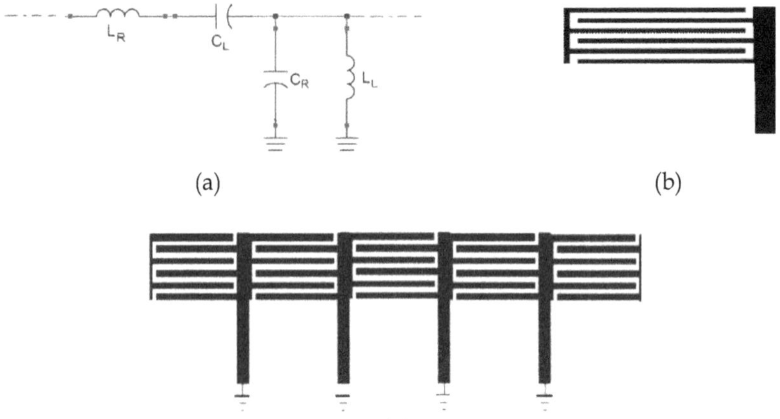

(a) (b)

(c)

Fig. 1. (a) Equivalent circuit model of a CRLH TL for one cell. (b) Layout of a CRLH TL by using interdigital capacitor and shorted stub inductor for one cell. (c) Microstrip implementation of a CRLH TL.

An interdigital capacitor is a multifinger periodic structure which, as mentioned, can be used as a series capacitor in microstrip transmission lines technology (Bahl, 2003). This capacitor uses the capacitance that occurs across a narrow gap between thin-film conductors. Fig. 2 shows an interdigital capacitor and its equivalent circuit model. As seen in this figure, an interdigital capacitor is made of some gaps. The gap meanders back and forth in a rectangular area forming two sets of fingers that are interdigital. These gaps are essentially very long and folded to use a small amount of area. By using a long gap in a small area, compact single-layer small-value series capacitors can be realized. Typically, its capacitance values range from 0.05 pF to about 0.5 pF. The capacitance can be increased by increasing the number of fingers, or by using a thin layer of high dielectric constant material such as a ferroelectric between the conductors and the substrate (Bahl, 2003).

The value of series capacitance of an interdigital structure can be expressed as (Bahl, 2003):

$$C_L = \frac{\varepsilon'_{re}}{18\pi}(N-1)\frac{K(\kappa)}{K'(\kappa)} l_s \qquad (pF) \tag{1}$$

where ε'_{re} is effective permittivity of a strip with width W, N is the number of fingers and $\frac{K(k)}{K'(k)}$ is a constant that has been presented in (Bahl, 2003).

As is well-known, the characteristic impedance of a CRLH TL (Z_c) with equivalent circuit model of Fig. 1(a) is given by (Caloz & Itoh, 2005):

$$Z_c = Z_L \sqrt{\frac{\left(\frac{\omega}{\omega_{se}}\right)^2 - 1}{\left(\frac{\omega}{\omega_{sh}}\right)^2 - 1}} \tag{2}$$

where

$$Z_L = \sqrt{\frac{L_L}{C_L}}, \quad \omega_{se} = \frac{1}{\sqrt{L_R C_L}}, \quad \omega_{sh} = \frac{1}{\sqrt{L_L C_R}} \tag{3}$$

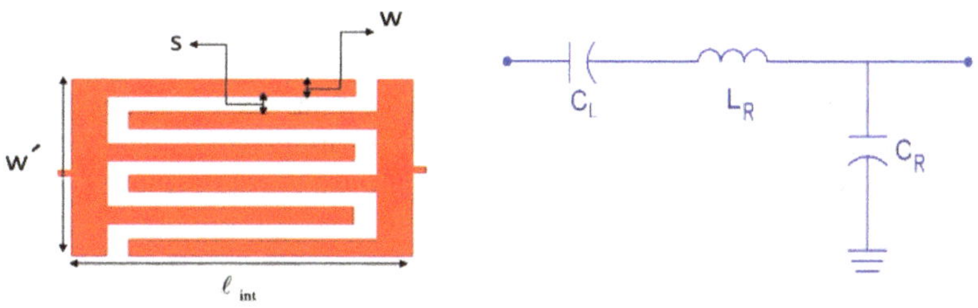

(a) (b)

Fig. 2. (a) Interdigital capacitor. (b) Its equivalent circuit model.

According to Fig. 2, the equivalent circuit model of an interdigital capacitor is similar to the equivalent circuit model of one cell of CRLH TL when $L_L \to \infty$. Inserting $L_L \to \infty$ into (2) results the characteristic impedance (Z_c^{int}) of a TL consists of cascaded interdigital capacitors as:

$$Z_c^{\text{int}} = \sqrt{\frac{\left(\frac{\omega}{\omega_{se}}\right)^2 - 1}{\omega^2 C_L C_R}} = \sqrt{\frac{L_R}{C_R} - \frac{1}{\omega^2 C_L C_R}} \tag{4}$$

It is seen from above equation that Z_c^{int} is real for $\omega > \omega_{se}$. From TL theory, it is clear that $\sqrt{L_R/C_R}$ is the characteristic impedance of a microstrip TL consists of a strip with width $W' = (4N-1)W$.

Similarly, the propagation constant for this TL is obtained as (Caloz & Itoh, 2005):

$$\beta^{\text{int}} = \sqrt{\omega^2 L_R C_R - \frac{C_R}{C_L}} \tag{5}$$

So, in a transmission line composed of interdigital capacitors which can be named "interdigital transmission line", for $\omega > \omega_{se}$, the propagation constant (β^{int}) is real and positive. It means that in this frequency interval, the interdigital transmission line operates in the right-handed (RH) band.

3. Coupled-Line Couplers (CLCs)

3.1 Conventional CLCs

Coupled line couplers are indispensable components in radio frequency (RF)/microwave communication systems. In these structures two unshielded transmission lines are close together, as indicated typically in Fig. 3, and power can be coupled between the lines. Such lines are referred to as coupled transmission lines (Mongia et al., 1999). The coupler is frequently utilized in a variety of circuits including modulators, balanced amplifiers, balanced mixers, and phase shifters. Rapidly expanding applications such as modern wireless technology continue to challenge couplers with extremely stringent requirements—high performance, broad bandwidth, and small size (Pozar, 2004).

In general, two types of CLCs have been proposed; backward and forward CLCs. When the coupled port is located on the same side of the structure as the input port and power is subsequently coupled backward to the direction of the source, this coupler is conventionally called a backward coupler and otherwise the CLC is called forward coupler (Mongia et al., 1999).

On the other hand, two types of edge-coupled backward CLCs have been presented. The first is the symmetrical coupler. When the two lines constituting a CLC are the same, the structure is called symmetric. In the symmetric structures, coupling mechanism is based on the difference between the characteristic impedances of the even and odd modes. The second one is the asymmetrical coupler. This coupler is asymmetrical as it is constituted of two different transmission lines. In this case, decomposition in even and odd modes is not possible anymore. The analysis becomes more difficult and the even/odd modes have to be replaced by the more general c and π modes, which are two fundamental independent modes, as described in (Mongia et al., 1999).

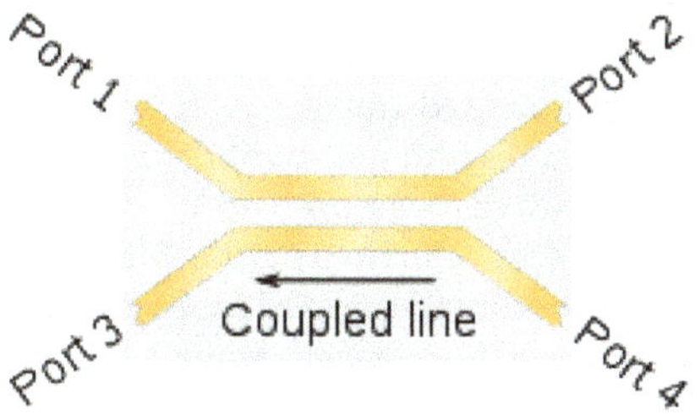

Fig. 3. Typical structure of a coupled-line coupler (CLC).

Symmetrical coupled lines represent a very useful but restricted class of couplers. In many practical cases, it might be more useful or even necessary to design components using asymmetrical coupled lines. For example, in some situations, the terminal impedance of one of the coupled lines may be different from those of the other. It may then be more useful to choose two coupled lines with different characteristic impedances. Also, an asymmetrical coupled-line coupler has usually broader bandwidth than symmetrical one (Mongia et al., 1999).

3.2 CRLH CLCs

The conventional CLC has several intrinsic drawbacks. First, their operating bandwidths are usually limited. Second, to raise the coupling level of a coupler, a very small space between the coupled lines is required and it is usually difficult to obtain due to fabrication constraints (Mongia et al., 1999).

As mentioned, in the past few years there has been a great interest in the field of metamaterials, especially composite right/left-handed structures (e.g. interdigital/stub configurations), and the microwave circuits based on the unusual properties of them (Caloz & Itoh, 2005). By closely placing two identical CRLH lines in parallel, such as configuration shown in Fig. 4, a strong contrast exists between the impedances of two fundamental modes of propagation (i.e. the even and odd mode impedances), which would result in high coupling-level.

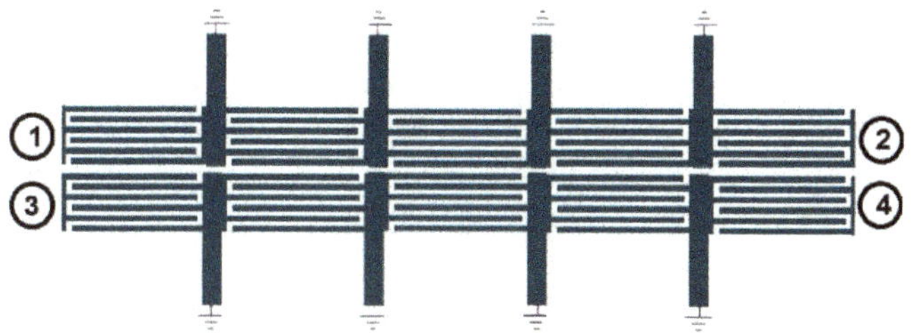

Fig. 4. Prototype of a CRLH edge-coupled directional coupler constituted of two interdigital/stub CRLH TLs.

For the first time, a novel composite right/left-handed coupled-line directional coupler composed of two CRLH TLs was proposed in (Caloz et al., 2004) and an even/odd-mode theory was used to analyze the phenomenon of complete backward coupling. Then, an asymmetric RH-CRLH coupler was introduced and studied in (Caloz & Itoh, 2004b). It was composed of a conventional right-handed transmission line and a CRLH TL. That coupler showed the advantage of broad bandwidth and tight coupling characteristics, and coupled-mode theory based on traveling waves was used to discuss these interesting features. In (Islam & Eleftheriades, 2006), it was shown that the formation of a stop-band and the excitation of complex modes occurred in the case of coupling between a forward wave and a backward-wave mode for a range of frequencies around the tuning frequency. Moreover, authors in (Wang et al., 2007) presented the conditions for tight coupling and detailed formulas were given to define the edges of the coupling range.

Moreover, some CLCs based on the CRLH TLs with arbitrary coupling levels have been developed, recently (Fouda et al., 2010; Hirota et al., 2009; Hirota et al., 2011; Kawakami et al., 2010; Mocanu et al., 2010). In these couplers, the backward coupling depends on the difference between even and odd modes characteristic impedances and length of the coupled lines (Caloz & Itoh, 2005).

The interdigital/stub CLCs have been typically adapted to increasing coupling level, but these couplers increase in size (Caloz et al., 2004; Caloz & Itoh, 2004b; Islam & Eleftheriades, 2006), band width of them is narrow (Hirota et al., 2011; Mocanu et al. 2010; Wang et al. 2007) and the multiconductors of the interdigital construction complicate the design procedure (Caloz & Itoh, 2005).

It is considerable that the microstrip CRLH TL structures have been mostly implemented in the form of interdigital capacitors and stub inductors. In the other hand, using shorted stub inductors with large sizes to achieve the required inductances can cause the structure width to be also enlarged. For instance, the length and width of 3-dB microstrip coupled-line coupler proposed in (Caloz et al., 2004) are approximately $\lambda g/3$ and $\lambda g/6$, respectively. Also, bandwidth of the CRLH CLCs which presented in (Mocanu et al., 2010) and (Fouda et al., 2010) are 25% and 30%, respectively.

Also, forward coupling level in CRLH coupled line couplers is low (nearly -10 dB in (Fouda et al., 2010)).

3.3 Proposed CLCs

In this section, some of the authors' proposed CLCs based on the CRLH concepts to reach new couplers with better specifications, such as smaller size, broader bandwidth and more simplicity in fabrication are presented. In these new CLCs one has been trying to eliminate some drawbacks and disadvantages of conventional CRLH CLCs mentioned in previous section.

3.3.1 Backward symmetrical CLC

The proposed backward-wave directional coupler is shown in Fig. 5 (Keshavarz et al., 2011a). It is a coupled-line coupler consisting of an interdigital capacitor with one finger as a CRLH TL in each coupled-line. It is seen that using only one interdigital capacitor to realize the interdigital TLs is more suitable to reach a coupler with better matching and wider bandwidth. As it was mentioned, for $\omega > \omega_{se}$, these interdigital TLs will be operating completely in their RH range for the presented coupler application. So in this coupler, similar to the conventional edge-coupled couplers, the coupling coefficient is (Pozar, 2004):

$$S_{31} = \frac{jk\sin\theta}{\sqrt{1-k^2}\cos\theta + j\sin\theta}, \quad k = \frac{Z_{ce} - Z_{co}}{Z_{ce} + Z_{co}} \tag{6}$$

where, $\theta=(2\pi l/\lambda g)$ is electrical length and ℓ is the length of CLC. Therefore, setting the interdigital capacitor length as $l=\lambda g/4$ or $\theta=\pi/2$ results in maximum coupling level. On the other hand, selection of $l=\lambda g/4$ preserves the homogeneity condition in CRLH structure (i.e., $p \le \frac{\lambda_g}{4}$, where p is structural cell size) (Caloz & Itoh, 2005).

The equivalent circuits model of the even and odd modes of Fig. 5 for one cell have been presented in Fig. 6. In this figure, L is the inductance for a strip with width W' and C_e, C_o are the distributed capacitances for the even and odd modes, respectively.

Even and odd mode characteristic impedances (Z_{ce}, Z_{co}) of the coupled-lines composed of interdigital TLs are obtained from (Caloz & Itoh, 2005) with setting $L_L \to \infty$ as:

$$Z_{ce} = \sqrt{\frac{L}{C_e} - \frac{1}{\omega^2 C_e C_L}}, \quad Z_{co} = \sqrt{\frac{L}{C_o} - \frac{1}{\omega^2 C_o C_L}} \tag{7}$$

and

$$Z'_{ce} = \sqrt{\frac{L}{C_e}}, \quad Z'_{co} = \sqrt{\frac{L}{C_o}} \tag{8}$$

Z'_{ce} and Z'_{co} are even and odd mode characteristic impedances of a conventional microstrip CLC with strips of width W' for each TL, where $W'(=(4N-1)W)$ is total width of the interdigital capacitor.

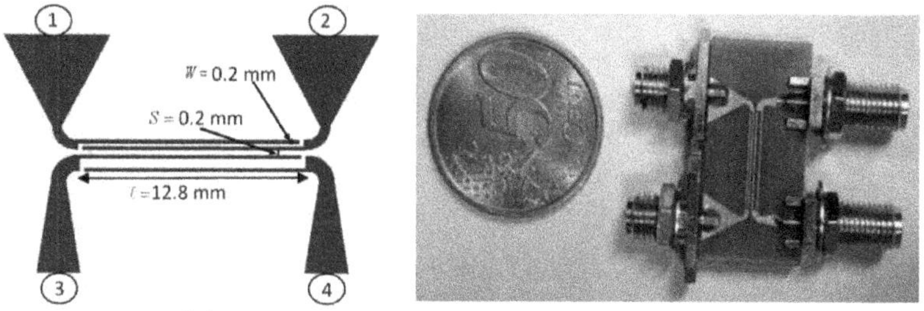

(a) (b)

Fig. 5. Structure of the proposed microstrip coupled-line backward coupler on FR4 substrate, $\varepsilon_r = 4.7$, thickness of 1.6 mm. (a) Structure layout. (b) Fabricated coupler (Keshavarz et al., 2011a).

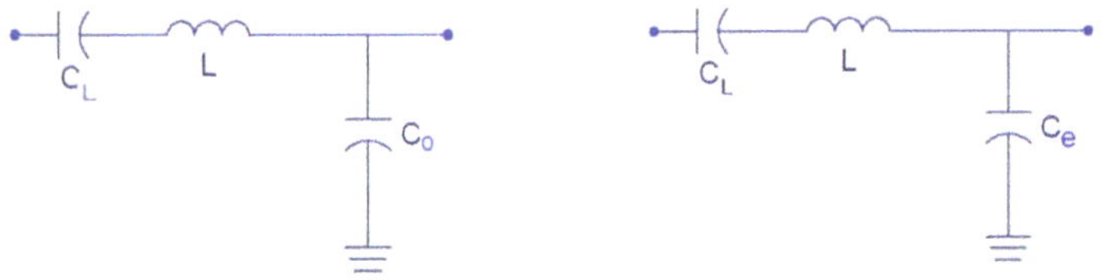

Fig. 6. (a) Odd and (b) even modes equivalent circuit models of proposed coupler in Fig. 5 (Keshavarz et al., 2011a).

In the proposed coupler for given even and odd mode characteristic impedances, according to (7), selection of a small C_L leads to larger values of Z_{ce} and Z_{co}. This situation is very suitable for elimination of the fabrication restrictions in CLCs with tight coupling level. Consequently, to decrease the value of C_L in the proposed structure, interdigital capacitors with only one finger (i.e., $N = 1$) are used.

In design procedure, for an indicated coupling-level (c) and characteristic impedance (Z_c^{int}), Z_{ce} and Z_{co} can be obtained from conventional expressions as (Pozar, 2004):

$$Z_{ce} = Z_c^{\text{int}} \sqrt{\frac{1+c}{1-c}}, \quad Z_{co} = Z_c^{\text{int}} \sqrt{\frac{1-c}{1+c}} \tag{9}$$

With setting $N = 1$, $\ell = \lambda_g/4$ and the substrate profile being determined, C_L and C_R can be calculated using expressions presented in (Bahl, 2003) and Z'_{ce} and Z'_{co} are obtained from (7) and (8). Then, W' and S can be determined by using achieved Z'_{ce}, Z'_{co} and relative design graphs for conventional coupled microstrip lines.

For instant, Fig. 7(a) illustrates the required width of the interdigital TL (W) in the proposed coupler realized on FR4 substrate, with $\varepsilon_r = 4.7$ and thickness of 1.6 mm, for different values of Z'_{ce}. In addition, the necessary spacing between two coupled interdigital TLs (S) for the presented structure versus Z_m, where $Z_m = \dfrac{2Z'_{ce}Z'_{co}}{Z'_{ce} - Z'_{co}}$, has been provided in Fig. 7(b).

As it was mentioned, since Z'_{ce} and Z'_{co} would be larger than Z_{ce} and Z_{ce}, for constant coupling-level (c) and characteristic impedance (Z_c^{int}) in comparison with the conventional CLCs, W' decreases and S increases. Therefore in the proposed coupler, the fabrication constrains in conventional edge-coupled couplers to get a tight coupling-level caused very small spacing between two coupled lines (i.e., S) can be removed.

To validate the proposed technique, a 3-dB coupled line coupler based on the design procedure and presented expressions has been designed on FR4 substrate with $\varepsilon_r = 4.7$, thickness of 1.6 mm and $\tan\delta = 0.021$. Fig. 5 shows the designed coupler layout and the fabricated structure. A 3-dB coupled line coupler with nearly 60% bandwidth (from 2.3 to 4 GHz) around the design frequency $f_c = 3.2$ GHz is achieved in the measured prototype. The spacing between two TLs (S) and width of the interdigital capacitor fingers (W) are 0.2

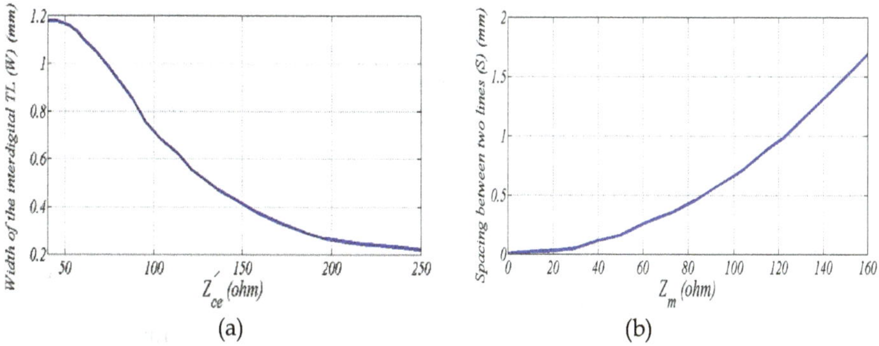

Fig. 7. a) Width of the interdigital TL (W) versus Z'_{ce}, b) Spacing between two coupled interdigital TLs (S), in the proposed coupler on FR4 substrate, $\varepsilon_r = 4.7$, thickness of 1.6 mm (Keshavarz et al., 2011a).

mm. Also, the length of the TLs is 12.8 mm (see Fig.5). For better matching and wider bandwidth, we use only one interdigital capacitor, i.e. one cell, in every interdigital TL. Moreover, to reach a large isolation parameter, spacing between the fingers in the lower interdigital capacitor is set larger than the upper one, when ports 1 and 4 are the input and isolated ports, respectively. As shown in the layout of the coupler in Fig. 5, at the all four ports of the structure, tapered microstrip TLs have been used for the impedance matching to 50 Ω, as well as to fit the ports size to the inner conductors of the coaxial-to-microstrip transitions.

Fig. 8 presents the full-wave simulated (by using Agilent ADS software) and measured S-parameters for the coupler of Fig. 5. Excellent agreement can be observed between simulated and experimental results. There is only a small difference between S_{11} parameter of simulated and measurement results. Due to small distance between coupler connectors, we could not connect network analyzer ports to adjacent coupler connectors, directly. Therefore, two interface cables were connected to the coupler connectors and then S-parameters were measured. This drawback shows its bad effect on S_{11} parameter more strongly than other S-parameters.

Using these figures, a amplitude balance of ±2 dB over a bandwidth of 60% (2.3–4 GHz), a matching (10 dB bandwidth) and an isolation at least −20 dB over a bandwidth of 80% (2.2–4.6 GHz) are observed. Fig. 9 illustrates the phase difference between ports 2 and 3 of the coupler. This phase difference is 90^0 at design frequency and exhibits a phase-balance ($\pm 10^0$) bandwidth of 1.3 GHz.

In comparison with the conventional CRLH CLCs, the electrical length of the proposed CLC is more compact than the CRLH CLCs presented in (Islam et al., 2004; Mao & Wu, 2007; Nguyen & Caloz, 2006; Zhang et al., 2008). Moreover, due to the elimination of the stubs in the structure, its width is also smaller. For instance, the width of the coupler is nearly 11 times smaller than CRLH CLC reported in (Caloz et al., 2004) and its coupled-line electrical length is shortened to 60% of the 3-dB CRLH coupler electrical length presented in

(Keshavarz et al., 2011a). Moreover, the bandwidth of the proposed CLC is wider than CRLH CLCs presented in (Islam et al., 2004) and (Nguyen & Caloz, 2006).

In comparison with the conventional planar microstrip CLC realized in the same substrate material and similar spacing between coupled TLs, this CLC achieves higher coupling level. The high coupling level (8 dB or higher) is extremely difficult to achieve in the conventional CLC due to the present limit in fabrication (Pozar, 2004). Also, simulation results show that in the proposed structure if the spacing between the coupled lines increases, the bandwidth increases up to 85% for 7-dB coupling factor. Moreover, this coupler exhibits much higher design simplicity than the existing CRLH CLCs. Due to the wide bandwidth and compact size, the proposed coupler is well suitable for microwave and millimeter-wave integrated circuits, wideband communication systems and many kinds of antenna arrays.

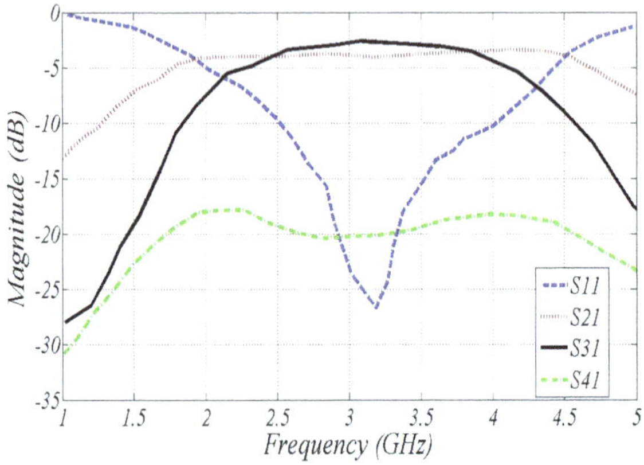

(a)

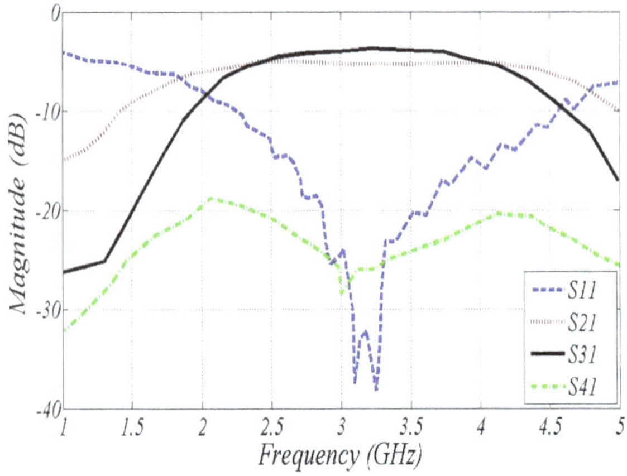

(b)

Fig. 8. *S*-parameters of the proposed coupler have shown in Fig.5 (a) Full-wave simulation results. (b) Measurement results (Keshavarz et al., 2011a).

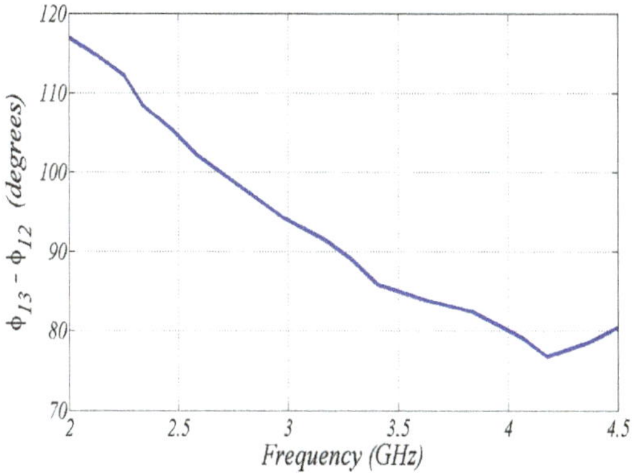

Fig. 9. Measured phase difference between the through port and the coupled port for the proposed coupler of Fig. 5 (Keshavarz et al., 2011a).

3.3.2 Backward asymmetrical CLC

In this section, an asymmetrical coupled-line coupler based on the interdigital TL is presented. Fig. 10 shows layout and circuit model of the interdigital TL and conventional TL which are adjacent to each other as asymmetrical backward coupled-line coupler. As depicted in Fig. 10(b), C_m represents the mutual capacitance between interdigital and strip of the microstrip conductors in the absence of the structure ground conductor while C_1 and C_2 represent the capacitance between interdigital or microstrip strip conductors and ground, respectively. Moreover, the circuit model includes mutual inductance (L_m) and self-inductances of interdigital (line 1) and conventional microstrip (line 2) conductors, i.e. L_1 and L_2, respectively. C_{int} is series interdigital capacitor of line 1. It should be stated that all of parameters in the circuit model are per unit length quantities. Also, Fig. 11 shows the capacitance representation for quasi-TEM mode of cross section of the proposed asymmetrical coupler. For structure analysis, it is assumed that lines 1 and 2 are terminated to impedances Z_a and Z_b, respectively.

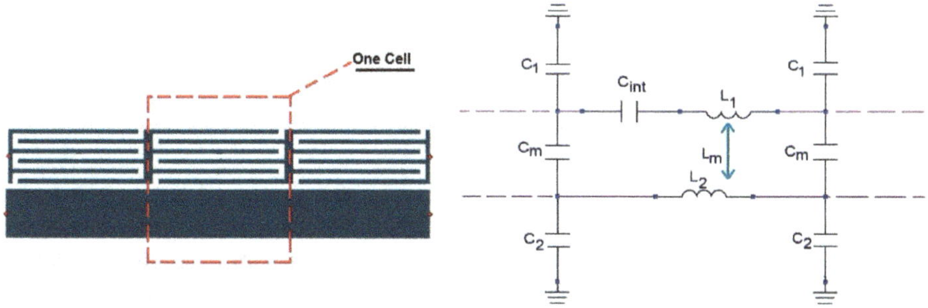

(a) (b)

Fig. 10. Proposed asymmetrical coupled-line coupler consisted of interdigital TL and microstrip conventional TL. a) Its layout and b) lumped equivalent circuit model.

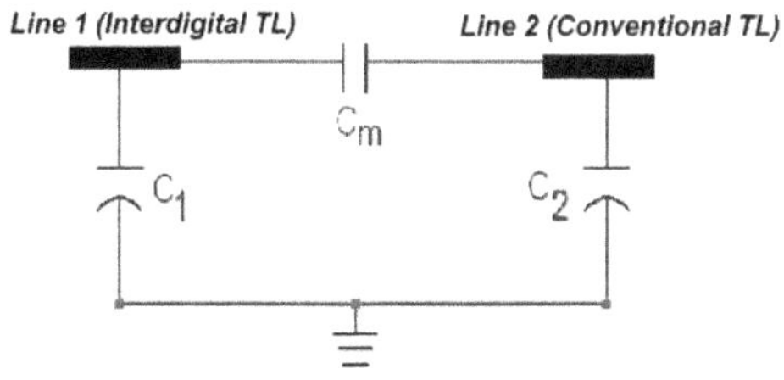

Fig. 11. Capacitance representation for cross section of the asymmetrical coupler presented in Fig. 10.

Characteristics of the proposed coupled transmission line can be described by a superposition of characteristics of c and π modes. A set of two coupled lines can support two fundamental independent modes of propagation (called normal modes). For asymmetrical coupled lines, the two normal modes of propagation are known as c and π modes (Mongia et al., 1999). Both c and π modes are composed of two traveling waves in the backward and forward directions. The c mode is characterized by four parameters: γ_c, Z_{c1}, Z_{c2} and R_c which are the propagation constant of the mode, the characteristic impedances of lines 1 and 2 and the ratio of the voltages on the two lines of the c mode, respectively. Similarly, the π mode is also characterized by four parameters: γ_π, $Z_{\pi 1}$, $Z_{\pi 2}$ and R_π which are propagation constant of the mode, characteristic impedances of lines 1 and 2 and the ratio of the voltages on the two lines of the π mode, respectively (Mongia et al., 1999).

As it has been shown in (Mongia et al., 1999), the relation between the characteristic impedances, i.e. Z_{c1}, Z_{c2}, $Z_{\pi 1}$ and $Z_{\pi 2}$, and also the ratio parameters, i.e. R_c and R_π, are as:

$$\frac{Z_{c2}}{Z_{c1}} = \frac{Z_{\pi 2}}{Z_{\pi 1}} = -R_c R_\pi \tag{10}$$

So, a total number of only six quantities, i.e. γ_c, γ_π, Z_{c1} or Z_{c2}, $Z_{\pi 1}$ or $Z_{\pi 2}$, R_c and R_π are required to characterize asymmetrical coupled lines. For a lossless TEM-mode coupled-line, the propagation constants of both c and π modes are the same, and are given by (Cristal, 1966):

$$\gamma_c = \gamma_\pi = j\beta \tag{11}$$

As special case for asymmetrical coupled lines, symmetrical coupled line are completely characterized by four parameters, the even and odd modes characteristic impedances of any lines (as both lines are identical) and even and odd modes propagation constants. In symmetrical coupled lines, R_c and R_π are equal to 1 and -1, respectively (Mongia et al., 1999).

By assuming the quasi-TEM mode for proposed structure and according to equations (10), (11) and (Cristal, 1966) for above asymmetrical coupler (Fig. 10), it is obtained that:

$$R_c = -R_\pi = \sqrt{\frac{Z_2}{Z_1}} \tag{12}$$

where

$$Z_1 = \sqrt{\frac{L_1}{C_1} - \frac{1}{\omega^2 C_{int} C_1}}\,, \qquad Z_2 = \sqrt{\frac{L_2}{C_2}} \tag{13}$$

where Z_1 and Z_2 are characteristic impedances of uncoupled lines 1 (interdigital TL) and 2 (conventional microstrip TL), respectively.

Moreover, the capacitance matrix of the coupled lines (Fig. 10) can be expressed as (Cristal, 1966):

$$[C] = \begin{bmatrix} C_1 & C_{12} \\ C_{21} & C_2 \end{bmatrix} = \begin{bmatrix} C_1 + C_m & -C_m \\ -C_m & C_2 + C_m \end{bmatrix} \tag{14}$$

According to equations (13) and (14), c and π mode characteristic impedances of interdigital transmission line ($Z_{0c}{}^a, Z_{0\pi}{}^a$) and the conventional microstrip transmission line ($Z_{0c}{}^b, Z_{0\pi}{}^b$) are obtained as (Cristal, 1966):

$$\begin{cases} Z_{0c}{}^a = \sqrt{\dfrac{L_1}{C_1} - \dfrac{1}{\omega^2 C_{int} C_1}} \\ Z_{0\pi}{}^a = \sqrt{\dfrac{L_1}{(C_1 + 2C_m)} - \dfrac{1}{\omega^2 C_{int} (C_1 + 2C_m)}} \\ Z_{0c}{}^b = \sqrt{\dfrac{L_2}{C_2}} \\ Z_{0\pi}{}^b = \sqrt{\dfrac{L_2}{C_2 + 2C_m}} \end{cases} \tag{15}$$

and

$$Z'_{0c}{}^a = \sqrt{\frac{L_1}{C_1}}\,, \qquad Z'_{0\pi}{}^a = \sqrt{\frac{L_1}{(C_1 + 2C_m)}} \tag{16}$$

$Z'_{0c}{}^a$ and $Z'_{0\pi}{}^a$ are c and π mode characteristic impedances of a conventional microstrip TL with a strip of width W', where W' (= $(2N-1)S' + 2NW$) is the total width of the interdigital capacitor.

In coupler design procedure, for an indicated coupling-level (k) and impedance ports Z_a and Z_b of lines 1 and 2, respectively, $Z_{0c}{}^a, Z_{0\pi}{}^a, Z_{0c}{}^b$ and $Z_{0\pi}{}^b$ can be calculated from following equations (Cristal, 1966):

$$\begin{cases} Z_{0c}{}^{a} = \dfrac{Z_a Z_b \sqrt{1-k^2}}{Z_b - k\sqrt{Z_a Z_b}} \\ Z_{0\pi}{}^{a} = \dfrac{Z_a Z_b \sqrt{1-k^2}}{Z_b + k\sqrt{Z_a Z_b}} \\ Z_{0c}{}^{b} = \dfrac{Z_a Z_b \sqrt{1-k^2}}{Z_a - k\sqrt{Z_a Z_b}} \\ Z_{0\pi}{}^{b} = \dfrac{Z_a Z_b \sqrt{1-k^2}}{Z_a + k\sqrt{Z_a Z_b}} \end{cases} \tag{17}$$

In order the values of $Z_{0c}{}^{a}$ and $Z_{0c}{}^{b}$ to be positive, it is necessary that:

$$\frac{1}{k^2} \geq \frac{Z_a}{Z_b} \quad \text{and} \quad \frac{1}{k^2} \geq \frac{Z_b}{Z_a} \tag{18}$$

where k^2 denotes the power coupling coefficient between two coupled lines.

As it was mentioned, for indicated coupling level (k) and ports impedance (Z_a,Z_b) in the proposed coupler, the c and π characteristic impedances, i.e. $Z_{0c}{}^{a}$, $Z_{0\pi}{}^{a}$, $Z_{0c}{}^{b}$ and $Z_{0\pi}{}^{b}$, can be determined using (17). It is clear from (15) that selecting a small C_{int} in the introduced coupler, increases values of $Z'_{0c}{}^{a}$ and $Z'_{0\pi}{}^{a}$ which can lead to smaller value for C_m. It means that in this situation, the required spacing between two coupled-lines can be increased in comparison with the conventional microstrip coupled-lines. It is due to the inverse relationship between mutual capacitance value and spacing between coupled lines. Therefore, it is suitable for realizing high coupling-level coupled-line couplers with relatively larger spacing between two lines than conventional coupled-line couplers.

Fig. 12 illustrates the layout and fabrication of the proposed asymmetrical coupler that above considerations have been considered in its design (Keshavarz et al., 2011b).

For an asymmetric coupled microstrip line of the type shown in Fig. 12, the design graphs presented in Figs. 13, 14 and 15 can be used to determine the necessary interdigital and microstrip strip widths and spacing for a given set of characteristic impedances, $Z_{oc}{}^{a}$, $Z_{0c}{}^{b}$ and Z_m on FR-4 substrate with $\varepsilon_r = 4.6$ and thickness of 1.6 mm. In Fig. 15, Z_m is defined as:

$$Z_m = \frac{2 Z_{0c}{}^{b} Z_{0\pi}{}^{b}}{Z_{0c}{}^{b} - Z_{0\pi}{}^{b}} \tag{19}$$

The asymmetrical coupled line coupler presented in this study is a 3-dB coupler at center frequency of 3 GHz which is simulated on FR-4 substrate with 1.6 mm substrate thickness and a dielectric constant of 4.6. Impedances of all four ports have been considered equal to 50 Ω $(Z_a = Z_b = 50\ \Omega)$. The final structure of designed coupler has been presented in Fig. 12(b) with $W_1 = 0.6\ mm, W_2 = 1\ mm$ and the spacing between two coupled lines (s) is 0.2 mm.

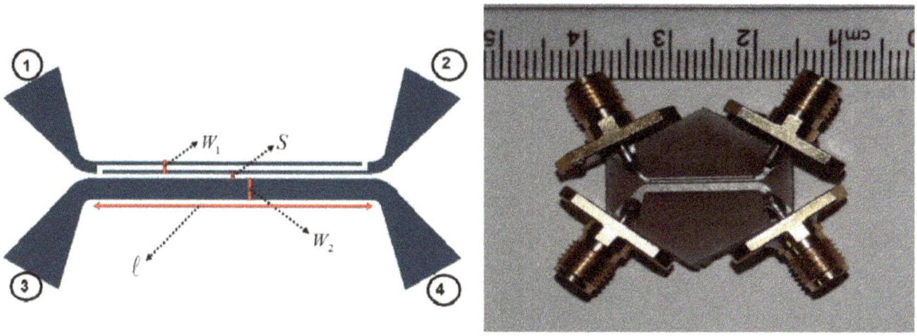

(a) (b)

Fig. 12. Proposed asymmetrical backward coupler based on the interdigital and conventional microstrip coupled TLs. (a) Structure layout. (b) Fabricated coupler. (Keshavarz et al., 2011b).

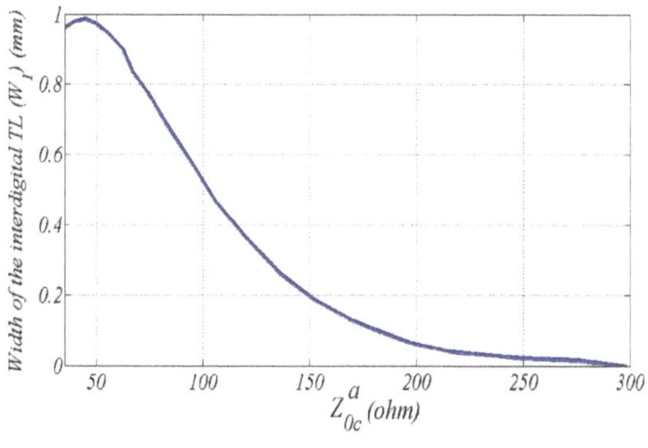

Fig. 13. Design graph for width of the interdigital TL (W_1) on FR-4 substrate versus *c* mode characteristic impedance. (Keshavarz et al., 2011b).

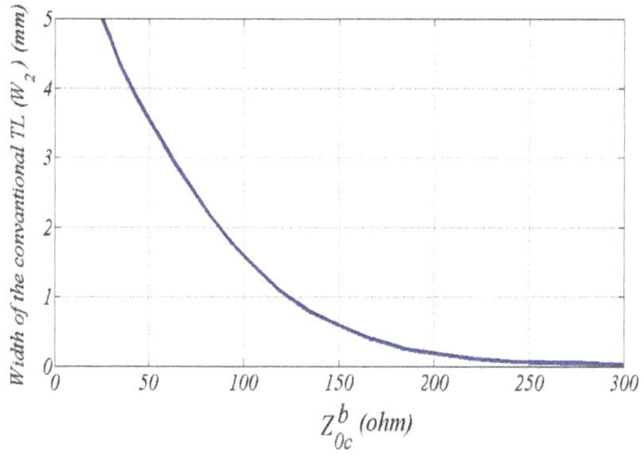

Fig. 14. Design graph for width of the conventional microstrip TL (W_2) on FR-4 substrate versus *c* mode characteristic impedance. (Keshavarz et al., 2011b).

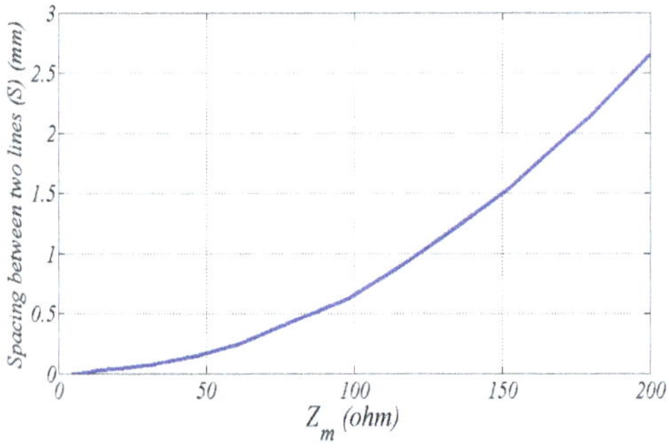

Fig. 15. Design graph for two lines separation (S) on FR-4 substrate versus Z_m . (Keshavarz et al., 2011b).

So, this coupler is more compact than CRLH coupled line couplers reported in (Abdelaziz et al., 2009; Garcia-Perez et al., 2010; Joon-Boom et al., 2001), due to the elimination of the stubs. The structure coupled-line length (ℓ) is equal to 12 mm, which is approximately $\lambda_g/4$ at center frequency of 3 GHz and is smaller than the CRLH microstrip CLC with the coupled line length around $\lambda_g/3$ (Caloz et al., 2004).

In addition to the equivalent circuit model which is used to simulate the designed coupler, a full-wave electromagnetic simulator (ADS) is also used to examine the structure. Fig. 16 illustrates the full-wave and equivalent circuit model analysis results of the proposed asymmetric backward coupler along with its measured S-parameters. Excellent agreement can be observed between full-wave simulated and experimental results. The elements of the equivalent circuit model are obtained using equations (13) and (15) and for this example are equal to $L_1 = 7.33\,nH$, $C_1 = 0.7\,pF$, $C_{\text{int}} = 1.82\,pF$, $L_2 = 6.18\,nH$, $C_2 = 0.86\,pF$. Using this figure, performance of the introduced 3-dB edge-coupled coupled-line coupler can be stated as the following: the power which is coupled to port 3 is approximately -3 dB, the return loss is less than -14 dB and the isolation is better than -13 dB over the bandwidth of 66% from 2.2 GHz to 4.2 GHz. Moreover, Fig. 17 shows the phase difference between the ports 2 and 3 of the coupler. As it is seen, this difference is equal to $90 \pm 10°$ for a frequency range from 2.2 GHz to 3.5 GHz. Proposed asymmetrical backward coupler exhibits reachable dimension, broad bandwidth and smaller size than the conventional and CRLH couplers.

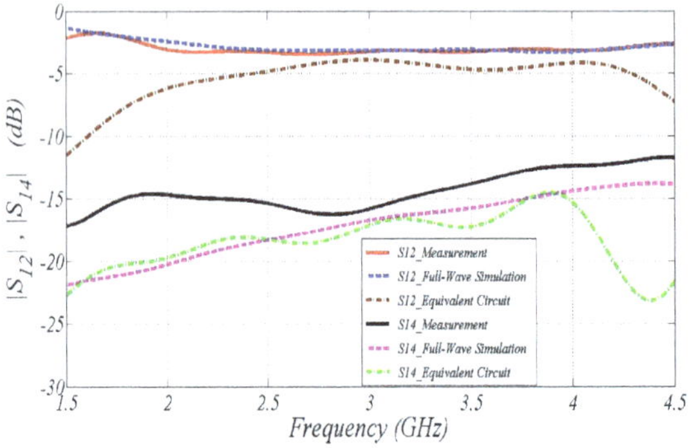

(a)

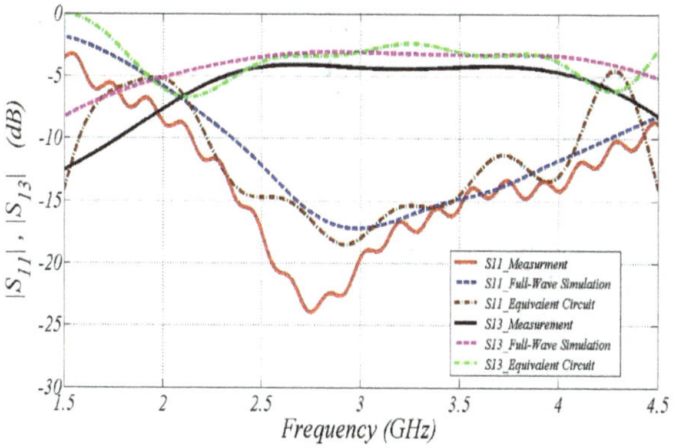

(b)

Fig. 16. Magnitude of the S-parameters for the proposed coupler obtained by full-wave simulation, equivalent circuit model and measurement results. (a) $|S_{12}|,|S_{14}|$ (b) $|S_{11}|,|S_{13}|$ (Keshavarz et al., 2011b).

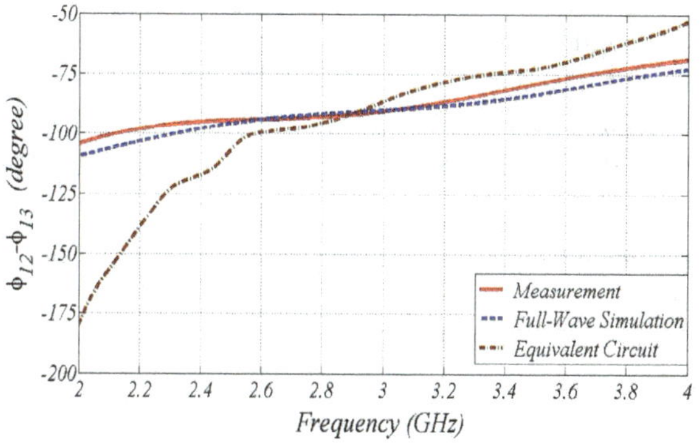

Fig. 17. Phase difference between the through and the coupled ports for the proposed coupler of Fig. 12. (Keshavarz et al., 2011b).

3.3.3 Forward symmetrical CLC

The scattering parameters of an ideal forward-wave directional coupler, as shown in Fig. 3, are given by (Mongia et al., 1999):

$$\begin{cases} S_{11}=0 \\ S_{12}=-je^{\frac{-j(\beta_e+\beta_o)l}{2}}\cos[\frac{(\beta_e-\beta_o)l}{2}] \\ S_{13}=0 \\ S_{14}=-je^{\frac{-j(\beta_e+\beta_o)l}{2}}\sin[\frac{(\beta_e-\beta_o)l}{2}] \end{cases} \tag{20}$$

where β_e and β_o are even and odd mode propagation constants of coupled lines, respectively. Also, l is length of the coupled line. As it was mentioned, forward-wave directional couplers cannot be realized using TEM mode transmission lines such as coaxial lines. It is due to this fact that for the TEM mode, the propagation constant of the even and odd modes are equal, and as shown in (20), there is no coupling between ports 1 and 4. Therefore, forward-wave coupling mechanism can only be appeared in non-TEM coupled TLs such as metallic waveguides, fin lines, dielectric waveguides and also quasi-TEM mode TLs like microstrip lines at high operating frequencies. In these transmission line structures, in general, the phase velocities of the even and odd modes are not equal (Mongia et al., 1999).

From (20), it is clear that complete power can be transferred between lines if the length l of the coupled line is chosen as:

$$l = \frac{\pi}{|\beta_e - \beta_o|} \tag{21}$$

Above result is significant in the sense that even for arbitrarily small values of difference in the propagation constants of even and odd modes, complete power can be transformed between the lines if the length of the coupler is chosen according to (21). In this situation, the directivity and isolation of the coupler are thus infinite. Also, the phase difference between ports 1 and 4 (S_{41} and S_{21}) is 90°. However, in general, situation (21) cannot be completely satisfied. Hence, some finite amount of backward-wave coupling always exists between coupled lines.

Our proposed forward-wave coupled-line coupler is shown in Fig. 18(a), where the coupled-lines have the same width of W and periodic stubs have been loaded between these coupled-lines (Keshavarz et al., 2010). In this structure, W_s and ℓ_s are the width and length of the periodic stubs, respectively, and d_s is a period of the stubs. The mid plane (red line in Fig. 18(a)) between the coupled-lines remains two different equivalent circuits for the even and odd modes. The even and odd modes are associated with a magnetic wall (open-circuit) and an electric wall (short-circuit), respectively. These two equivalent circuit models have been presented in Figs. 18(b) and 18(c) for one period. In these circuits, C_e and C_o are even and odd mode capacitances per unit length, respectively, and L is inductance per unit length of the coupled-lines. C_e and C_o are equal to:

$$C_e = C_{11} = C_{22}, \quad C_o = C_{11} + 2C_{12} + C_{\text{int}} \tag{22}$$

where C_{11} and C_{22} represent the capacitance between one strip conductor and ground in absence of the other strip conductor, in planar structures. Because of the strip conductors of the coupled lines are identical in size and location relative to the ground conductor, C_{11} will be equal to C_{22} or $C_{11} = C_{22}$. From transmission line theory, it is well known that the value of C_{11} is (Pozar, 2004):

$$C_{11} = \frac{\sqrt{\varepsilon_{re}}\, Z}{c} \tag{23}$$

where ε_{re} is effective permittivity of a microstrip transmission line with a strip with width W, Z is characteristic impedance of the transmission line and c is the speed of light. Also, C_{12} represents the capacitance between the two coupled lines without stubs and ground conductor. C_{int} is capacitance per unit length of the interdigital capacitor formed between the two coupled lines.

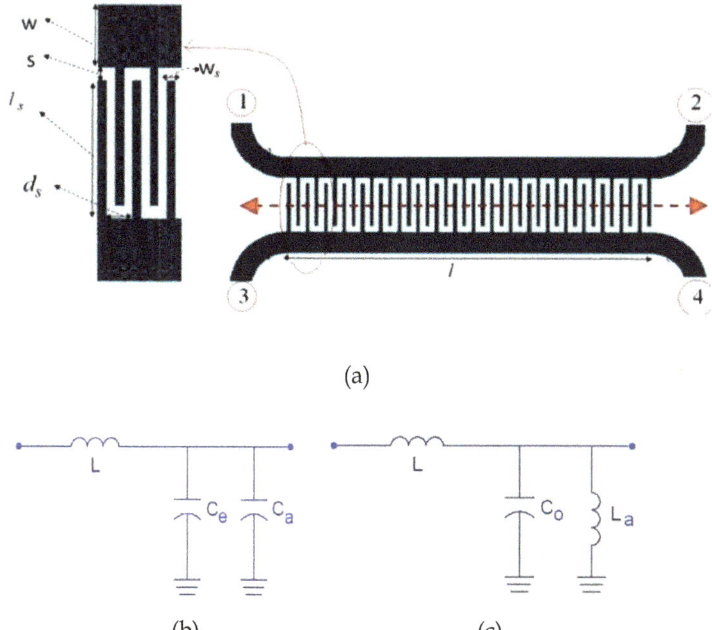

Fig. 18. (a) Proposed forward-wave coupled-line coupler with periodic stubs. (b) Even mode, and (c) odd mode equivalent circuit models of each coupled line for one period (Keshavarz et al., 2010).

Some extra distributed shunt capacitance and inductance per unit length are added to the equivalent circuit models for the even and odd modes, respectively, which are given based on the TL theory as (Pozar, 2004):

$$\begin{aligned} L_a &= \frac{1}{d_s}\left(\frac{Z_s}{\omega}\tan\beta_s\left(\frac{l_s+s}{2}\right)\right) \approx \frac{Z_s\beta_s(l_s+s)}{2\omega d_s} \\ C_a &= \frac{1}{d_s}\left(\frac{1}{\omega Z_s}\tan\beta_s\left(\frac{l_s+s}{2}\right)\right) \approx \frac{\beta_s(l_s+s)}{2\omega Z_s d_s} \end{aligned} \tag{24}$$

where Z_s and β_s represent characteristic impedance and phase constant of the shunt stubs, respectively.

Series impedance and shunt admittance of these equivalent circuit models in even and odd modes are given by:

$$\begin{aligned} Z_e &= j\omega L, \quad Y_e = j\omega(C_e + C_a) \\ Z_o &= j\omega L, \quad Y_o = j\omega C_o + 1/j\omega L_a \end{aligned} \tag{25}$$

According to the TL theory, the propagation constants and the characteristic impedances of the transmission coupled-lines in even and odd modes are:

$$\begin{aligned} \gamma_e &= \sqrt{Z_e Y_e} = j\omega\sqrt{L(C_e + C_a)} = j\beta_e \\ \gamma_o &= \sqrt{Z_o Y_o} = j\omega\sqrt{L(C_o - 1/\omega^2 L_a)} = j\beta_o \end{aligned} \tag{26}$$

and

$$\begin{aligned} Z_{ce} &= \sqrt{\frac{Z_e}{Y_e}} = \sqrt{\frac{j\omega L}{j\omega(C_e + C_a)}} = \sqrt{\frac{L}{(C_e + C_a)}} \\ Z_{co} &= \sqrt{\frac{Z_o}{Y_o}} = \sqrt{\frac{j\omega L}{j\omega(C_o - 1/\omega^2 L_a)}} = \sqrt{\frac{L}{(C_o - 1/\omega^2 L_a)}} \end{aligned} \tag{27}$$

Since, the length of the stubs is relatively large, the value of C_{12} would be very smaller than C_{11} and C_{int} . So, (22) can be approximated as:

$$C_o \cong C_{11} + 2C_{\text{int}} \tag{28}$$

As it is seen in (26), the difference between β_e and β_o in proposed structure becomes larger than conventional structures without stubs in coupled line couplers. Moreover, this difference can be controlled by stub length, so that for a fixed coupling-level, increasing length of stubs (ℓ_s) results reduction of structure length (Fig. 19).

In the coupled-line couplers, input matching condition for termination of impedance $Z_c (Z_{in} = Z_c)$ is achieved under condition which is given by (12).

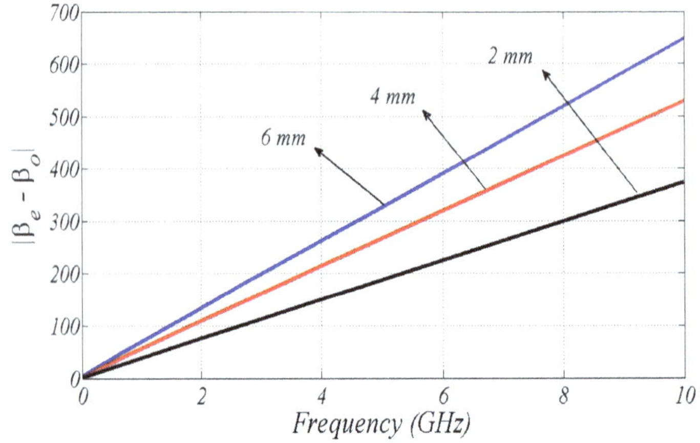

Fig. 19. $|\beta_e - \beta_o|$ for three lengths of the stubs ($l_s = 2, 4$ *and* 6 *mm*) (Keshavarz et al., 2010).

Fig. 20 presents some curves for selecting dimension of the proposed coupler for three coupling-levels (0-dB, 3-dB and 6-dB) with $W_s = 0.2mm$, $d_s = 0.6mm$ and $S = 0.2mm$ on FR-4 substrate ($\varepsilon_r = 4.6$, $h = 1.6\ mm$). These curves illustrate that with increasing the coupling-level, dimension of the coupler increase. But, it is interesting to note that for a fixed coupling-level, the area of the coupler (product of the stub length by the structure length) will remain constant, approximately.

The proposed structure of the forward-wave CLC in this section is fabricated on FR4 substrate with 1.6 mm thickness and dielectric constant of 4.6, as shown in Fig. 21. The full-wave simulator Agilent Technologies Advanced Design System (ADS) is used to examine the structure. For good matching, the width of the microstrip transmission lines for 50Ω port impedances is selected equal to 1 mm (i.e. $W = 1mm$). To have a coupling level of 0-dB, according to the derived relations and Fig. 20, the length (l) and width ($l_s + 2W$) of the structure in Fig. 18 have been chosen equal to 26 mm and 4 mm, which are approximately $\lambda_g/2$ and $\lambda_g/13$ at center frequency of 3 GH, respectively.

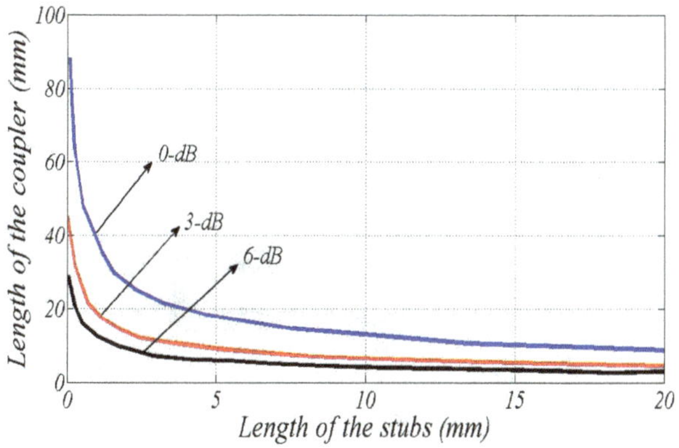

Fig. 20. Data for designing dimension of the proposed coupler on FR-4 substrate ($\varepsilon_r = 4.6, h = 1.6\ mm$) (Keshavarz et al., 2010).

Therefore, the proposed CLC is more compact than the microstrip coupler with the coupled-line length around $0.75\lambda_g$ presented in (Fujii & Ohta, 2005). Also, the width (W_s) and period distance (d_s) of the stubs are considered as: $W_s = 0.2mm$, $d_s = 0.6mm$ and the space between the stubs and transmission lines is 0.2 mm (i.e. $S = 0.2mm$).

The measured and simulated S-parameters of the proposed coupler are shown in Fig. 22. This figure shows the measured amplitude balance of ±2 dB over a bandwidth of 66% (2-4 GHz). In this figure, full-wave simulation and equivalent circuit model results have also been presented for verification. A good agreement between measurement, full-wave simulation and equivalent circuit model results is obtained and thus the usefulness of the presented equivalent circuit model is validated. The element values of the equivalent circuit model (Fig. 18) for the layout are: $L = 1.8nH, L_a = 3.2nH, C_a = 0.1pF, C_e = 0.2pF$ and $C_o = 1.8pF$.

In comparison with the conventional forward CLCs, the electrical length of the proposed CLC is more compact than CLCs presented in (Chang et al., 2001; Deng et al., 2002; Lauro et al., 2009). For instance, the coupled-line electrical length of the coupler is shortened to 50% of the conventional CLC electrical length reported in (Deng et al., 2002). Moreover, the bandwidth of the proposed CLC is wider than forward CLCs presented in (Deng et al., 2002; Lauro et al., 2009; Chang et al., 2001; Sen-Kuei & Tzong-Lin, 2010). For example, compared with the forward couplers reported in (Deng et al., 2002) and (Lauro et al., 2009), the proposed structure is capable of producing 65% bandwidth enhancement for the amplitude and a 0-dB coupling level with a smaller coupled-line length. Moreover, the proposed structure exhibits broader bandwidth than couplers presented in (Chang et al., 2001; Huang & Chu, 2010; Ikalainen & Matthaei, 1987; Sen-Kuei & Tzong-Lin, 2010; Lauro et al., 2009).

Fig. 23 shows the even- and odd-mode characteristic impedances computed using full-wave simulation. This result indicates that the proposed structure is matched to 50 Ω port impedance over the operating bandwidth, such that the additional tapered structure at each port for impedance matching can be eliminated. Hence, the proposed forward coupler would be more compact in size. As it was mentioned, for the proposed forward CLC, the coupler area is approximately constant. It means that reduction of the structure length results width increasing, proportionally (Fig. 20).

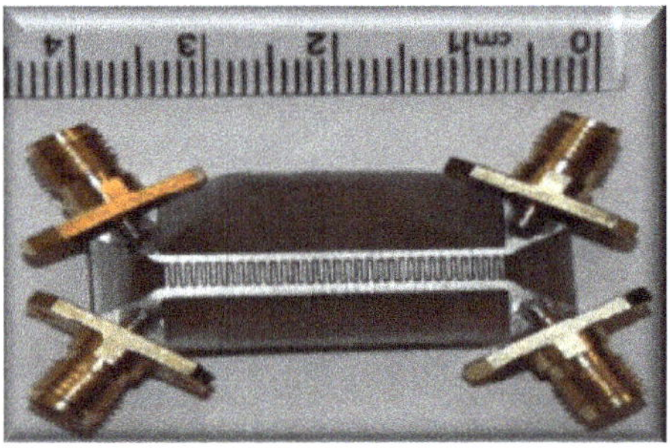

Fig. 21. Proposed forward symmetrical coupler which realized on FR-4 substrate ($\varepsilon_r = 4.6$, $h = 1.6\,mm$) (Keshavarz et al., 2010).

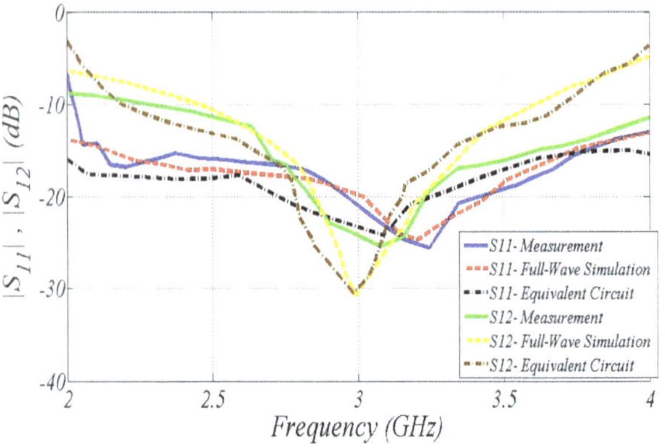

(a)

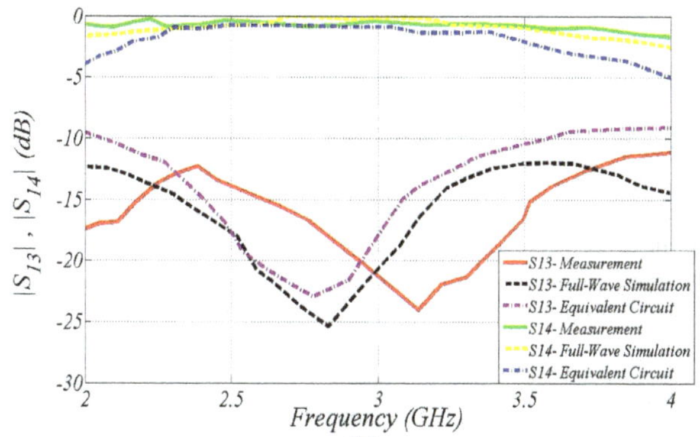

(b)

Fig. 22. Magnitude of the S-parameters, (a) S_{11},S_{12} (b) S_{13},S_{14} for the proposed coupler in Fig. 21 obtained by the full-wave simulation, equivalent circuit model and measurement results (Keshavarz et al., 2010).

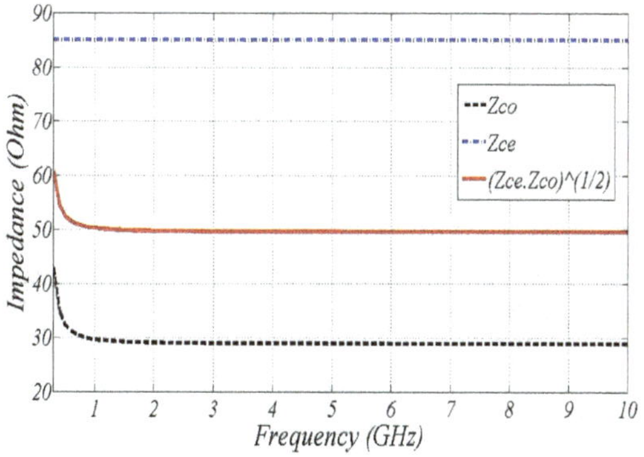

Fig. 23. Even and odd modes characteristic impedances of the coupler presented in Fig. 21 (Keshavarz et al., 2010).

4. Conclusion

In this chapter, some new techniques for realizing compact and tight coupling microstrip backward and forward CLCs with obtainable dimension, broad bandwidth and smaller size than the most conventional microstrip and CRLH couplers have been introduced. We presented three CLCs based on the concept of CRLH CLCs; a symmetrical backward CLC, an asymmetrical backward CLC and a symmetrical forward CLC.

New symmetrical backward coupler structure consists of only one interdigital capacitor in each coupled TL without shorted stubs as the CRLH TL. Designed and fabricated 3-dB microstrip coupler at center frequency about f_c= 3.2 GHz exhibits a matching (10-dB) bandwidth of over 2 GHz, a phase-balance (±10°) bandwidth of 1.3 GHz and at least 20-dB isolation between adjacent ports. The coupled-line length and the width of the proposed structure are approximately, $\lambda_g/4$ and $\lambda_g/36$, respectively. Also, this coupler exhibits higher design simplicity than the existing CRLH CLCs.

Moreover, a new type of backward CLC composed of two different coupled lines, i.e. interdigital and conventional microstrip TLs has been proposed, fabricated, and investigated theoretically and experimentally. In this structure, an interdigital capacitor with only one finger is used as interdigital TL. This interdigital TL is coupled with a conventional microstrip TL and achieves an asymmetrical backward CLC. The proposed backward-wave coupler with 0.2 mm spacing between two coupled lines exhibits the amplitude balance of ±2 dB from 2.2 GHz to 4.2 GHz and the phase balance of 90°±10° from 2.2 GHz to 3.5 GHz.

Finally, a forward CLC composed of two identical microstrip TLs and periodic shunt stubs between them has been proposed and investigated experimentally and theoretically. Using loaded stubs between two microstrip coupled-lines forms the proposed 0-dB forward CLC which exhibits the amplitude balance of ±2 dB around center frequency of 3 GHz from 2 GHz to 4 GHz (66% bandwidth). A matching ($|S_{11}|$<15-dB) bandwidth of over 4 GHz (1-5 GHz) bandwidth and at least 15 dB isolation between adjacent ports have been seen in measurement results. In this forward-wave CLC, by increasing the length of the stubs, the coupler length decreases, proportionally.

5. References

Abdelaziz, A. F., Abuelfadl, T. M., & Elsayed, O. L. (2009). Realization of composite right/left-handed transmission line using coupled lines. *Progress In Electromagnetics Research*, Vol. 92, pp. 299–315.

Bahl, I. (2003). *Lumped Elements for RF and Microwave Circuits*, ArtechHouse, Boston.

Caloz, C. & Itoh, T. (2002). Application of the transmission line theory of left-handed (LH) materials to the realization of a microstrip LH transmission line. *In Proc. IEEE-AP-S USNC/URSI National Radio Science Meeting*, vol. 2, pp. 412–415.

Caloz, C. & Itoh, T. (2004). A novel mixed conventional microstrip and composite right/left handed backward-wave directional coupler with broadband and tight coupling characteristics. *IEEE Microwave Wireless Component Letter*, vol. 14, no. 1, pp. 31-33.

Caloz, C. & Itoh, T. (2004). Transmission line approach of left-handed (LH) structuresand microstrip realization of a low-loss broadband LH filter. *IEEE Trans. Antennas Propagation*, vol. 52, no. 5, pp. 1159–1166.

Caloz, C., & Itoh, T. (2005). *Electromagnetic Metamaterials: Transmission Line Theory and Microwave Applications*, Wiley, New York.

Caloz, C., Sanada, A., & Itoh, T. (2004). A novel composite right/lefthanded coupled-line directional coupler with arbitrary coupling level and broad bandwidth. *IEEE Trans. Microwave Theory Technique*, vol. 52, pp. 980–992.

Chang, C., Qian, Y. & Itoh, T. (2001). Enhanced Forward Coupling Phenomena Between Microstrip Lines on Periodically Patterned Ground Plane. *IEEE MTT-S International Microwave Symposium Digest*, pp. 2039–2042.

Cristal, E.G. (1966). Coupled-Transmission-Line Directional Couplers with Coupled Lines of Unequal Characteristic Impedances. *G-MTT International Symposium Digest*, vol. 66 , no. 1, pp. 114 - 119.

Deng, J.D.S., Feng-ka H., Kuo, J.I., Kuo Y. H., Ching-Yuan L., & Chung-Sen W. (2002). Tightly coupling LTCC microwave coupled lines: analysis, modeling and realization Electronic Materials and Packaging. *Proceedings of the 4th International Symposium on Digital Object*, pp. 391 - 396.

Fouda, A.E., Safwat, A.M.E., & El-Hennawy, H. (2010). On the Applications of the Coupled-Line Composite Right/Left-Handed Unit Cell. *IEEE Transactions on Microwave Theory and Techniques*, vol. 58, no. 6, pp. 1584 - 1591.

Fujii, T. & Ohta, I. (2005). Size-Reduction of Coupled-Microstrip 3-dB Forward Couplers by Loading with Periodic Shunt Capacitive Stubs. *IEEE MTT-S International Microwave Symposium Digest*, pp. 1235–1238.

Garcia-Perez, O., Garcia Munoz, L. E., Segovia-Vargas, D., & Gonzalez-Posadas, V. (2010). Multiple order dual-band active ring filters with composite right/left-handed cells. *Progress In Electromagnetics Research*, vol. 104, 201–219.

Hirota, A., Tahara, Y., & Yoneda, N. (2009). A compact coupled-line forward coupler using composite right-/left-handed transmission lines. *IEEE MTT-S International Microwave Symposium Digest*, pp. 617 - 620.

Hirota, A., Tahara, Y., & Yoneda, N. (2011). A wide band forward coupler with balanced composite right-/left-handed transmission lines. *IEEE MTT-S International Microwave Symposium Digest*, pp. 1 - 4.

Huang, J.Q., & Chu, Q.X. (2010). Compact UWB band-pass filter utilizing modified composite right/left-handed structure with cross coupling. *Progress In Electromagnetics Research*, vol. 107, pp. 179–186.

Ikalainen, K., & Matthaei, L. (1987). Wide-Band, Forward-Coupling Microstrip Hybrids with High Directivity. *IEEE Trans. on Microwave Theory and Techniques*, vol. 35, no. 8, pp. 719–725.

Islam, R. & Eleftheriades, G. V. (2006). Printed high-directivity metamaterial MS/NRI coupled-line coupler for signal monitoring applications. *IEEE Microwave and Wireless Component Letter*, vol. 16, no. 4, pp. 164-166.

Islam, R., Elek, F., & Eleftheriades, G. V. (2004). Coupled line metamaterial coupler having co-directional phase but contradirectional power flow. *Electronics Letters*, vol. 40, no. 5, pp. 315–317.

Joon-Bum K., Chul-Soo K., Kwan-Sun C., Jun-Seok P., & Dal A. (2001). A design mapping formula of asymmetrical multi-element coupled line directional couplers. *IEEE MTT-S International Microwave Symposium Digest*, vol. 2, pp. 1293 - 1296.

Kawakami, T., Inoue, N., Horii, Y., & Kitamura, T. (2010). A super-compact 0dB/3dB forward coupler composed of multi-layered CRLH transmission lines with double left-handed shunt-inductors. *European Microwave Conference*, pp. 1409 - 1412.

Keshavarz, R., Movahhedi, M., Hakimi, A., & Abdipour, A. (2010). A Compact 0-dB Coupled-Line Forward Coupler by Loading with Shunt Periodic Stubs. *Asia Pacific Microwave Conference*, pp. 1248-1251.

Keshavarz, R., Movahhedi, M., Hakimi, A., & Abdipour, A. (2011). A novel broad bandwidth and compact backward coupler with high coupling-level. *Journal of Electromagnetic Waves and Applications*, vol. 25, no. 2/3, pp. 283-293.

Keshavarz, R., Movahhedi, M., Hakimi, A., & Abdipour, A. (2011). A broadband and compact asymmetrical backward coupled-line coupler with high coupling level. *Submitted to International Journal of Electronics and Communication.*

Lauro, S., Toscano, E. & Vegni, L. (2009). Symmetrical Coupled Microstrip Lines With Epsilon Negative Metamaterial Loading, *IEEE Transactions on Magnetics*, vol. 45, no. 7, pp. 1182-1189.

Mao, S. G. & Wu, M. S. (2007). A novel 3-dB directional coupler with Bandwidth and coupler with coupling-level, broad bandwidth and compact size using composite right/lefthanded coplanar waveguides. *IEEE Microwave Wireless Components Letter*, vol. 17, no. 5, pp. 331–333.

Mocanu, I.A., Militaru, N., Lojewski, G., Petrescu, T., & Banciu, M.G. (2010). Backward couplers using coupled composite right/left-handed transmission lines. *8th International Conference on Communications*, pp. 267 - 270.

Mongia, R., Bahl, I., & Bhartia, P. (1999). *RF and Microwave Coupledline Circuits*, Artech House, Norwood, MA.

Nguyen, H. V., & Caloz, C. (2006). Simple-design and compact MIM CRLH microstrip 3-dB coupled-line coupler. *Proc. IEEE International Microwave Symposium Digest*, pp. 1733–1736.

Pozar, M. D. (2004). *Microwave Engineering*, John Wiley & Sons.

Sen-Kuei, H. & Tzong-Lin W. (2010). A novel microstrip forward directional coupler based on an artificial substrate. *European Microwave Conference*, pp. 926-930.

Veselago, V. (1968). The electrodynamics of substances with simultaneously negative valuesof ε and μ. *Soviet Physics Uspekhi*, vol. 10, no. 4, pp. 509–514.

Wang, Y., Zhang, Y., Liu, F., He, L., Li, H., Chen, H., & Caloz, C. (2007). Simplified Description of Asymmetric Right-Handed Composite Right/Left-Handed Coupler in Microstrip-Chip Technology. *Microwave and Optical Technology Letters*, vol. 49, no. 9, pp. 2063-2068.

Zhang, Q., Khan, S. N., & Sailing, H. (2008) Coupled-line directional coupler based on Composite Right / Left-Handed coplanar waveguides. *International Workshop on Metamaterials*, pp. 301 - 304.

11

Magnetic Refrigeration Technology at Room Temperature

Houssem Rafik El-Hana Bouchekara[1,2] and Mouaaz Nahas[1]
[1]*Department of Electrical Engineering, College of Engineering and Islamic Architecture, Umm Al-Qura University, Makkah,*
[2]*Electrical Laboratory of Constantine "LEC", Department of Electrical Engineering, Mentouri University - Constantine, Constantine,*
[1]*Saudi Arabia*
[2]*Algeria*

1. Introduction

Modern society largely depends on readily available refrigeration methods. Up till now, the conventional vapor compression refrigerators have been mainly used for refrigeration applications. Nonetheless, the conventional refrigerators - based on gas compression and expansion - are not very efficient because the refrigeration accounts for 25% of residential and 15% of commercial power consumption (Tishin, 1999). Moreover, using gases such as chlorofluorocarbons (CFCs) and hydrochlorofluorocarbons (HCFCs) have detrimental effects on our environment. Recently, the development of new technologies - such as magnetic refrigeration - has brought an alternative to the conventional gas compression technique (Manh, 2007).

The magnetic refrigeration at room temperature is an emerging technology that has attracted the interest of researchers around the world (Bouchekara, 2008). Such a technology applies the magnetocaloric effect which was first discovered by Warburg (Bohigas, 2000; Zimm, 2007). In 1881, Warburg noticed an increase of temperature when an iron sample was brought into a magnetic field and a decrease of temperature when the sample was removed out of it. Thus, the magnetocaloric effect is an intrinsic property of magnetic materials; where it is defined as the response of a solid to an applied magnetic field which appears as a change in its temperature (Bohigas, 2000; Zimm, 2007). Such materials are called magnetocaloric materials. The magnetocaloric effect is present in all transition metals and lanthanide-series elements, which may have ferromagnetic behaviour. When a magnetic field is applied, the magnetic moments of these metals tend to align parralel to it, and the thermal energy released in this process produces the heating of the sample. The magnetic moments become randomly oriented when the magnetic field is removed, thus the ferromagnet cools down (Gschneidner, 1998).

The ultimate goal of this technology would be to develop a standard refrigerator for home use. The use of magnetic refrigeration has the potential to reduce operating and maintenance costs when compared to the conventional method of compressor-based refrigeration. By eliminating the high capital cost of the compressor and the high cost of electricity to operate the compressor, magnetic refrigeration can efficiently (and economically) replace compressor-based refrigeration technology. Some potential advantages of the magnetic refrigeration technology over the compressor-based refrigeration are: [1] green technology (no toxic or antagonistic gas emission); [2] noiseless technology (no compressor); [3] higher energy efficiency; [4] simple design of machines; [5] low maintenance cost; and [6] low (atmospheric) pressure (this is an advantage in certain applications such as in air-conditioning and refrigeration units in automobiles).

This chapter is concerned with the magnetic refrigeration technology form the material-level to the system-level. It provides a detailed review of the magnetic refrigeration prototypes available until now. The operational principle of this technology is explained in depth by making analogy between this technology and the conventional one. The chapter also investigates the study of the magnetocaloric materials using the molecular field theory. The thermal and magnetic study of the magnetic refrigeration process using the finite difference method (FDM) is also explained and are presented and discussed in detail.

The chapter is organized as follows. Section 2 introduces the magnetocaloric effect and its application to produce cold. It also introduces active magnetic regenerative refrigeration. Section 3 reviews ten various magnetic refrigeration systems and highlights their pros and cons. In Section 4 and 5, the thermal and magnetic study of the magnetic refrigeration process using the finite difference method are explained and the results from the thermal study are also presented and discussed in detail. Finally, the conclusions are drawn in Section 6.

2. The magnetocaloric effect

2.1 Definition

The magnetocaloric effect (MCE) is an intrinsic property of magnetic materials; it consists of absorbing or emitting heat by the action of an external magnetic field (Tishin, 1999). This results in warming or cooling (both reversible) the material as shown in Fig. 1.

Fig. 1. Magnetocaloric effect (the arrows symbolize the direction of the magnetic moments).

2.2 Thermodynamic approach

The absolute entropy, which is a function of temperature and induction in the magnetocaloric material is a combination of the magnetic entropy, the entropy of the lattice and the entropy of the conduction electrons (assumed negligible). It is given by the following equation:

$$S(T,B) = S_m + S_l \tag{1}$$

where S (J K-1) is the entropy (subscripts m and l are respectively for magnetic and lattice entropies), T (K) is the temperature and B (T) is the magnetic filed induction.

In magnetocaloric materials, a significant variation of the entropy can be observed by the application or removal of an external magnetic field. For a given material, MCE depends only on its initial temperature and the magnetic field. The MCE can be interpreted as the isothermal entropy change or the adiabatic temperature change.

The separation of entropy into three terms given in (1) is valid only for second order phase transition materials characterized by a smooth variation of the magnetization as a function of temperature. For first order transitions (abrupt change of magnetization around the transition temperature), this separation is not accurate (Kitanovski, 2005). For most applications, it is sufficient to work with the total entropy which - in its differential form - can be given as:

$$dS(T,B)=\left(\frac{\partial S}{\partial T}\right)_B dT+\left(\frac{\partial S}{\partial B}\right)_T dB \tag{2}$$

The specific heat capacity C_B (J m-3 K) of the material is given as:

$$C_B=\left(\frac{\partial S}{\partial T}\right)_B T \tag{3}$$

This gives:

$$\left(\frac{\partial S}{\partial T}\right)_B=\frac{C_B}{T} \tag{4}$$

From (2) and (4) we can write:

$$dS(T,B)=\frac{C_{\mathbf{B}}}{T}dT+\left(\frac{\partial S}{\partial B}\right)_T dB \tag{5}$$

In the case of an adiabatic process (no entropy change $\Delta S=0$) the temperature variation can be written as:

$$dT=-\frac{T}{C_B}\left(\frac{\partial S}{\partial B}\right)_T dB \tag{6}$$

Using the Maxwell relation given as:

$$\left(\frac{\partial S}{\partial B}\right)_T=\left(\frac{\partial M}{\partial T}\right)_B \tag{7}$$

where M (A m-1) is the magnetization.

We can write:

$$dT=-\frac{T}{C_B}\left(\frac{\partial M}{\partial T}\right)_B dB \tag{8}$$

The magnetocaloric effect (the adiabatic variation in temperature) can then be expressed as follows:

$$\Delta T_{ad} = -\int_{B_i}^{B_f} \frac{T}{C_B}\left(\frac{\partial M}{\partial T}\right)_B dB = MCE \tag{9}$$

In the case of an isothermal process, the temperature does not change during the magnetization and we can express the entropy as:

$$dS(T,B) = \left(\frac{\partial S}{\partial B}\right)_T dB \tag{10}$$

Using the Maxwell relation given by (7), the magnetic entropy change can be expressed as:

$$\Delta S = \int_{B_i}^{B_f} \left(\frac{\partial M}{\partial T}\right)_B dB \tag{11}$$

and the heat saved in this way is transferred to the lattice thermal motion.

2.3 Theoretical approach of MCE: molecular field theory

The theoretical calculation of the MCE is based on the model of Weiss (MFT: Molecular Field Theory) and the thermodynamic relations (Huang, 2004). To interpret quantitatively the ferromagnetism, Weiss proposed a phenomenological model in which the action of the applied magnetic field **B** was increased from that of an additional magnetic field proportional to the volume magnetization density B_v as:

$$B_v = \lambda \mu_0 M \tag{12}$$

The energy of a magnetic moment is then:

$$E = -\mu\left(B + B_v\right) \tag{13}$$

The magnetic moments will tend to move in the direction of this new field. Adapting the classical Weiss-Langevin classical calculations to a system of quantum magnetic moments, one finds:

$$M(x) = n g_J \mu_B B_J(x) \tag{14}$$

where:

$$x = \frac{J g_J \mu_B \left(B + \lambda \mu_0 M(x)\right)}{k_B T} \tag{15}$$

and

$$B_J(x) = \frac{2J+1}{2J}\coth\left(\frac{2J+1}{2J}x\right) - \frac{1}{2J}\coth\left(\frac{1}{2J}x\right) \tag{16}$$

where: J (N m s) is the total angular momentum, n (mol^{-1}) is the Avogadro number, g_J is the Landé factor, μ_B (J T^{-1}) is the Bohr magnetron, k_B (J K-1) is the Boltzmann constant, $B_J(x)$ is the Brillouin function, λ is the Weiss molecular field coefficient and μ_0 (T m A^{-1}) is the Permeability of vacuum.

The magnetic entropy is given by the relationship of Smart (Allab, 2008):

$$S_m(x) = R\left(\ln\left(\sinh\left(\frac{2J+1}{2J}x\right)\Big/\sinh\left(\frac{1}{2J}x\right)\right) - xB_J(x)\right) \tag{17}$$

The lattice contribution can be obtained using the Debye model of phonons (Allab, 2008). It is given by the following equation:

$$S_r = R\left(-3\ln\left(1 - e^{\frac{-T_D}{T}}\right) + 12\left(\frac{T}{T_D}\right)^3 \int_0^{\frac{T_D}{T}} \frac{y^3}{e^y - 1}dy\right) \tag{18}$$

where: T_D (K) is the Temperature of Debye and R (J K^{-1} mol^{-1}) is the universal gas constant.

2.3.1 Application of MFT to gadolinium (Gd)

In this section the theoretical study based on the MFT developed in the previous section is applied to the gadolinium. Table 1 gives the parameters used to calculate the magnetocaloric properties.

J	n	g_J	μ_B	k_B	μ_0	T_C	T_D
3.5	$6.023\ 10^{23}$	2	$9.2740154\ 10^{-24}$	$1.380662\ 10^{-23}$	$4\pi\ 10^{-7}$	293	184

Table 1. Parameters used for applying MFT to the gadolinium.

The numerical solution of equations (14), (15) and (16) allows getting the isotherms of magnetization and its evolution as a function of temperature calculated by the method of Weiss as shown in Fig. 2 (a) and Fig. 2 (b). Fig. 2 (c) represents the total heat capacity calculated from the equation (3) for different levels of induction. The magnetic entropy and its variation with temperature are shown respectively in Fig. 2 (d) and Fig. 2 (e). Finally, Fig. 2 (f) shows the magnetocaloric effect calculated by the MFT.

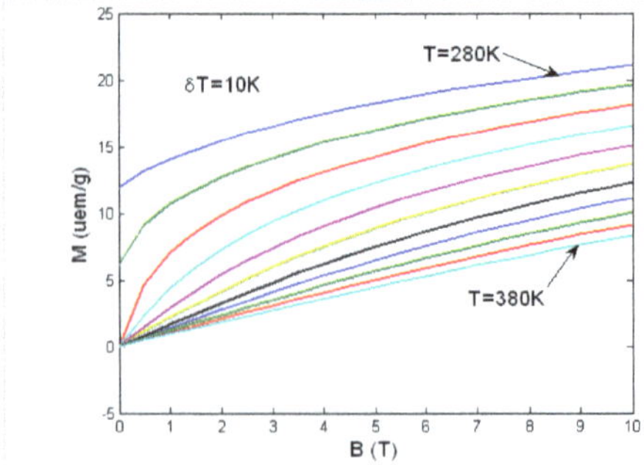

(a) Variation of magnetization versus the Magnetic field.

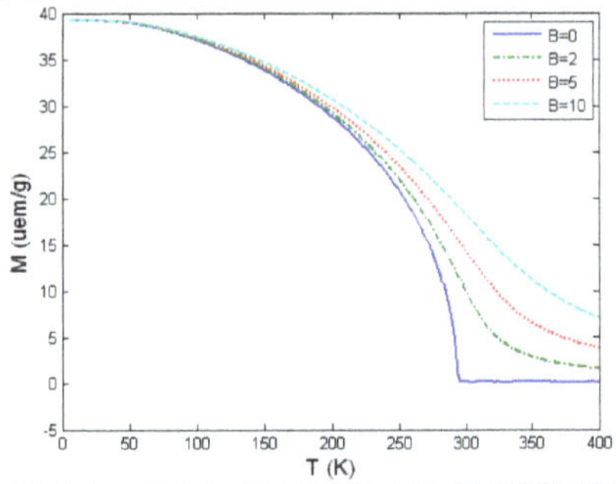

(b) Variation of magnetization versus the temperature.

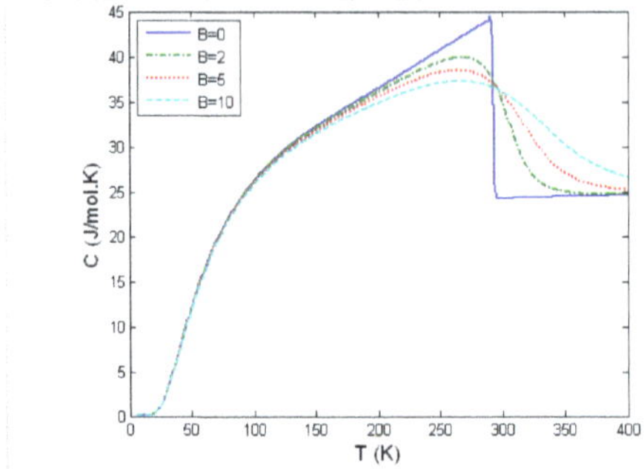

(c) Evolution of the total heat capacity as a function of temperature.

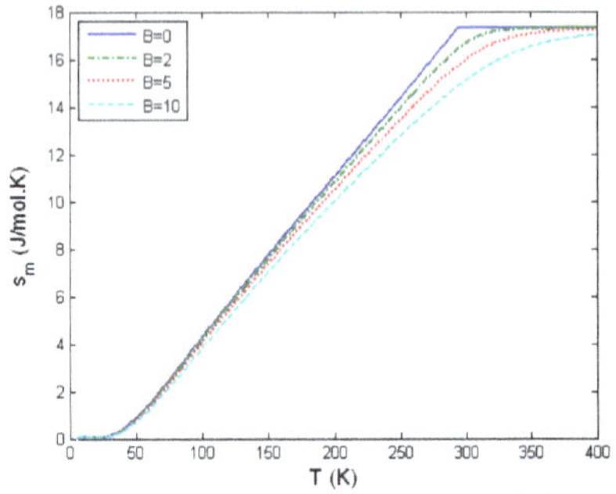

(d) Magnetic entropy.

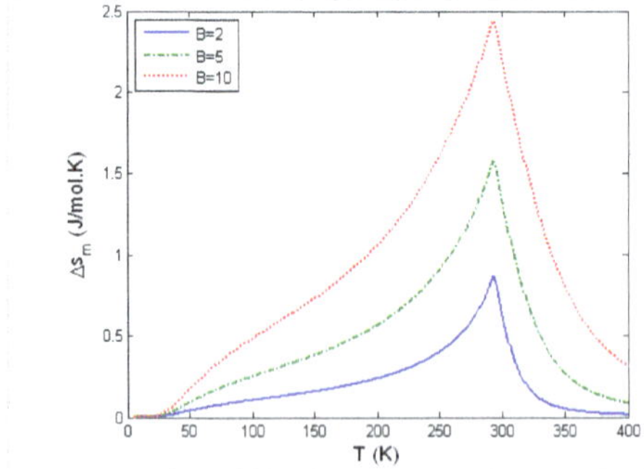

(e) Variation of the magnetic entropy.

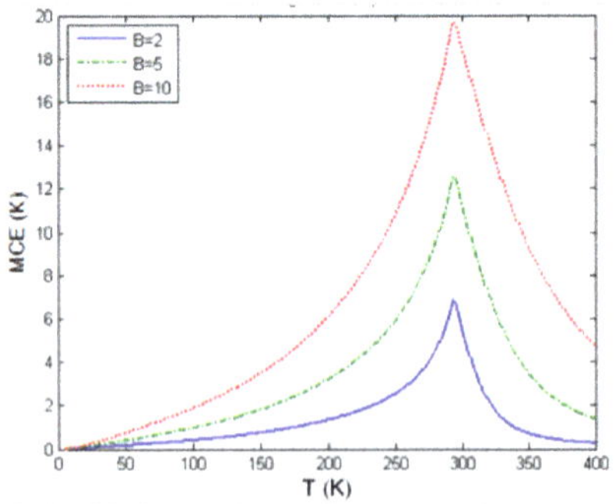

(f) MCE of Gd calculated by the MFT.

Fig. 2. Results of the theoretical study applied to Gd.

2.4 Application of MCE to produce cold

The magnetic cycles are generally composed of the process of magnetization and demagnetization, in which heat is discharged or absorbed in four steps as depicted by Fig. 3. From thermodynamic point of view, the magnetic cooling can be realized by: Carnot, Stirling, Ericsson and Brayton, where the Ericsson and Brayton cycles are believed to be the most suitable for such medium or room temperature cooling. Such cycles are predisposed to yield high cooling efficiency of the magnetic materials (Bouchekara, 2008).

Fig. 3(a) shows the conventional gas compression process that is driven by continuously repeating the four different basic processes shown while Fig. 3 (b) shows the magnetic refrigeration cycle comparison. The steps of the magnetic refrigeration process are analogous to those of the conventional refrigeration. By comparing (a) with (b) in Fig. 3, one can see that the compression and expansion are replaced by adiabatic magnetization and demagnetization, respectively. These processes change the temperature of the material and heat may be extracted and injected just as in the conventional process.

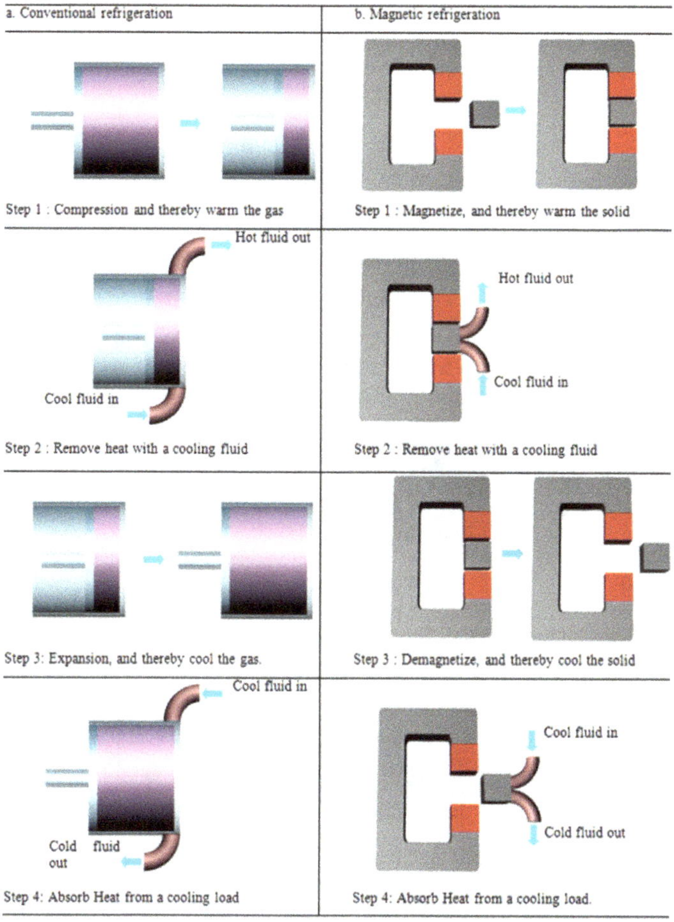

Fig. 3. Analogy between magnetic refrigeration and conventional refrigeration.

2.5 The Active Magnetic Regenerative Refrigeration (AMRR)

The direct exploitation of the giant MCE around the room temperature is limited by the fact that existing MCE materials do not achieve high temperature differences (Lebouc, 2005). For example, a sample of gadolinium around room temperature produces an MCE of approximately 10 K in a magnetic field of 5 T.

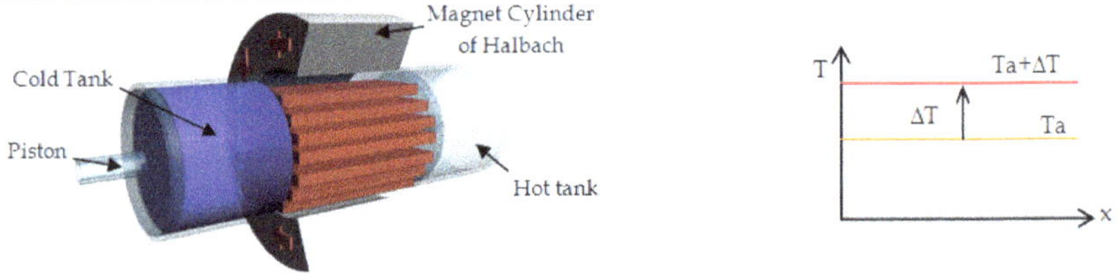

Step 1: Magnetization of the material from an initial state where the entire system is at temperature T_a. Each point of the regenerator material sees its temperature increase by ΔT following the application of the magnetic field.

Step 2: Flow of the fluid from the cold source to the hot source. The heat produced by the magnetization step is removed by the fluid flowing from the cold source T_c to the hot source T_h. This creates a temperature gradient along the bed.

Step 3: demagnetization of the material. The temperature of Each point of the regenerator decreases by Δ T due to the demagnetization.

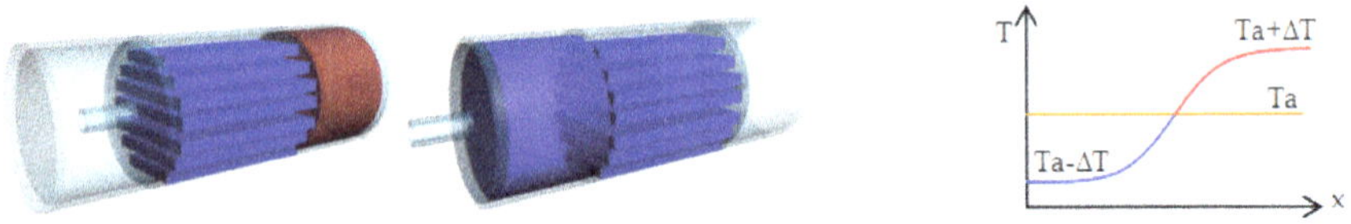

Step 4: Flow of fluid from the hot source to cold source. The flow of the fluid from the hot source T_h to the cold source T_f transfers its heat to the regenerator. The temperature gradient is amplified.

Fig. 4. Representation of AMRR cycle and temperature profile along the MCE material.

Since the gadolinium is considered as one of the best magnetocaloric materials currently available (Lebouc, 2005), the MCE corresponds to the absolute maximum value that can be obtained between the hot tank and cold tank. Thus it is obviously hard to imagine the exploitation of the MCE in most refrigeration applications (Engelbrecht, 2005).

This technical barrier has been overcome by the application of the Active Magnetic Regenerative Refrigeration (AMRR) (Engelbrecht, 2005; Lebouc, 2005; Tura, 2002). Regeneration in magnetic refrigeration systems allows the heat rejected by the network in any step of the cycle to be restored and returned to the network in another step in the same cycle (Yu, 2003). Thus, the capacity used for cooling the network load can be used effectively to increase the actual change of entropy and the obtained temperature difference (Yu, 2003).

AMRR cycles are illustrated in Fig. 4. The regenerative bed consists of plates of MCE material that initially have a quasi-linear temperature profile between the hot and cold tanks.

The bed itself acts as a regenerator. The different solid parts of the regenerator are connected by the fluid, so the heat does not need to be transferred between two solid parts separated, but on the same block.

Each particle of the bed undergoes a regenerative Brayton cycle and the entire bed undergoes a cascade Brayton cycle (Yu, 2003). This cycle is repeated 'n' times and the ΔT generated is amplified at each cycle to reach the temperatures limits of hot and cold sources (steady state). This ΔT is higher than the adiabatic temperature change of refrigerant material (MCE). In addition, the regenerator bed can be achieved by superposing different materials of different composition to expand the temperature's range of variation and thus to extend the utilization range of the system.

3. Magnetic refrigeration systems

Since the first magnetic refrigeration system manufactured by Brown in 1976, many researchers around the world have paid considerable attention to the magnetic refrigeration around room temperature and consecutively developed some interesting systems (Bouchekara, 2008) (Yu, 2010) (Bjørk, 2010). This section reviews - in detail - some of the magnetic refrigeration systems available until now.

3.1 The magnetic system of Brown

The system of Brown is a rotating system and employs an Ericsson cycle (Yu, 2003). The magnetic field is produced by an electromagnet (water cooled) with a maximum magnetic field of 7 T. The MCE material used is the Gd in the form of plates with 1 mm thickness, separated by stainless steel wires with 1 mm intervals to allow the regenerator fluid to flow vertically. The fluid is composed of 80% of water and 20% of alcohol. Without load and after 50 cycles, the temperatures reached were 46 ° C for the heat source and -1 ° C for the cold source, thus $\Delta T = 47°C$. However, the cooling power obtained was not rely important, this is due to the large ΔT obtained. Moreover, the cycle can operate only at low frequencies; the temperature gradient is reduced because both warm and cold sides have time to interact.

3.2 The magnetic system of Steyert

An alternative system with a rotating refrigerant, implementing a Brayton's cycle has been designed by Steyert (Yu, 2003). In this system, the porous magnetocaloric material has a form of rings. This wheel (the regenerator with a ring form) rotates through a first area of low magnetic field and a second area of high magnetic field as shown in Fig. 5. The exchange fluid enters the wheel (regenerator) at the temperature T_{hot} and exits at the temperature T_{cold}, having transferred its heat to the coolant located in the area of weak field. After receiving the heat of the load to cool Q_{cold} the fluid enters the wheel again at a temperature $T_{cold}+\Delta$ due to heat exchange with the wheel which is at this instant at the temperature $T_{hot}+\Delta$. The temperature of the fluid increases to $T_{hot}+\Delta$. Finally, the fluid transfers heat Q_{hot} to the reservoir of the hot source completing one cycle at the same time. Fig. 5 describes schematically the magnetic system of Steyert.

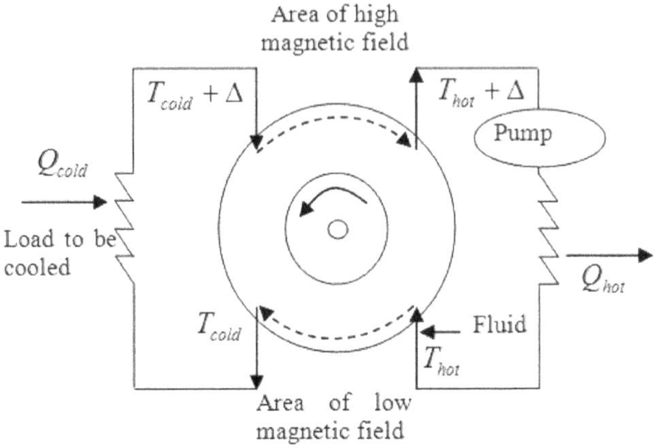

Fig. 5. Schematic representation of the Steyert's magnetic system.

3.3 The magnetic system of Kirol

This system was designed by Kirol (Yu, 2003) on the principle of a rotating machine and Ericsson's cycle. The magnetic field is produced by permanent magnets NdFeB and reaches a maximum value of 0.9 T in the air-gap. The refrigerant rotor is composed of a flat disk of 270 g of gadolinium as magnetocaloric material. During one rotation of the rotor, the four thermodynamic cycles are operated and a ΔT of 11 K is obtained.

3.4 The Spanish device

The device shown in Fig. 6 was developed by the team of the Polytechnic University of Catalonia in Barcelona (Allab, 2008). The magnetocaloric material is a ribbon of gadolinium (Gd 99.9%) fixed on a plastic disc and immersed in a fluid (olive oil). The magnetic cycle of magnetization / demagnetization is provided by the rotation of the plastic disc and its interaction with a magnet. The temperature span is obtained respectively: 1.6 and 5 K for a magnetic field of 0.3 T and 0.95 T. This corresponds to 2.5 times the MCE of Gd. Even if obtained performances of this system were weak, this device is the first that has shown the feasibility of magnetic refrigeration with fields accessible by permanent magnets.

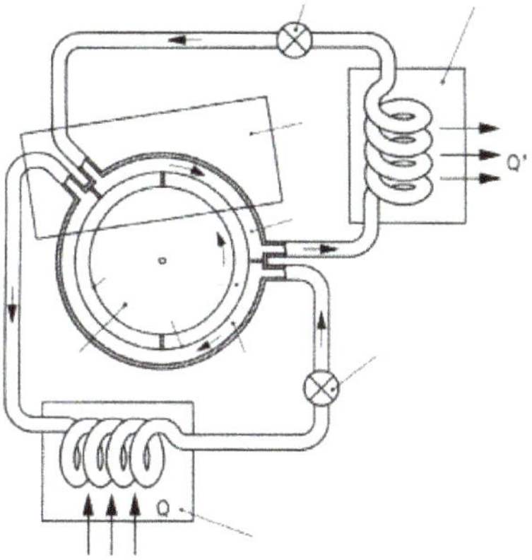

Fig. 6. The Magnetic device made in Spain (Bohigas, 2000).

3.5 Japanese system

Okamura et al. have constructed a magnetic refrigeration system, as shown in Fig. 7-a (Okamura, 2006). The yoke has an outer diameter of 27 cm and a length of 40 cm. The magnetic field is produced by rotating permanent magnets, producing a maximum field of 0.77 T. The bed regenerator is composed of 4 blocks. Each block is composed of a different alloy GdDy (sphere shaped) to enhance the range of variation of temperature. The fluid circulation is ensured by a pump and a rotary valve. The power obtained is about 60 W. The initial system has been improved as shown in Fig. 7-b (Okamura, 2007). The stator used was a laminated yoke and the magnetic field source was improved (the maximum field is 0.9 T). This helped to obtain a power of 100 W (using Gd as MCE material).

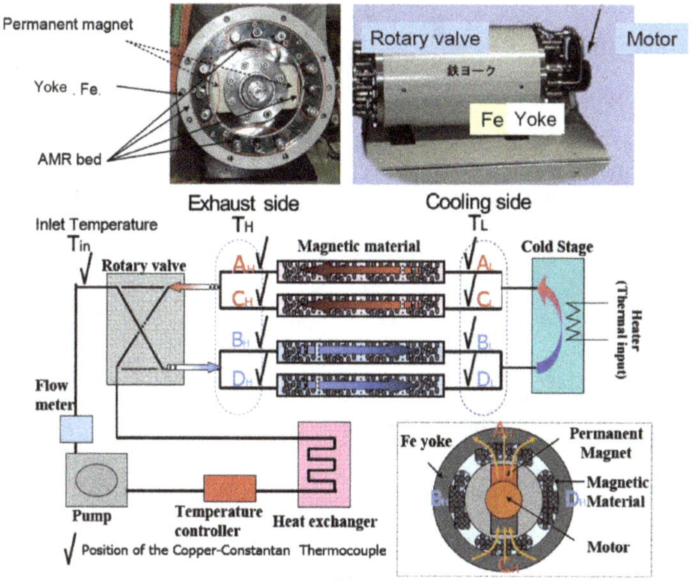

(a)

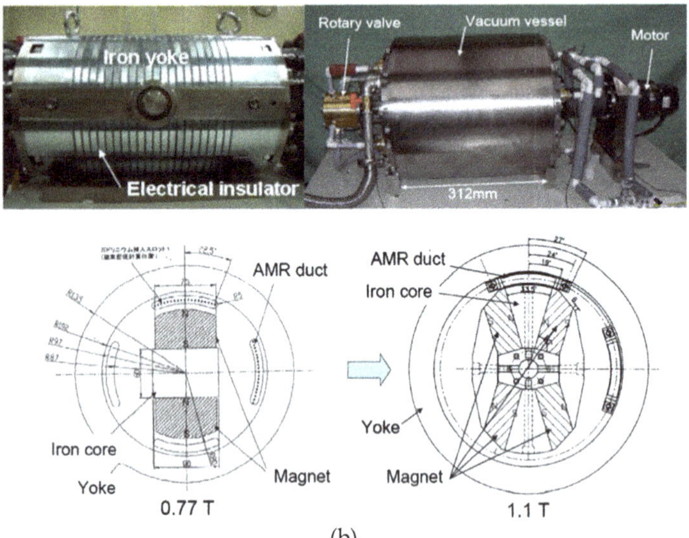

Fig. 7. The Japanese Device: (a) The initial one, (b) The improved one (Hirano et al., 2007; Okamura et al., 2006, 2007).

3.6 The magnetic system of Zimm

The ACM (Astronautics Corporation in Madison) has led many researches on magnetic refrigeration and achieved several patents in this field (Engelbrecht, 2005). In this corporation, an AMRR system was designed; it consists of a wheel with 6-bed regenerators composed of gadolinium powder. This wheel is rotating inside an area of high magnetic field of 1.5 T. The regenerative beds exchange with the fluid. The flow of the fluid is adjusted according to the relative position of each bed inside the magnetic field. Fig. 8 shows the photography of this prototype.

For cycles rotating with a frequency varying from 0.16 to 2 Hz and water rate flows varying from 0.4 to 0.8 l/min, the temperature spans obtained between the heat source and the cold source are from 4 to 20°C and the cooling power values are from 50 to 100W.

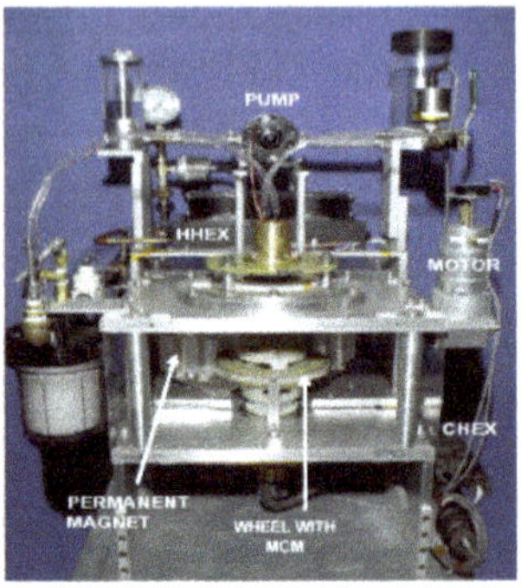

Fig. 8. ACM prototype (Zimm, 2002).

3.7 Canadian system

At the University of Victoria in Canada, Tura and Rowe (2007) constructed a magnetic refrigerator containing permanent magnets for a testing of all sorts of magnetic refrigerants in different configurations. This machine is shown in Fig. 9. A nested Halbach array of NbFeB permanent magnets was applied and led to a magnetic field of 0.1-1.47 T strength. Water was the heat transfer fluid with a heat rejection temperature range of 253-311 K, and the operation frequency was between 0 Hz and 4 Hz. The prototype showed cylindrical magnetocaloric regenerators (with a porosity of 57%) whose volume, diameter and length were 20 cm³, 16 mm, and 110 mm, respectively. The void in the regenerator of the hot heat exchanger and the cold heat exchanger was 0.83 cm³ and 0.4 cm³, and the parallel flow paths in the heat exchangers were optimized with a computational fluid dynamics (CFD) approach. The system which is designed to be flexible showed many advantages: for example, a simple design, easy accessibility to all the components and very low heat leakages. This machine reached a maximum temperature span of 13.2 K (Yu, 2010).

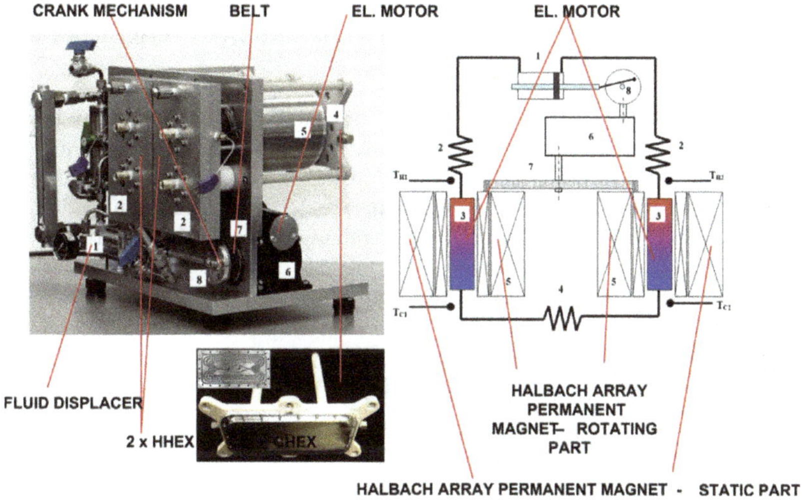

Fig. 9. Rotary magnetic refrigerator with permanent magnets as presented by researchers of the University of Victoria in Canada (Tura and Rowe, 2007).

3.8 Cooltech systems

The company Cooltech Applications in France built a rotary magnetic refrigerator composed of eight pieces of supporting discs positioned in synthetic material (see Fig. 10), which were mechanically stable and thermally isolated (Vasile and Muller, 2005, 2006).. These inserts were interchangeable for the test of different magnetocaloric materials, different sensors for temperature, pressure, air velocity, hydrometry and electrical power. Each insert was packed with 165 g Gd. The rotating axes were made of stainless steel, where four pieces of NdFeB permanent magnets were rotating to provide a magnetic field of 1 T. However, the authors reported on a new type (open Halbach) magnetic assembly, which yielded a magnetic field between 1 to 2.4 T. The flow of fluid was controlled to improve the cooling capacity, which was obtained in the range of 100W to 360 W (Yu, 2010).

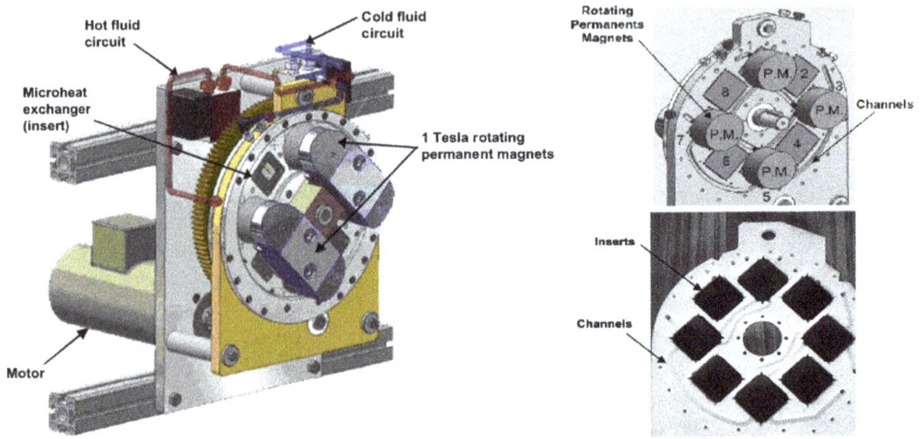

Fig. 10. 3D structure form of the Cooltech magnetic refrigerator, some details in further pictures, photography of the assembly and the open Halbach type of magnet (Vasile and Muller, 2005, 2006).

A rotary magnetic refrigerator prototype was developed in collaboration between the National Institute of Applied Sciences INSA of Strasbourg and the company Cooltech Applications in France (Muller et al., 2007). The system was composed by a rotary magnet assembly and of four static blocks of magnetocaloric material performed by gadolinium. The maximum magnetic field was 1.3 T and water was the working fluid. Unfortunately, there is no more information available. However, from Fig. 11, one may easily verify the manner how the magnetic field is produced in the magnetocaloric material (Yu, 2010).

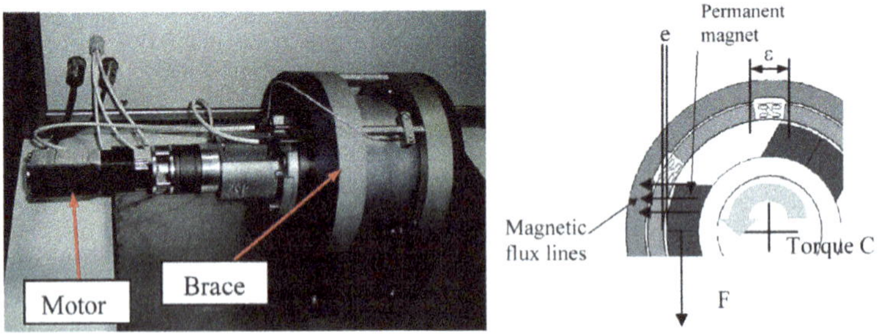

Fig. 11. A rotary magnetic refrigerator of the joint collaboration action between Cooltech Applications and INSA in France (Muller et al., 2007).

In France at Cooltech Applications, Bour et al. built a reciprocating prototype as it is shown in Fig. 12. The AMR bed was composed of 37 parallel plates of Gadolinium of 0.6 mm thickness, showing a spacing of the heat transfer fluid channels of 0.1 mm and 0.2 mm, respectively. The Halbach arrays, which produced a magnetic field intensity between 0.8 T and 1.1 T in the air gap, consisted of an assembly of three sets of NdFeB magnets of 50 mm thickness. The French experts obtained experimentally the evolutions of the average temperatures at the hot end and the cold end reservoirs for different initial temperatures and operation frequencies. The device led to a maximum temperature span of 16.1 K (Yu, 2010).

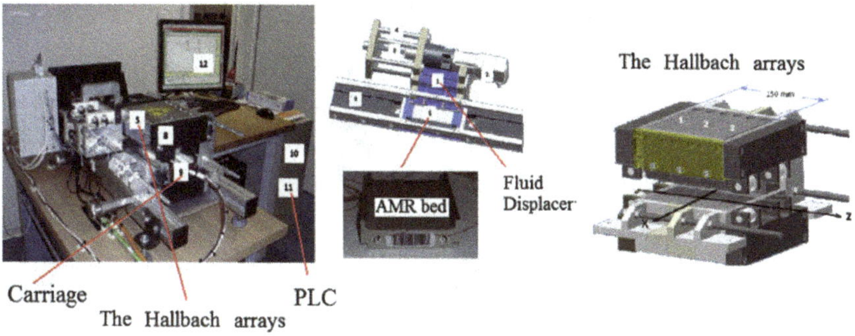

Fig. 12. The reciprocating prototype built at Cooltech Application in France (Bour et al., 2009).

3.9 The G2Elab prototypes

The first device constructed at G2Elab (Grenoble Electrical Engineering Laboratory) is an alternating device type as shown in Fig. 13. The regenerator is composed of parallel plates of gadolinium with 1 mm in thickness and 50 mm in length. The magnetic field is produced by a permanent magnet (Halbach cylinder) creating a magnetic field of 0.8 T. The fluid used is water. Its circulation is ensured by a peristaltic pump operating in both directions (Clot, 2002). The pneumatic actuator produces the movement of the refrigerant blocs and provides magnetization / demagnetization phases. The controller is programmed to manage the Halbach cylinder and the flow of fluid to perform the four phases of the cycle. The system is closed and there is no exchange with the outside. It was designed to study the Active Magnetic Regenerative Refrigeration (AMRR) cycles and exploit different materials.

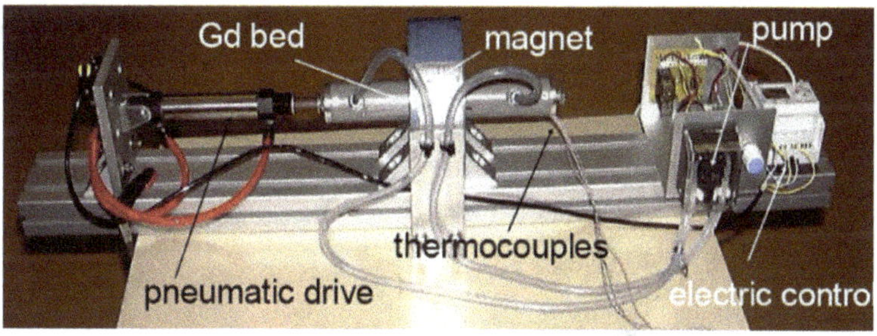

Fig. 13. The G2Elab first device (Clot, 2002).

A second prototype was developed at G2Elab (Allab, 2008), (Bouchekara, 2008), (Dupuis, 2009). This structure is quite similar to a rotating machine. It is also similar to some existing prototypes (Okamura, 2005, 2007). It consists of a permanent magnet which forms the rotor and of a stator made of magnetic yoke and four refrigerant beds (see Fig. 14).

The yoke is composed of four poles which are aimed to better conduct the magnetic flux within the refrigerant bed. The magnetization and demagnetization phases are obtained by a simple rotation of the permanent magnet. The beds undergo an active magnetic regenerative refrigeration AMRR cycle and operate two by two in the opposite way.

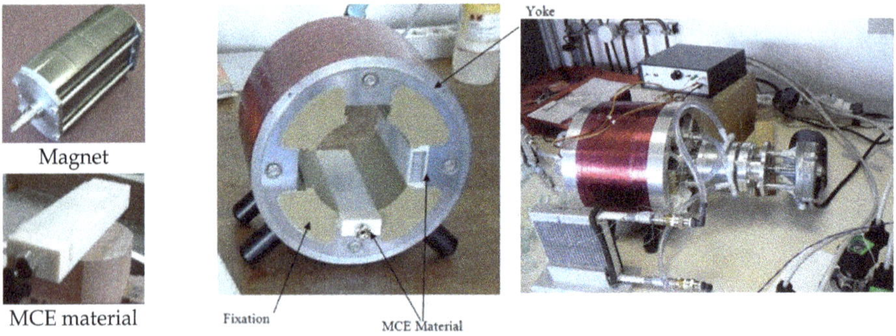

Fig. 14. The second prototype of the G2Elab, Components of the prototype (left) and the Prototype in its actual environment (right) (Bouchekara, 2008).

3.10 The Slovenian system

An interesting prototype of a rotary magnetic refrigerator (Tusek et al., 2009) has been built on the basis of permanent magnets at the University of Ljubljana in Slovenia. Their rotary magnetic refrigerator consisted of a rotating drum (cylinder) that rotated around an internally positioned stationary soft iron core and externally positioned stationary permanent magnets. As shown in Fig. 15, the magnetic structure was composed of four NdFeB permanent magnets and low carbon 1010 steel used as a soft ferromagnetic material, and two magnetic circuits existed to allow the rotary movement of the AMR's. After optimization of the magnet structure geometry, a range of magnetic field intensities from 0.05 T to 0.98 T was obtained in the air gaps. There were 34 AMR's in the rotary drum and each AMR had the dimensions 10 mm × 10 mm × 50 mm. Gd plates, with a thickness of 0.3 mm, were filled in the AMR's and the total mass of Gd was approximately 600 g. The prototype could operate up to a frequency of 4 Hz. This reference mainly focused on the experience in development of such a rotary magnetic refrigeration prototype and no experimental results were reported. However, first predictions according to the researchers are that approximately a 7 K temperature difference will be achieved (Yu, 2010).

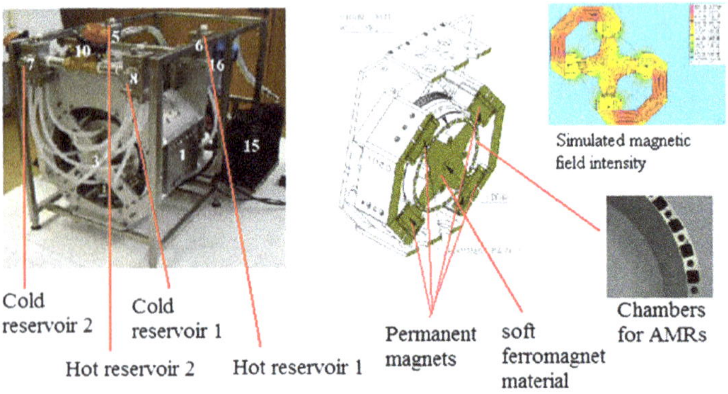

Fig. 15. The rotary magnetic refrigerator developed at the University of Ljubljana in Slovenia (Tusek et al., 2009).

4. Thermal study

Heat exchanges play an important role in magnetic refrigeration systems, both in the cold production cycles, and in the interaction with external environments, including the substance to be cooled. Thus, a thermal study is needed to determine the performance of a magnetic refrigeration system and optimize it. The aim of this section is to focus on the thermal modeling of magnetic AMRR systems.

Most of heat exchanges operating in the magnetic refrigeration are via convection. The convection represents transfer processes performed by the motion of fluids (Bianchi, 2004). In a solid (index 's') in contact, with a fluid (index 'f'), the flow through the wall (index 'w') can be written as:

$$\lambda_s\left(\frac{\partial T}{\partial n}\right)_{ws} = \lambda_f\left(\frac{\partial T}{\partial n}\right)_f = \varphi_p \tag{19}$$

where : n is the normal to the wall and $\lambda\left[W/(m\ K)\right]$ is the thermal conductivity

whereas, the continuity of temperatures can be given by:

$$\left(T_s\right)_{wM} = \left(T_f\right)_{wM} \tag{20}$$

where : $\left(T_s\right)_{wM}$ is the temperature of the solid at a point 'M' of the wall and $\left(T_f\right)_{wM}$ represents the temperature of the fluid at this point.

According to Newton, there is a linear relationship between the density of heat flow φ and temperature difference $\Delta T = T_s - T_f$ between the solid (T_s) and the fluid (T_f):

$$\varphi = h\Delta T = h\left(T_s - T_f\right) \tag{21}$$

where: $h\left[W/Km^2\right]$ represents the coefficient of heat transfer by convection or simply the convection coefficient.

Using the first law of thermodynamics, by subtracting the mechanical energy, we get the balance of internal energy that gives us the heat equation governing the temperature field at any point in the domain (Janna, 2000)

$$\rho C_p\left(\frac{\partial T}{\partial t} + \text{V.grad}T\right) = \beta T\left(\frac{\partial p}{\partial T} + \text{V.grad}T\right) + P + \Phi + \lambda\ div\left(\text{grad}T\right) \tag{22}$$

where: $\rho\left[kg/m^3\right]$ is he volume density, $C_p\ \left[J/(kg\ K)\right]$ is the specific heat, $\text{V}\ \left[m/s\right]$ is the velocity of the fluid, $p\ \left[Pa\right]$ is the pressure, $\beta\ \left[1/K\right]$ is the coefficient of dilatation, $\Phi\ \left[W\right]$ is the dissipation function and $P\ \left[W\right]$ is the local thermal power produced or absorbed.

For low viscosity fluids and isochors (Janna, 2000), the energy equation reduces to:

$$\rho.C_p\left(\frac{\partial T}{\partial t} + \text{V.grad}T\right) = P + \lambda\ div\left(\text{grad}T\right) \tag{23}$$

if $a = \lambda / \rho C_p$ is defined as the thermal diffusivity of the fluid, thus :

$$\frac{\partial T}{\partial t} + \text{V.grad}T = \frac{P}{\rho C_p} + a.\Delta T \tag{24}$$

4.1 Application to AMRR

Governing equations for AMRR system have been developed throughout the years with the objective of to analytically or numerically describe the thermal behaviour at specific time and for a given set of boundary conditions. They consist of a system of two equations, one for the fluid and the other for the solid matrix. These equations are derived from the energy balance expression for each phase. Since they are coupled they must be solved simultaneously (Bouchekara, 2008). The model of an AMRR cycle has been developed in (Bouchekara, 2008). Fig. 16 illustrates the concept of an AMRR regenerator modelled using one dimensional (1D) approximation.

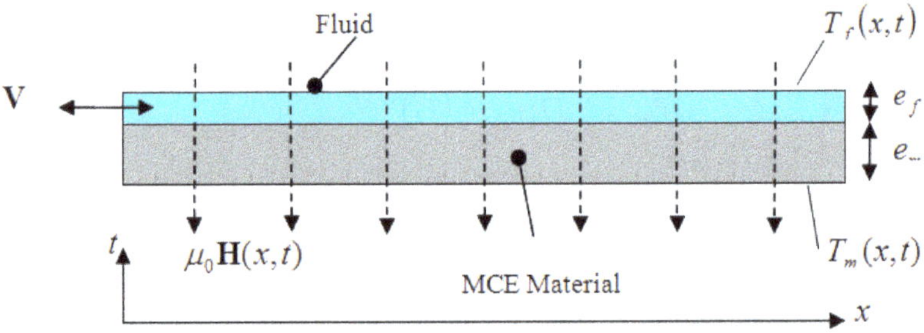

Fig. 16. Conceptual drawing of a 1D AMRR model.

The system of equations given by the energy balance (explained above) for both the magnetocaloric material and the fluid by neglecting the axial conduction (this approximation can be justified for different conditions: low thermal conductivity, very thin plates, etc.) can be summarized by the following system of equations:

$$\begin{cases} m_f C_f(T_f)\left(\frac{\partial T_f}{\partial t} + \dot{d}(t)\frac{\partial T_f}{\partial x}\right) = hS(T_m - T_f) \\ m_m C_m \frac{\partial T_m}{\partial t} = hS(T_f - T_m) \end{cases} \tag{25}$$

To solve this system we use the finite difference method. We use a grid of elements that range from 0 to L for the space and from 0 to τ for the time. Thus, the derivatives with respect to the time are calculated using forward formulas, and those with respect to the space are calculated using backward formulas. This gives a centered discretization scheme. Thus the system (25) becomes:

$$\begin{cases} T_{f_{(i+1,j)}} = A_{f1}T_{f_{(i,j)}} + A_{f2}T_{f_{(i,j-1)}} + A_{f3}T_{m_{(i,j)}} \\ T_{m_{(i+1,j)}} = A_{m1}T_{m_{(i,j)}} + A_{m2}T_{f_{(i,j)}} \end{cases} \tag{26}$$

where: $A_{f1} = \left(1 - \left(\dot{d}(t)\frac{\Delta t}{\Delta x} + \frac{hS}{m_f C_f}\Delta t\right)\right)$, $A_{f2} = \left(\dot{d}(t)\frac{\Delta t}{\Delta x}\right)$, $A_{f3} = \left(\frac{h.S}{m_f C_f}\Delta t\right)$,

$A_{m1} = \left(1 - \frac{h.S}{m_m C_m}\Delta t\right)$ and $A_{m2} = \left(\frac{h.S}{m_m.C_m}\Delta t\right)$

The AMRR model has been implemented using Matlab ® commercial software.

4.2 Results

We will now apply the model developed earlier to a regenerator in the form of plates, as shown in Fig. 17(a). The equivalent cell of the whole regenerator is given in Fig. 17 (b). This cell has the same parameters as the regenerator except the width that is $l_{eq} = N_p l$ (where l_{eq} represents the equivalent width, N_p represents the number of plates and l is the width of one plate).

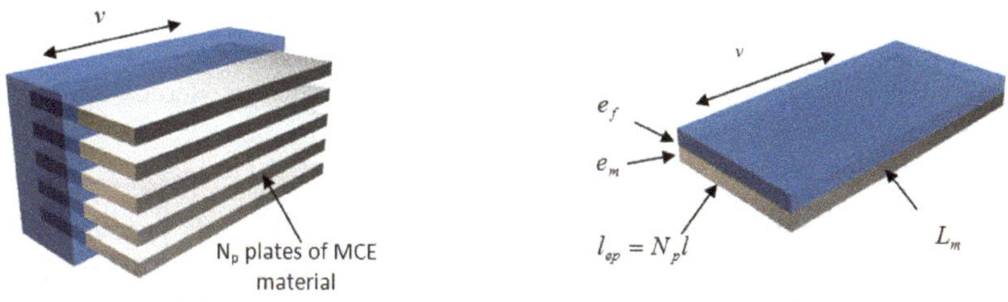

Fig. 17. (a) A regenerator in the form of plates, (b) Equivalent cell (plate + fluid).

The model parameters (for this simulation case) are shown in Table 2. The magnetocaloric material used is gadolinium, the coolant used is water and the magnetic field is generated by permanent magnets B = 1 T.

Parameters	D(t) [ml/s]	e_m [mm]	e_f [mm]	L_m [mm]	l_{eq} [mm]	ρ [kg/m³]	C_p [J/(kg K)]	MCE [K]
Values	5	1	0.157	50	573	1000	4185	1.75

Table 2. The parameters used in the simulation.

Fig. 18 (a) shows the temperature evolution of both sides (hot and cold) of the material versus time. After a transient phase, the two curves reach their steady state. In addition, we note that the final value is greater than the initial MCE. From this curve we can extract the evolution of the temperature at the end of each cycle (Fig. 18 (b)). The small delay between the two curves of this figure is due to programming constraints, i.e. the magnetization phase has been introduced (programmed) before the demagnetization phase.

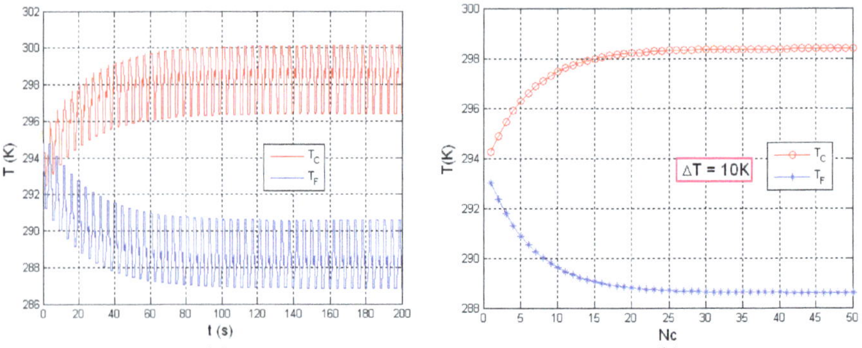

(a) (b)

Fig. 18. Temperature profiles given by the AMRR numerical model.

5. Magnetostatic study

From the definition of the MCE, it is obvious that the performance of a magnetic refrigeration system depends mainly on the efficiency of the MCE material and the strength of the magnetic field. Thus the study of magnetic field sources dedicated to magnetic refrigeration systems is of a paramount importance. In this section we will pay a particular attention to the design of these sources using the finite element method.

The field sources described throughout this section are built with permanent magnets (Neodymium Iron Boron) with a remanent magnetization of $B_r = 1.46$ T and a magnetic permeability of $\mu_r = 1.064$. The MCE material used is gadolinium with isotropic magnetic permeability $\mu_r = 2$. In this study, MCE solid blocks are considered. However, in actuality to ensure a better heat exchange between the MCE material and the exchanging fluid, other forms are considered (plates, powder, etc). The yoke, when it exists, is made of XC10 steel.

5.1 Structure A: monobloc linear system

This first structure is suitable for linear magnetic refrigeration systems with direct cycles. It consists of two magnets (to create the magnetic field), a soft magnetic material yoke (to canalize the magnetic flux) and a block of MCE material (to create the cold) as shown in Fig. 19 (a). The MCE material has a linear alternating movement along the 'y' axis (to achieve magnetization and demagnetization phases). The magnetic characteristics (induction and magnetic force profiles) are shown in Fig. 19 (b) and Fig. 19 (c) while Fig. 19 (d) represents the distribution of the magnetic induction B in Tesla.

5.2 Structure B: Halbach cylider

The structure B is a Halbach cylinder (Fig. 20). It is a magnetized cylinder composed of 'N' segments of ferromagnetic material producing (in the idealized case) an intense magnetic field confined entirely within the cylinder with zero field outside. This second structure can be used in an AMRR system. The MCE material, which can be in the form of plates stacked in a cylinder, is guided by a linear motor or actuator to create the phases of magnetization (material located inside the cylinder) and the demagnetization phase (material located

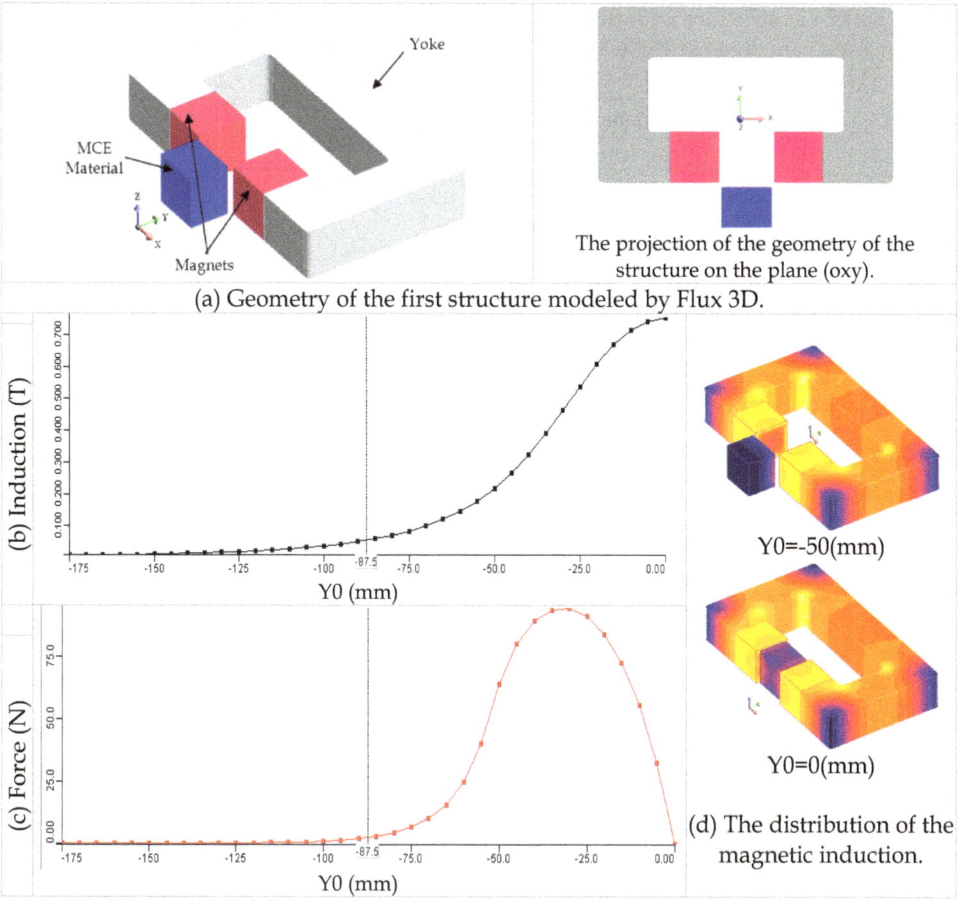

Fig. 19. Geometry and magnetic characteristics of the Structure A.

outside the cylinder). The opposite way works also; i.e. the material remains fixed and the magnet is connected to the linear motor. In this case, the magnetic behavior is the same as in the first case.

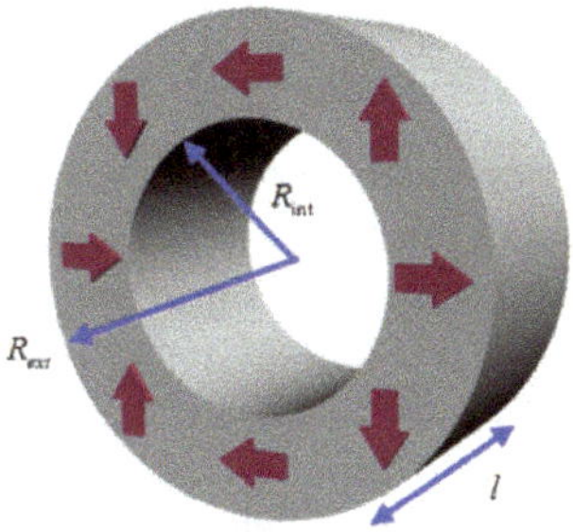

Fig. 20. Cylindre d'Halbach made of eight segments.

We modeled this structure with the commercial software 'Flux 3D'. The parameters used in this simulation are given in Table 3.

Parameters	R_{ext}	R_{int}	R_m	l
Values [mm]	65	25	22	50

Table 3. The dimensions of the Halbach cylinder used in the simulation.

Fig. 21 (a) shows the induction at the center of the material according to the movement. It shows clearly the phases of magnetization and demagnetization produced by this structure. Fig. 21 (b) represents the magnetic forces exerted by the cylinder on the block of MCE material. The distribution of the magnetic induction is shown in Fig. 21 (c).

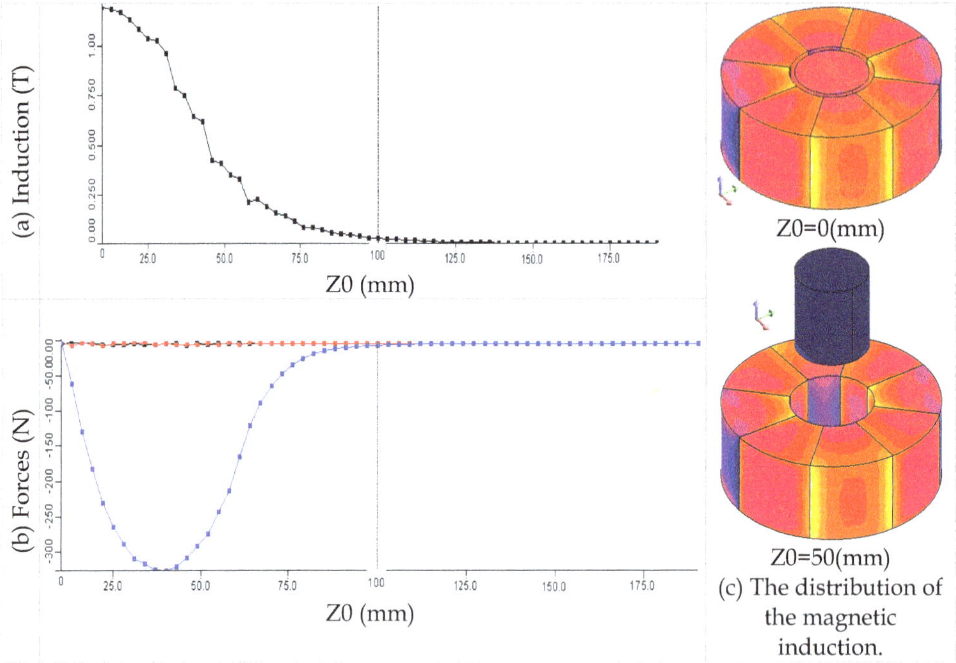

Fig. 21. Magnetic characteristics of the Structure B.

5.2.1 Structure C: double Halbach cylinder

Structure C is a double Halbach cylinder; two cylinders are concentric and have the same number of segments (Fig. 22). Using this structure for magnetic refrigeration systems allows having the phases of magnetization and demagnetization simply by rotating one of the cylinders while the active material remains stationary at the center of the structure. The magnetic field produced at the center is the sum of the two fields produced by each cylinder. When the magnetizations of the first cylinder segments are in the same direction of those of the second cylinder, the magnetic field produced is high and the magnetization phase is achieved (this position is taken as a reference, i.e. $\theta = 0°$). However, when the magnetizations of the first cylinder are in opposition with those of the second cylinder the magnetic field produced is low ($\theta = 180°$) and the demagnetization phase is achieved.

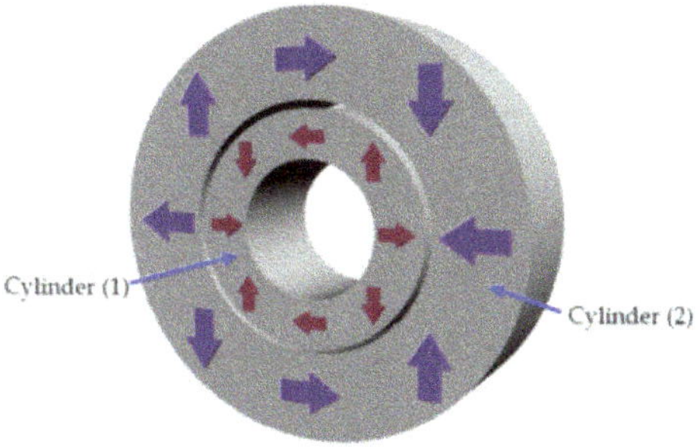

Fig. 22. Double Halbach cylinder (The two cylinders are opposed in this figure).

5.2.2 Structure D: rotating multiblock system

The structure D presented here has a configuration adapted to rotary magnetic refrigeration systems with a direct thermal cycle. It has two magnets to create the field; a yoke of soft material to canalize the magnetic flux and N MCE material blocks for the creation of the cold (Fig. 23). Table 4 below shows the parameter values used in our simulation for the rotating multiblock system.

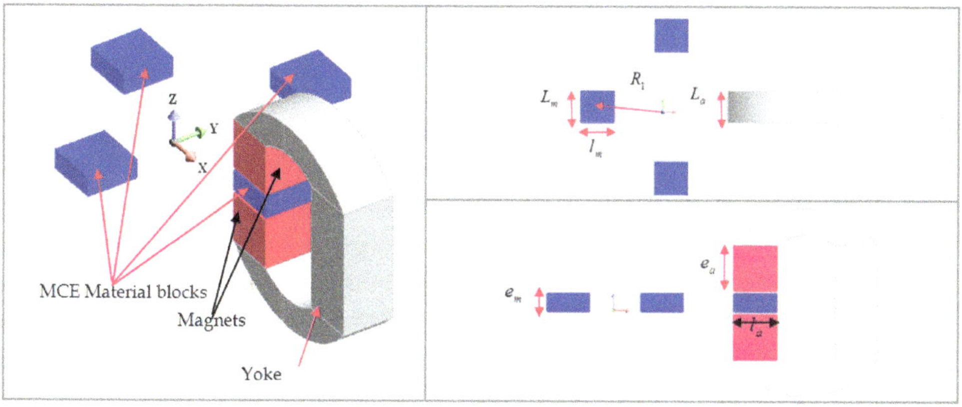

Fig. 23. Geometry of the rotating multiblock system.

Parameters	R_1	l_m	L_m	e_m	l_a	L_a	e_a	e
Values [mm]	100	50	50	20	50	50	20	3

Table 4. The dimensions used in the simulation of the rotating multiblock system.

Fig. 24 (b) shows the induction at the center of a block of material b_1. Fig. 24 (c) and Fig. 24 (d) represent the torque exerted on the blocks and the force exerted on the block b_1, respectively. Fig. 24 (e) represents the distribution of the magnetic induction for two positions $\theta = 0°$ and $\theta = 45°$.

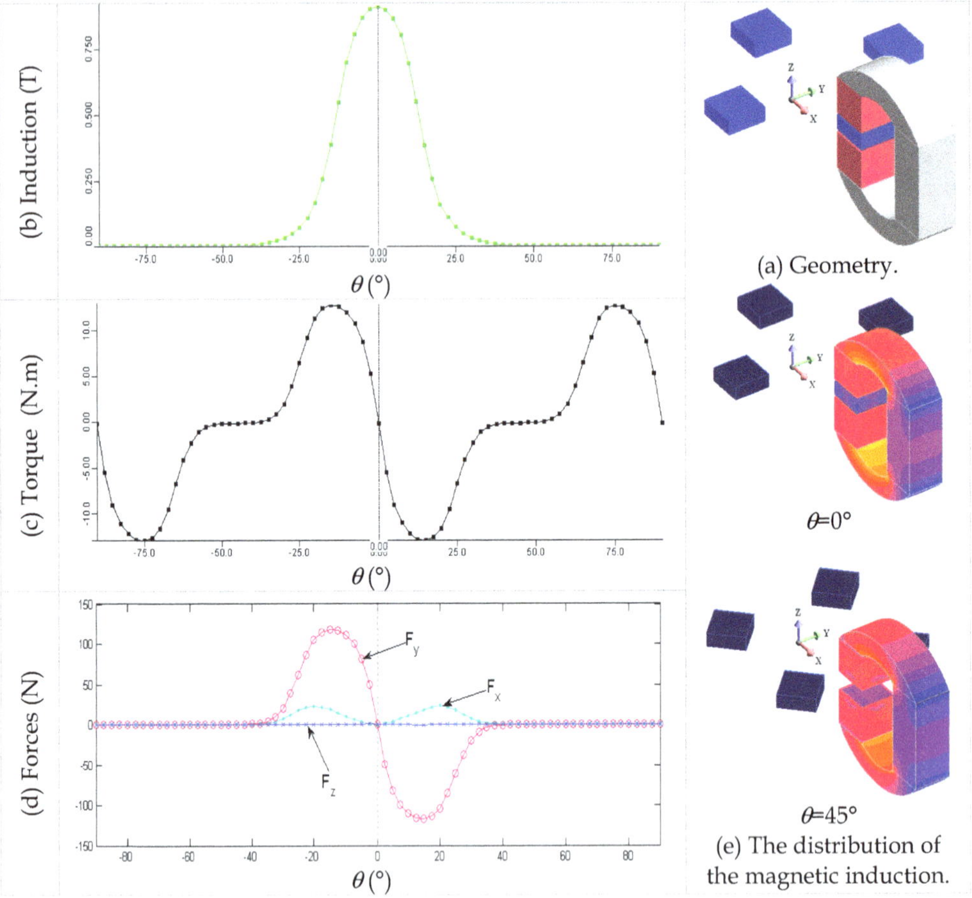

Fig. 24. Magnetic characteristics of the rotating multiblock system.

6. Conclusion

The conventional gas compression refrigerators have been mainly used for refrigeration applications. Generally, such refrigerators are not power-efficient. In addition, gases used in these refrigerators causes harmful effects on the environments. This has led to the development of magnetic refrigeration technology. Over the last decade or so, magnetic refrigeration at room temperature has become the subject of considerable attention. This technology is based on the use of magnetocaloric effect: that is the response of a solid to an applied magnetic field which emerges as a change in its temperature. This technology is ultimately aimed at developing a standard refrigerator for home use. Unlike conventional refrigerators, magnetic refrigerators are cost-effective and environmentally friendly. Moreover, it is noiseless and power-efficient and requires low maintenance cost and atmospheric pressure besides that its machines are easy to design.

This chapter introduced the magnetic refrigeration technology. It began by providing some essential definitions and provided a detailed review of ten magnetic refrigeration prototypes which are available until now. The operational principle of this technology was explained in depth through the comparison with the conventional one. The chapter then moved on to investigate the study of the magnetocaloric material using the Molecular Field Theory. The thermal study of the magnetic refrigeration process using the finite difference method (FDM) was then explained with providing some useful simulation results. Finally, the magnetic study of magnetic refrigerators using the finite element method (EFM) was presented with some practical results.

7. References

Allab, F. (2008). Conception et réalisation d'un dispositif de réfrigération magnétique basé sur l'effet magnétocalorique et dédié a la climatisation automobile, Thèse de doctorat, Grenoble, *Institut National Polytechnique de Grenoble.*

Bianchi, A.M., Fautrelle, Y., Etay, J. (2004). Transferts thermiques, première édition, *Presses polytechnique et universitaires romandes*, ISBN 2-88074-496-2, Lausanne.

Bjork, R., Bahl, C.R.H., Smith, A., Pryds, N. (2010). Review and comparison of magnet designs for magnetic refrigeration. *International. Journal of Refrigeration.* 33, 437-448.

Bohigas, X., Molins, E., Roig, A., Tejada, J., Zhang, X.X. (2000). Room-temperature magnetic refrigerator using permanent magnets. *IEEE Transactions on Magnetics*, 36 (3), 538-544.

Bouchekara H. (2008). Recherche sur les systèmes de réfrigération magnétique. Modélisation numérique, conception et optimisation, Thèse de doctorat, *Grenoble Institut National Polytechnique.*

Clot, P., Viallet, D., Kedous-Lebouc, A., Fournier, JM., Yonnet, JP., Allab, F. (2003). A magnetic device for active magnetic regeneration refrigeration", *IEEE Transactions on Magnetics*, vol. 39, n° 5, pp. 3349-3351.

Dupuis C. (2009). Matériaux à effet magnétocalorique géant et systèmes de réfrigération magnétique, Thèse de doctorat, Grenoble, *Institut National Polytechnique de Grenoble.*

Engelbrecht, K.L., Nellis, G.F., Klein, S.A., Boeder, A.M. (2005). Modeling Active Magnetic Regenerative Refrigeration Systems, *International Conference on Magnetic Refrigeration at Room Temperature*, Montreux, Switzerland.

Gschneidner, K., Pecharsky, V. (1998). The Giant Magnetocaloric Effect in Gd5 (SixGe1-x4) Materials for Magnetic Refrigeration" *Advances in Cryogenic Engineering*, Plenum Press, New York, pp. 1729.

Huang, W.N., Teng, C.C. (2004). A simple magnetic refrigerator evaluation model. *Journal of Magnetism and Magnetic Materials 282*, pp 311-316.

Janna, W.S. (2000).Engineering Heat Transfer'', ISBN 0-8493-2126-3, CRC Press USA.

Kitanovski, A.P., Egolf, W., Gender, F., Sari, O., Besson, CH. (2005). A Rotary Heat Exchanger Magnetic Refrigerator, *International Conference on Magnetic Refrigeration at Room Temperature*, Montreux, Switzerland.

Lebouc, A., Allab, F., Fournier, J.M., Yonnet, J.P. (2005). Réfrigération magnétique, *[RE 28], Techniques de l'Ingénieurs.*

Muller, C., Bour, L., Vasile, C., (2007). Study of the efficiency of a magnetothermal system according to the permeability of the magnetocaloric material around its Curie temperature. *Proceedings of the Second International Conference on Magnetic Refrigeration at Room Temperature,* Portoroz, Slovenia, pp. 323-330.

Okamura T., Yamada, K., Hirano, N. Nagaya S. (2006). Performance of a roomtemperature rotary magnetic refrigerator'' *International Journal of Refrigeration* 29, 1327-1331.

Okamura T., Yamada, K., Hirano, N. Nagaya S. (2007). Improvement of 100W class room temperature magnetic refrigerator. *Proceedings of the Second International Conference on Magnetic Refrigeration at Room Temperature,* Portoroz, Slovenia, 11-13 April. International Institute of Refrigeration, Paris.

Phan, M.H., Yu, S.C., 2007. Review of the magnetocaloric effect in manganite materials. Journal of Magnetism. and Magnetic. Materials. 308, 325-340.

Tishin, A.M. (1999), Hand Book of Magnetic Material, Vol.12, Ed. Buschow K.H.J., North Holland, Amsterdam.

Tura, A., Rowe, A. (2007). Design and testing of a permanent magnet magnetic refrigerator. *Proceedings of the Second International Conference on Magnetic Refrigeration at Room Temperature,* Portoroz, Slovenia, 11-13 April. International Institute of Refrigeration, Paris, pp. 363-371.

Vasile, C., Muller, C. (2005). A new system for a magnetocaloric refrigerator., *Proceedings of First International Conference on Magnetic Refrigeration at Room Temperature,* Montreux, Switzerland, 27-30 September, International Institute of Refrigeration, Paris, pp. 357-366.

Vasile, C., Muller, C., (2006). Innovative design of a magnetocaloric system. *International Journal of Refrigeration* 29 (8), 1318-1326.

Yu, B., Liu, M., Egolf, P.W., Kitanovski, A., 2010. A review of magnetic refrigerator and heat pump prototypes built before the year 2010. International. Journal of Refrigeration. 33, 1029-1060.

Yu, B.F., Gao, Q., Zhang, B., Meng, X.Z., Chen, Z., 2003. Review on research of room temperature magnetic refrigeration. International. Journal of Refrigeration. 26, 622-636.

Zimm, C., Auringer, J., Boeder, A., Chells, J., Russek, S., Sternberg, A., 2007. Design and initial performance of a magnetic refrigerator with a rotating permanent magnet. *Proceedings of the Second International Conference on Magnetic Refrigeration at Room Temperature,* Portoroz, Slovenia, pp. 341-347.

12

Theory and Applications of Metamaterial Covers

Mehdi Veysi, Amir Jafargholi and Manouchehr Kamyab
E.E. Dept. K. N. Toosi University of Technology, Tehran, Iran

1. Introduction

Metamaterial covers exhibit inimitable electromagnetic properties which make them popular in antenna engineering. Two important features of metamaterial covers are: (1) increasing of the transmission rate and (2) control of the direction of the transmission which enable one to design directive antennas. In this chapter, the possibility of increasing both bandwidth and directivity of the printed patch antenna using metamaterial covers is examined. The printed patch antennas are a class of low-profile antennas, which are conformable to planar surfaces, simple and inexpensive to manufacture using printed-circuit technology.

Furthermore, novel polarization-dependent metamaterial (PDMTM) covers, whose transmission phases for two principal polarizations are different, are presented (Veysi et al., 2011). A full-wave Finite Difference Time Domain (FDTD) numerical technique is adopted for the simulations. A schematic of the metamaterial cover with square holes is shown in Fig. 1. It consists of two planar layers with similar square lattices. It was demonstrated in (Pendry et al., 1996; Tsao & Chern, 2006) that in the frequency range, where the wavelength is very large compared to the period of the metamaterial cover, this structure acts as a homogenous medium. The equivalent refractive index of this medium, in the microwave domain, is given by:

$$n_{eff} = \sqrt{1 - \left({f_P}/{f} \right)^2} \tag{1}$$

where f_P denotes the plasma frequency and f denotes the operating frequency. If the operating frequency is selected slightly larger than the natural plasma frequency of the metamaterial cover, the equivalent refractive index will be extremely low. Consequently, the transmission phase at the plasma frequency is extremely low.

The ultra refraction phenomena, in which the transmitted rays are parallel to each other, can be expected where the transmission coefficient reaches its maximum value. In other words, the zero transmission phase occurs at the same frequencies where the magnitude of the transmission coefficient becomes maximum. Hence, it acts similar to an equally phase surface at its plasma frequency. It is evident from Eq.1, that the equivalent refractive index and thus the antenna directivity are very sensitive to the frequency.

As a starting point, we consider a two-layer metallic grid placed on top of the patch antenna backed by a ground plane. The simulations have been carried out to examine the transmission characteristic of the metamaterial cover, without the ground plane and without the antenna, using FDTD code developed by the authors.

Fig. 2 shows an effective unit cell model of the metamaterial cover which takes into account the image effect of the ground plane. This unit cell is a convenient method of computing of the transmission coefficient of a two layer metamaterial cover placed on top of the patch antenna backed by a ground plane. Here, Perfect Match Layers (PMLs) are applied to realize a medium with no reflection. The normalization in the code consists of choosing the peak magnitude of the transmission coefficient to be unity. Therefore, the magnitude of the transmitted field from the metamaterial cover has been normalized to that without the metamaterial cover. We have used the same methodology applied in the measurements (Enoch et al., 2002). The periodic boundary conditions (PBCs) have been also applied to model an infinite periodic replication. Since an infinite periodic structure has been simulated, the peak magnitude of the transmission coefficient is unity, unlike the results obtained in the measurements (Enoch et al., 2002).

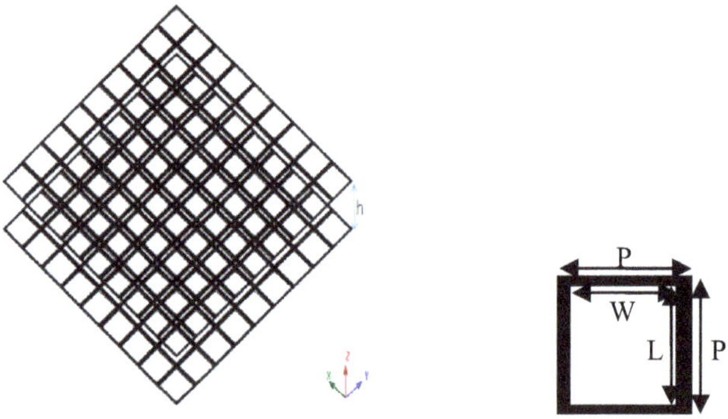

Fig. 1. Schematic view of two layer metamaterial cover together with its unit cell (Veysi et al., 2011).

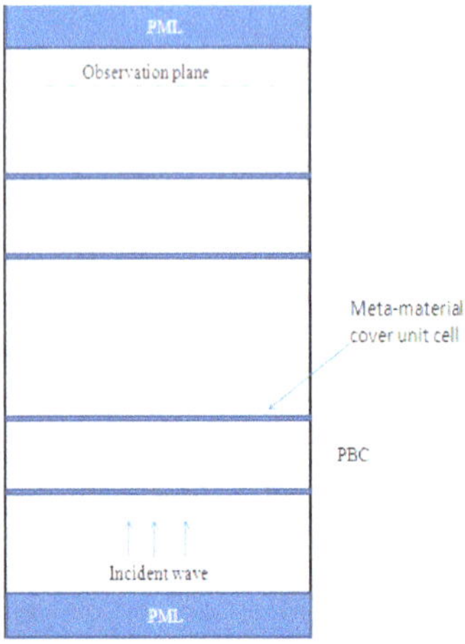

Fig. 2. FDTD model for metamaterial cover analysis (Veysi et al., 2011).

2. Directivity and bandwidth enhancement of proximity-coupled microstrip antenna

Directive patch antennas are very popular in electromagnetic community. Their attractive features, such as low profile, light weight, low cost and compatibility with Microwave Monolithic Integrated Circuits (MMICs), do not exist in other antennas.

Two distinctive types of directive antennas are parabolic antennas and large array antennas. Bulk and curved surface of parabolic antennas limits their use in many commercial applications. Also, complex feeding mechanism and loss in the feeding network are two major disadvantages associated with microstrip array antennas.

One solution to these problems is to use metamaterial cover over the patch antenna (Alu et al., 2006; Xu et al., 2008; Zhu et al., 2005; Huang et al., 2009). One of the first works was done by B. Temelkuaran in 2000, (Temelkuaran et al., 2000). In 2002, S. Enoch proposed a kind of metamaterial for directive emission, (Enoch et al., 2002). Another problem associated with microstrip antennas is their narrow bandwidth. The previous works so far (Xu et al., 2008; Zhu et al., 2005; Huang et al., 2009) have dealt only with the enhancement of the antenna directivity using metamaterial cover, but the effect of this cover on the antenna input impedance has not been investigated.

Recently, a new metamaterial cover has been proposed to enhance both the antenna bandwidth and directivity, (Ju et al., 2009). But, its directivity is significantly lower compared to the primary metamaterial cover, (Xu et al., 2008; Zhu et al., 2005; Huang et al., 2009).

In this section, it is demonstrated that both the impedance and directivity bandwidths of the proximity-coupled patch antenna can be enhanced using the metamaterial cover. It is known that proximity-coupled patch antennas are sensitive to the transverse feed point location. In the case at hand, a parasitic microstrip line has been used on the opposite side of the feed line to mitigate this drawback (Jafargholi et al., 2011). The dimensions of the analyzed metamaterial cover are:

$$P=0.41\lambda_{6GHz},\ t = 0.01\ \lambda_{6GHz},\ L= 0.31\lambda_{6GHz},\ h=0.49\lambda_{6GHz} \tag{2}$$

Where λ_{6GHz} (50mm) denotes the free space wavelength at 6GHz, P is the periodicity, t is the thickness of the metallic grids, L is the edge of the square holes and h is the distance between the two sheets which is the same as the distance between the patch antenna and the first sheet.

In the FDTD simulations, a uniform $0.01\lambda_{6GHz}$ grid size is used. The resulting transmission curve is plotted in Fig.3. As can be seen, this structure has three microwave plasma frequencies at about 5GHz, 5.81GHz and 8.1GHz which make it suitable for the antenna applications. When the aforementioned metamaterial cover is placed over the conventional proximity-coupled patch antenna, the final metamaterial antenna can be approximated by a homogenous medium terminated in a ground plane.

This approximation is similar to that used for the transmission coefficient calculations. It is a simple matter to obtain the surface impedance of this grounded slab as a function of metamaterial parameters. A surface impedance of the grounded slab of thickness h is:

$$Z_s = j\eta \tan(2\pi h / \lambda) \tag{3}$$

where η and λ are the wave impedance and wavelength in the slab, respectively. For the extremely low values of ε_{eff}, the surface impedance is inductive. In addition, the inductive reactance is $X_l = j\omega L$.

For equivalence we can equate them, leading to the following equation:

$$j\omega L = j\eta \tan(2\pi h / \lambda) \tag{4}$$

Since $\varepsilon_{eff} << 1$ we can apply the small-angle approximation, so that above equation then becomes $L = \mu_0 h$. Consequently, the operation mechanism of this metamaterial based cover can be explained using this equivalent inductance.

In addition, coupling between the feed line and the patch antenna is totally capacitive. And thus, one can expect another resonant frequency due to the reactive cancellation between the capacitive feeding structure and the inductive metamaterial cover. Consequently, an appropriate selection of the coupling capacitor value can result in a broadband operation. To this aim, the metamaterial cover described above is placed over the conventional proximity-coupled patch antenna.

A schematic of proposed metamaterial patch antenna is shown in Fig. 4. In general, the two dielectrics can be of different thicknesses and relative permittivity, but here both dielectrics are 0.762mm Duroid with, $\varepsilon_r = 2.2$. For the case discussed here, the patch of the antenna is rectangular with 12.45mm width and 16mm length.

The distance between the main microstrip line and the parasitic line is also 7mm. Each metamaterial cover composed of 9×9 unit cells, as shown in Fig.4. Consequently, the total size of the dielectric substrate and the metamaterial cover is 184.5mm×184.5mm. Furthermore, the working frequency of the conventional patch antenna is selected at 5.9GHz.

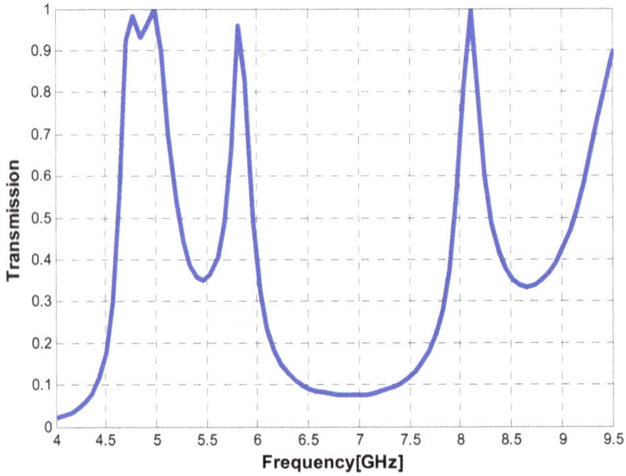

Fig. 3. FDTD simulated transmission of metamaterial cover.

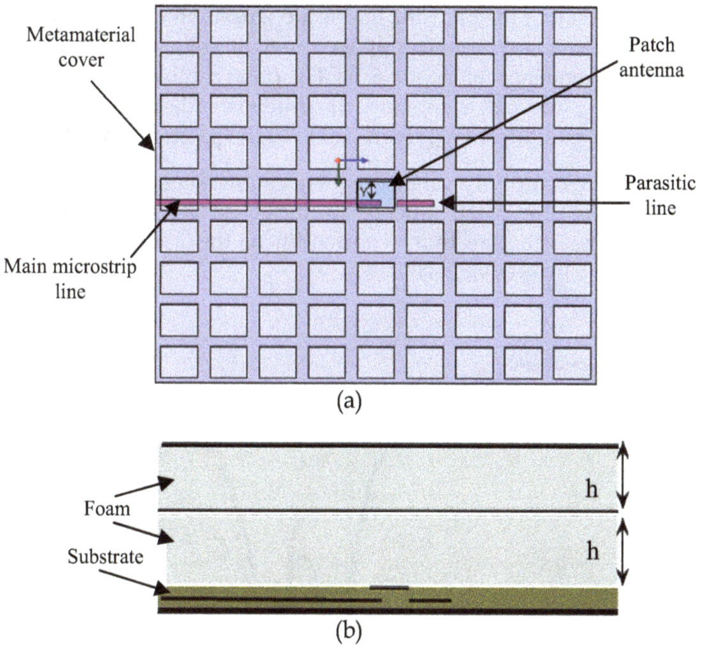

Fig. 4. Geometry of a metamaterial proximity-coupled patch antenna, (a) top view and (b) cross view.

Reflection coefficient of the proposed metamaterial patch antenna has been simulated and was compared to the one obtained for the conventional proximity-coupled patch antenna in Fig. 5. As revealed in the figure, the antenna return loss is significantly improved compared to the reference patch antenna without the metamaterial cover.

The impedance bandwidth of the patch antenna is increased from 2.9% to 5.23% (ranging from 5.649GHz to 5.952GHz). Using the usual formulas mentioned in (Garg et al., 2001), the conventional proximity-coupled patch antenna discussed here has a TM_{01} mode resonant frequency of approximately 5.9GHz. The second resonant frequency of the metamaterial patch antenna is obviously due to the TM_{01} mode of the conventional patch antenna. (See Fig.5)

Since the metamaterial superstrate disturbs the current distribution of the TM_{01} mode, this resonant frequency slightly shifts down to a lower frequency. An interested reader is recommended to refer to (Zhong et al., 1994) for more details. The first resonant frequency is the result of reactive cancellation between the capacitive feeding structure and the inductive metamaterial cover.

On the other hand, the first resonant frequency is close to the second resonant frequency, which results in broadband operation. The simulation results of Fig. 5 are in good agreement with the theoretical predictions discussed above, which serve to justify the approximations used to model the metamaterial patch antenna as a grounded homogenous medium.

It is necessary to mention that the parasitic line section, used on the opposite side of the feed line, stabilizes the antenna performance at its resonant frequency at the expense of an additional resonance frequency at about 6.24GHz (Jafargholi et al., 2011). By using the metamaterial cover over the patch antenna, the antenna radiation patterns in E- and H-planes are concentrated in a direction perpendicular to the patch antenna (θ=0).

The simulated broadside directivity versus frequency is shown in Fig. 6. As can be seen, the maximum directivity of the patch antenna is increased from 6.25dB to16.16dB using metamaterial cover. The 3dB directivity bandwidth of the metamaterial antenna is also between 5.685GHz and 5.91GHz, or 3.88%.

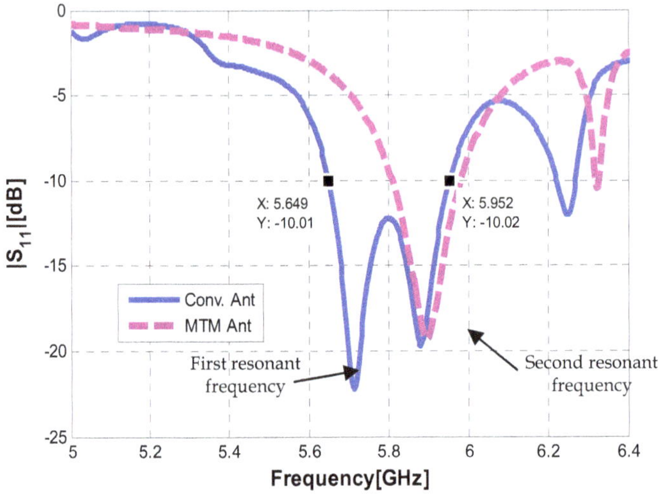

Fig. 5. Simulated reflection coefficient versus frequency.

The antenna radiation patterns within its bandwidth are also investigated. The E-plane and H-plane patterns of the metamaterial patch antenna at three frequencies (5.7GHz, 5.8GHz and 5.9GHz) have been simulated and were compared to the one obtained for the conventional antenna, at 5.9GHz, in Fig. 7.

Although the radiation pattern of the metamaterial antenna changes a bit at each frequency, the main lobe of the metamaterial antenna at all frequencies (ranging from 5.65 to 5.95GHz) is in the broadside direction and maximum directivity is reasonably good. The variation of the radiation pattern is mainly attributed to the nature of metamaterial cover.

The maximum directivities of the metamaterial antenna at 5.7GHz and 5.9GHz are 13.24dB and 14dB, respectively. The maximum directivity of an aperture antenna is calculated by $D_{max}=4\pi A/\lambda^2$. In the present case, the area of the aperture is A=and $\lambda=c_0/f_0$= 51.724mm, so that maximum directivity then becomes D_{max}= 22dB.

The maximum directivity of the metamaterial patch antenna, occurring at 5.81GHz, (16.16dB) has approached the maximal directivity obtained, theoretically, with the same aperture size.

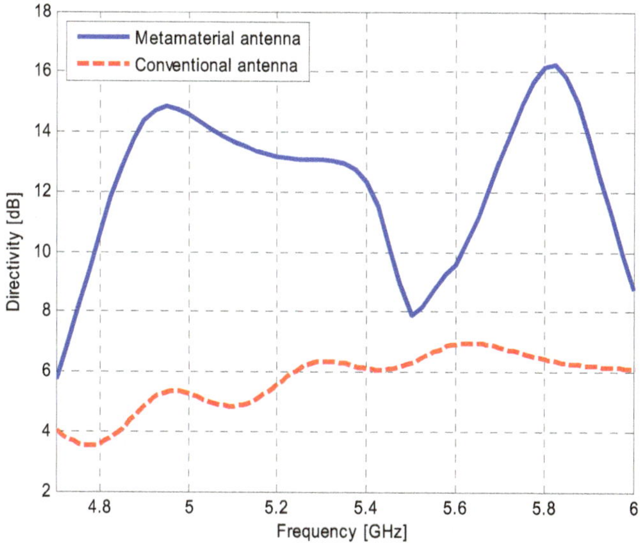

Fig. 6. Simulated broadside directivity versus frequency.

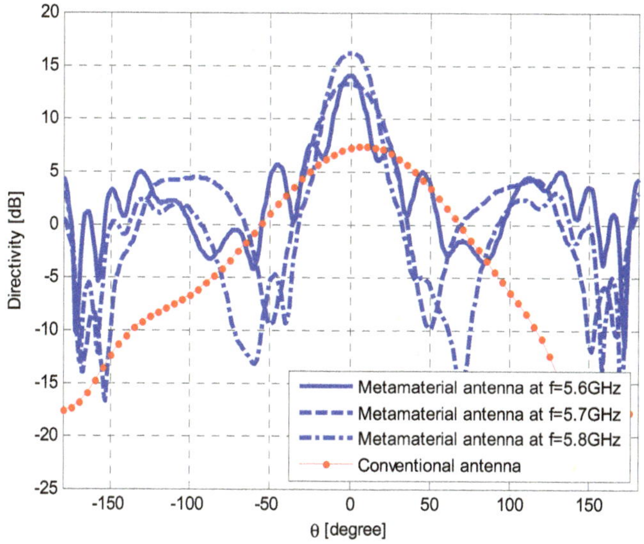

(a)

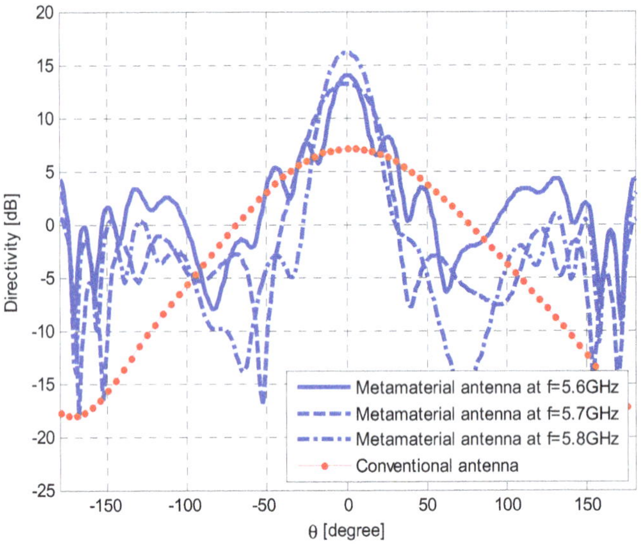

(b)

Fig. 7. CST simulated radiation patterns at different frequencies over the operating bandwidth, (a) E-plane, and (b) H-plane.

3. Polarization dependent metamaterial cover designs

A considerable part of research at microwave frequencies is focused on isotropic metamaterial covers that are independent on polarization states. Polarization-dependent surfaces have recently found useful applications in changing the polarization state of the incident wave (Yang & Rahmat-samii, 2005; Veysi et al., 2010).

For a traditional metamaterial cover, the transmission phase remains the same regardless of the x- or y-polarization state of the incident plane wave. In contrast, the transmission phase of a PDMTM cover is a function of both frequency and polarization state. Hence, when a PDMTM cover is employed as a director, the polarization state of the transmitted wave is fully characterized by the transmission phase difference between the x- and y-polarizations and the polarization state of the incident wave. And thus a proper phase difference between x- and y-polarized waves leads to a desired change in the polarization state of the transmitted wave.

Directive circularly polarized antennas are widely used in satellite communication systems. To obtain directive circularly polarized antenna, various types of metamaterial antenna have been proposed in the literature (Iriarte et al., 2006; Diblanc et al., 2005; Arnaud et al., 2007). It was demonstrated in (Iriarte et al., 2006) that the directive circularly polarized antenna can be realized by metamaterial antenna with a circular feed. A major limitation of this method is inability to tune mechanically. In other words, the polarization state of the antenna is only determined by the feed mechanism.

Directive circularly polarized antenna can be also realized using either metallic wire polarizer (Diblanc et al., 2005) or meander line polarizer (Arnaud et al., 2007) stacked on the top of the one-layer metamaterial cover. In this section, instead of using meander line or metallic wire polarizer, the geometry of the metamaterial cover is changed to provide circular polarization. Consequently, the second layer (polarizer layer) can be replaced with another metamaterial cover layer, which in turn results in higher directivity.

Moreover, in contrast to the previous directive circularly polarized antennas (Iriarte et al., 2006; Diblanc et al., 2005; Arnaud et al., 2007), polarization state of the directive antennas using our proposed metamaterial cover can be mechanically changed regardless of the feed mechanism.

In this section, a useful guideline has been established as how to use the magnitude and phase of the transmission coefficient to identify the operational frequency band of the directive circularly polarized antennas based on polarization dependent metamaterial covers.

As revealed in the previous sections, when a plane wave normally impinges upon a metamaterial cover, the phase and magnitude of the transmitted wave change with frequency. In order to illustrate the polarization feature of the PDMTM cover, we assume that a left-hand circularly polarized (LHCP) wave, namely,

$$\vec{E^i} = \vec{ax} e^{-jkz} + j\vec{ay} e^{-jkz}$$

where k is the free-space wavenumber, is normally impinged upon a director placed in X-Y plane. The field transmitted through the director can be easily calculated from the following equation:

$$\vec{E^t} = e^{-jkz} e^{j\theta_x} (\vec{ax} + j\vec{ay} e^{j(\theta_y - \theta_x)}) \quad (5)$$

The above field can be decomposed into two circularly polarized components

$$\vec{E^t} = e^{-jkz} e^{j\theta x} [\vec{e_r}(\frac{1 - e^{j(\theta y - \theta x)}}{\sqrt{2}}) + \vec{e_l}(\frac{1 + e^{j(\theta y - \theta x)}}{\sqrt{2}})] \quad (6)$$

Where

$$\vec{e_l} = \frac{(\vec{ax} + j\vec{ay})}{\sqrt{2}}, \quad \vec{e_r} = \frac{(\vec{ax} - j\vec{ay})}{\sqrt{2}}$$

and θ_x and θ_y denote the transmission phases for the x- and y-polarized waves, respectively. For a traditional metamaterial cover (θ_y- θ_x=0), the transmitted wave is purely LHCP and thus the polarization does not change. In order to change the polarization state of the antenna, the PDMTM cover can be used as a director. At a certain frequency where phase difference is 180° and transmission is considerable, the transmitted wave is purely right-hand circularly polarized (RHCP).

The left-hand circularly polarized incident wave can be also converted to the linearly polarized (LP) wave where the phase difference is 90° and the transmission is also considerable. One can follow the same procedure for the linearly polarized incident wave,

$$\vec{E^i} = \vec{ax} e^{-jkz} + \vec{ay} e^{-jkz},$$

so that the transmitted field then becomes:

$$\vec{E^t}=\vec{a_x}e^{-j(kz-\theta x)}+\vec{a_y}e^{-j(kz-\theta y)}=$$
$$e^{-jkz}e^{j\theta x}[\vec{e_r}(\frac{1+e^{j(\theta y-\theta x+\pi/2)}}{\sqrt{2}})+\vec{e_l}(\frac{1-e^{j(\theta y-\theta x+\pi/2)}}{\sqrt{2}})] \tag{7}$$

Consequently the radiation mechanism of the linearly polarized antenna with metamaterial cover is conceptually described by Eq. 7. For an isotropic metamaterial cover, the phase difference between two orthogonal polarizations is zero and thus the polarization state does not change.

An interesting feature of the PDMTM covers can be revealed by a closer investigation. When a PDMTM cover with 90° transmission phase difference is used as a director, the polarization state of the transmitted wave becomes LHCP. Moreover, when the phase difference is -90° the polarization state of the transmitted wave is RHCP. Consequently we can easily switch between LHCP and RHCP using a rotatory cover, which can be rotated smoothly with a 90° steps.

Based on above discussion, one can conclude that the metamaterial cover can be used as a changing polarization plane. The operational frequency band of an antenna with PDMTM cover is defined as the frequency region within which the magnitudes of the transmission coefficients for both x- and y-polarized waves are close to their maximum values and transmission phase difference takes the desired value.

This interesting feature has been realized by changing the unit cell geometry, such as cutting rectangular holes instead of square holes and changing the relative height difference between the x- and y-directed strips of each layer (Veysi et al., 2011).

3.1 Rectangular hole metamaterial cover

The traditional metamaterial cover uses symmetric square holes so that its transmission phase for normal incidence remains the same regardless of the x- or y-polarization state of the incident plane wave. Therefore, the logical step is to replace the square holes by rectangular ones (Veysi et al., 2011).

First, the design parameters of the metamaterial cover are selected to have a reasonable transmission at a specified frequency. The effect of different design parameters of the metamaterial cover on the magnitude of the transmission coefficient can be found in (Huang et al., 2009).

After the successful design of the isotropic metamaterial cover, the width or/and length of the square holes are changed to obtain both the desired transmission phase difference and the maximum transmission within the specified frequency band. When the hole width is increased, the plasma frequencies shift down to the lower frequencies.

Thus, by adjusting the width and length of the rectangular hole, the polarization sense of the transmitted wave can be changed. An example design for these parameters is as follows:

h= 24.5mm, P=20.5mm, L=17.5mm, W=16.5mm

For a linearly polarized antenna, namely

$$\vec{E^i} = \vec{a_x} e^{-jkz} + \vec{a_y} e^{-jkz},$$

The axial ratio of the transmitted wave is plotted in Fig. 8. Also, Fig. 9 shows the transmission curve for both x- and y-polarized incident plane waves. The frequency band inside which the axial ratio of the RHCP transmitted wave is below 6dB and the magnitudes of the transmission coefficients for both the x- and y-polarized incident waves are more than 90% ranges from 4.75GHz to 5GHz (5.12%).

3.2 Metamaterial cover with nonplanar strips

Another approach to realize PDMTM covers is to add space between the x- and y-directed strips of each layer (Veysi et al., 2011), as shown in Fig. 10. For a traditional metamaterial cover, the x-directed strip is located on the same plane as the y-directed strip.

For the nonplanar case discussed here the dimensions are chosen as follows: L=W=13.5mm, P=18.5mm, h=27.5mm, h_{r1}=3.5mm, and h_{r2}=7mm, where h_{r1} denotes the relative height between the x- and y-directed strips of the first layer and h_{r2} denotes the same height for the second layer.

Fig. 11 shows the magnitudes of the transmission coefficients for both x- and y-polarized waves. The axial ratio of the wave radiated from linearly polarized antenna is also plotted in Fig. 12. It can be seen from Figs. 11-12, that the operating frequency of the proposed structure is around 9.1GHz where the metamaterial cover has both the desired transmission phase difference and the remarkable transmission.

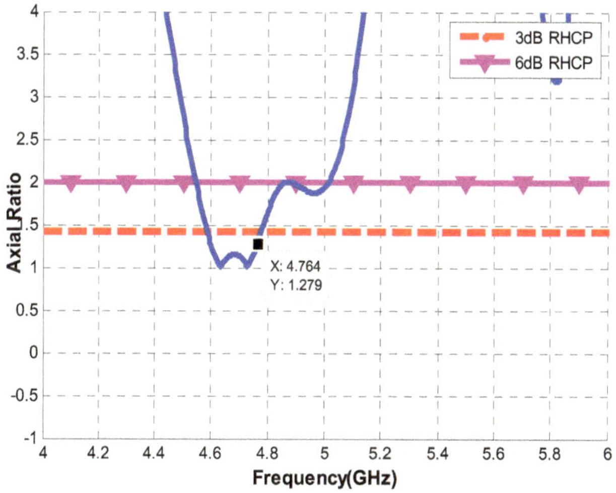

Fig. 8. Axial ratio of the transmitted plane wave from the rectangular hole metamaterial cover (Veysi et al., 2011).

The FDTD simulated results presented in this section confirm the concepts of the proposed approach to control both the direction and the polarization of the transmitted wave. The authors believe that the proposed cover can find many applications in the broad electromagnetic areas such as antenna engineering and optical sources.

However, a rigorous characterization should take into account the complex interactions between the antenna and the metamaterial cover, such as finite size of the ground plane and the antenna height. Consequently, it is indispensable to use full wave analysis method, such as the finite difference time domain (FDTD), in the antenna designs in order to obtain accurate results.

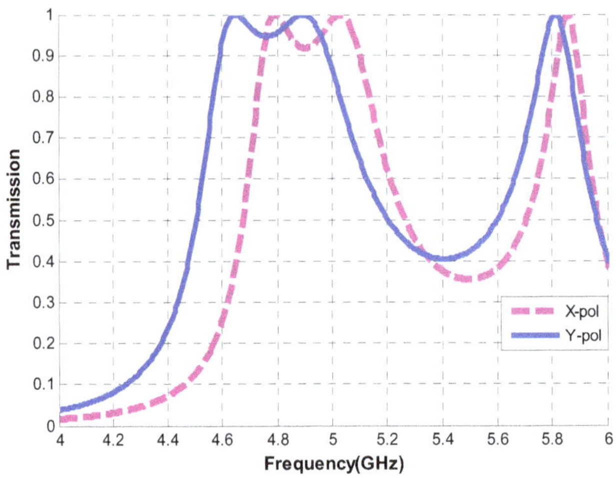

Fig. 9. FDTD simulated transmission of the rectangular hole metamaterial cover (Veysi et al., 2011).

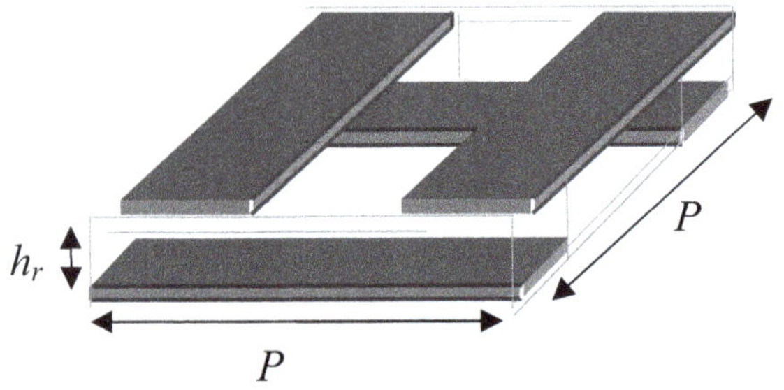

Fig. 10. A unit cell of metamaterial cover with offset strips (Veysi et al., 2011).

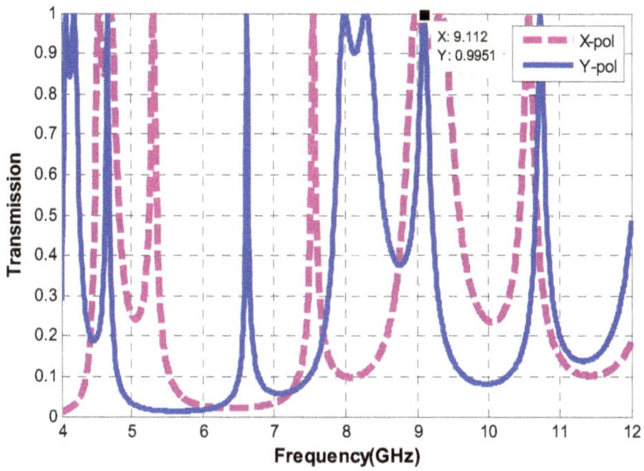

Fig. 11. FDTD simulated transmission of the metamaterial cover with offset strips (Veysi et al., 2011).

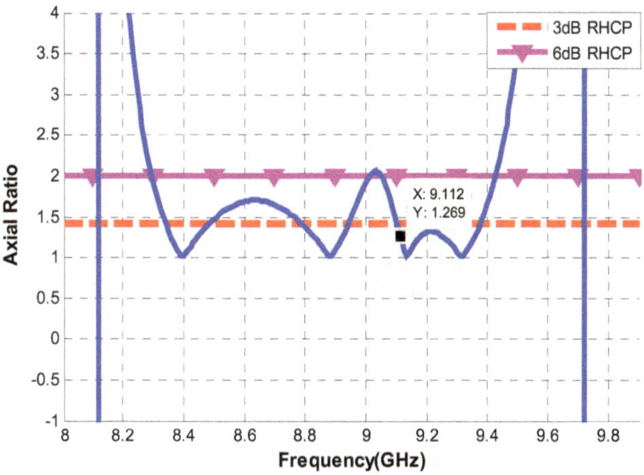

Fig. 12. Axial ratio of the transmitted plane wave from the metamaterial cover with nonplanar strips (Veysi et al., 2011).

4. Conclusions

Metamaterial covers can be applied to conventional antenna to improve their performance. These include conventional metamaterial covers to increase both the impedance and directivity bandwidths of the proximity coupled microstrip patch antenna and polarization dependent metamaterial covers to change the polarization state of the antenna. Thin lattices of ungrounded metal plates can behave as a metamaterial cover and can be analyzed using a simple FDTD code. These surfaces have two important properties: (1) increasing of the transmission rate and (2) control of the direction and polarization of the transmission. Polarization dependent metamaterial covers can be realized by cutting rectangular holes instead of square holes and changing the relative height difference between the x- and y-directed strips of each layer.

5. References

Alù, A., Bilotti, F., Engheta, N. & Vegni, L. (2006). Metamaterial Covers Over a Small Aperture, *IEEE Trans. Antennas Propag.*, Vol. 54, No. 6, pp. 1632–1643

Arnaud, E., Chantalat, R., Koubeissi, M., Menudier, C., Monediere, T., Thevenot, M., & Jecko, B. (2007). New Process of Circularly Polarized EBG Antenna by using meander Lines, *IET, EuCAP*, pp. 1-6

Diblanc, M., Rodes, E., Amaud, E., Thevenot, M., Monediere, T., & Jecko, B. (2005). Circularly Polarized Metallic EBG Antenna, *IEEE Microwave and Wireless Components Letters.*, Vol. 15, No. 10, pp. 638-640

Enoch, S., Tayeb, G., Sabouroux, P., Guérin, N., & Vincent, P. (2002). A Metamaterial for Directive Emission, *Physical Review Letters*, Vol. 89, pp. 213902

Garg, R., Bhartia, P., Bahl, I., Ittipiboon, A. (2001). Microstrip Antenna Design Handbook, *Artech House*

Huang, C., Zhao, Z., Wang, W., & Luo, X. (2009). Dual Band Dual Polarization Directive Patch AntennaUsing Rectangular Metallic Grids Metamaterial, *Int J Infrared Milli Waves*, Vol. 30, pp. 700–708

Iriarte, J. C., Ederra, I., Gonzalo, R., Gosh, A., Laurin, J., Caloz, C., Brand, Y., Gavrilovic, M., & Demers, Y. P. (2006). EBG Superstrate for Gain Enhancement of a Circularly Polarized Patch Antenna", *IEEE Trans. Antennas Propag.*, pp. 2993-2996

Jafargholi A, Kamyab, M., Veysi, M. & Nikfal Azar, M. (2011), Microstrip gap proximity fed-patch antennas, analysis, and design, *Int J Electron Commun* (AEÜ), doi:10.1016/j.aeue.2011.05.011

Ju, J., Kim, D., Lee, W. J., & Choi, J. I. (2009). Wideband High-Gain Antenna Using Metamaterial Superstrate with the Zero Refractive Index, *Microwave and Optical Tech. Lett.*, Vol. 51, No. 8, pp. 1973–1976

Pendry, J. B., Holden, A. J., Stewart, W. J., & Youngs, I. (1996). Extremely Low Frequency Plasmas in Metallic Mesostructures, *Physical Review Letters*, Vol. 76, pp.4773-4776

Temelkuaran, B., Bayindir, Ozbay, M., E., Biswas, R., Sigalas, M., Tuttle, G., & Ho, K. M. (2000). Photonic Crystal-Based Resonant Antenna with a Very High Directivity, *Journal of Applied Physics*, Vol. 87, pp. 603–605

Tsao, C. H. & Chen, J. L. (2006). Field Propagation of a Metallic Grid Slab That Act as a Metamaterial, *Physics Letters A*, Vol. 353, pp. 171–178

Veysi, M., Kamyab, M., Mousavi, S. M., & Jafargholi, A. (2010). Wideband Miniaturized Polarization-Dependent HIS Incorporating Metamaterials, *IEEE Antennas And Wireless Propagation Letters*, Vol. 9, pp. 764-766

Veysi, M., Kamyab, M., Moghaddasi, J., & Jafargholi, A. (2011). Transmission Phase Characterizations of Metamaterial Covers for Antenna Application, *Progress In Electromagnetics Research Letters*, Vol. 21, pp. 49-57

Xu, H., Zhao, Z., Lv, Y., Du, C., & Luo, X. (2008). Metamaterial Superstrate and Electromagnetic Band-Gap Substrate for High Directive Antenna, *Int J Infrared Milli Waves*, Vol. 29, pp. 493–498

Yang, F. & Rahmat-Samii, Y. (2005). A low profile single dipole antenna radiating circularly polarized waves, *IEEE Trans. Antennas Propag.*, Vol. 53, No. 9, pp. 3083–3086

Zhong, S.-S., Liu, G. & Qasim, G. (1994). Closed Form Expressions for Resonant Frequency of Rectangular Patch Antennas With Multidielectric Layers, *IEEE Trans. Antennas Propag.*, Vol. 42, pp.1360-1363

Zhu, F., Lin, Q., & Hu, J. (2005). A Directive Patch Antenna with a Metamaterial Cover, *Proceedings of Asia Pacific Microwave Conference*

Permissions

The contributors of this book come from diverse backgrounds, making this book a truly international effort. This book will bring forth new frontiers with its revolutionizing research information and detailed analysis of the nascent developments around the world.

We would like to thank Dr. Victor Barsan and Prof. Radu P. Lungu, for lending their expertise to make the book truly unique. They have played a crucial role in the development of this book. Without their invaluable contribution this book wouldn't have been possible. They have made vital efforts to compile up to date information on the varied aspects of this subject to make this book a valuable addition to the collection of many professionals and students.

This book was conceptualized with the vision of imparting up-to-date information and advanced data in this field. To ensure the same, a matchless editorial board was set up. Every individual on the board went through rigorous rounds of assessment to prove their worth. After which they invested a large part of their time researching and compiling the most relevant data for our readers. Conferences and sessions were held from time to time between the editorial board and the contributing authors to present the data in the most comprehensible form. The editorial team has worked tirelessly to provide valuable and valid information to help people across the globe.

Every chapter published in this book has been scrutinized by our experts. Their significance has been extensively debated. The topics covered herein carry significant findings which will fuel the growth of the discipline. They may even be implemented as practical applications or may be referred to as a beginning point for another development. Chapters in this book were first published by InTech; hereby published with permission under the Creative Commons Attribution License or equivalent.

The editorial board has been involved in producing this book since its inception. They have spent rigorous hours researching and exploring the diverse topics which have resulted in the successful publishing of this book. They have passed on their knowledge of decades through this book. To expedite this challenging task, the publisher supported the team at every step. A small team of assistant editors was also appointed to further simplify the editing procedure and attain best results for the readers.

Our editorial team has been hand-picked from every corner of the world. Their multi-ethnicity adds dynamic inputs to the discussions which result in innovative outcomes. These outcomes are then further discussed with the researchers and contributors who give their valuable feedback and opinion regarding the same. The feedback is then collaborated with the researches and they are edited in a comprehensive manner to aid the understanding of the subject.

Apart from the editorial board, the designing team has also invested a significant amount of their time in understanding the subject and creating the most relevant covers. They scrutinized every image to scout for the most suitable representation of the subject and create an appropriate cover for the book.

The publishing team has been involved in this book since its early stages. They were actively engaged in every process, be it collecting the data, connecting with the contributors or procuring relevant information. The team has been an ardent support to the editorial, designing and production team. Their endless efforts to recruit the best for this project, has resulted in the accomplishment of this book. They are a veteran in the field of academics and their pool of knowledge is as vast as their experience in printing. Their expertise and guidance has proved useful at every step. Their uncompromising quality standards have made this book an exceptional effort. Their encouragement from time to time has been an inspiration for everyone.

The publisher and the editorial board hope that this book will prove to be a valuable piece of knowledge for researchers, students, practitioners and scholars across the globe.

List of Contributors

Masao Kitano
Department of Electronic Science and Engineering, Kyoto University, Japan

Paul van Kampen
Centre for the Advancement of Science and Mathematics Teaching and Learning & School of Physical Sciences, Dublin City University, Ireland

Wei-Tou Ni
Center for Gravitation and Cosmology, Department of Physics, National Tsing Hua University, Hsinchu, Taiwan, ROC
Shanghai United Center for Astrophysics, Shanghai Normal University, Shanghai, China

Radu Paul Lungu
Department of Physics, University of Bucharest, Magurele-Bucharest, Romania

Manuel Arrayás and José L. Trueba
Universidad Rey Juan Carlos, Spain

Antonio F. Rañada
Universidad Complutense de Madrid, Spain

Victor Barsan
Department of Theoretical Physics, National Institute of Physics and Nuclear Engineering, Bucharest, Magurele, Romania

Bruno Carpentieri
University of Groningen, Institute of Mathematics and Computing Science, Groningen, The Netherlands

Hiroshige Kumamaru, Kazuhiro Itoh and Yuji Shimogonya
University of Hyogo, Japan

Ibrahim El Baba, Sébastien Lalléchère and Pierre Bonnet
Clermont University, Blaise Pascal University, BP 10448, F-63000, Clermont-Ferrand, France
CNRS, UMR 6602, LASMEA, F-63177, Aubière, France

Masoud Movahhedi and Rasool Keshavarz
Electrical Engineering Department, Shahid Bahonar University of Kerman, Kerman, Iran

Mouaaz Nahas
Department of Electrical Engineering, College of Engineering and Islamic Architecture, Umm Al-Qura University, Makkah, Saudi Arabia

Houssem Rafik El-Hana Bouchekara
Department of Electrical Engineering, College of Engineering and Islamic Architecture, Umm Al-Qura University, Makkah, Saudi Arabia
Electrical Laboratory of Constantine "LEC", Department of Electrical Engineering, Mentouri University - Constantine, Constantine, Algeria

Mehdi Veysi, Amir Jafargholi and Manouchehr Kamyab
E.E. Dept. K. N. Toosi University of Technology, Tehran, Iran

www.ingramcontent.com/pod-product-compliance
Lightning Source LLC
Chambersburg PA
CBHW070737190326
41458CB00004B/1210